Large Deviations for Gaussian Queues

Large Deviations for Gaussian Queues

Modelling Communication Networks

Michel Mandjes
Korteweg-de Vries Institute for Mathematics, University of Amsterdam, The Netherlands

John Wiley & Sons, Ltd

Telephone (+44) 1243 779777

Email (for orders and customer service enquiries): cs-books@wiley.co.uk
Visit our Home Page on www.wileyeurope.com or www.wiley.com

Other Wiley Editorial Offices

John Wiley & Sons Inc., 111 River Street, Hoboken, NJ 07030, USA

Jossey-Bass, 989 Market Street, San Francisco, CA 94103-1741, USA

Wiley-VCH Verlag GmbH, Boschstr. 12, D-69469 Weinheim, Germany

John Wiley & Sons Australia Ltd, 42 McDougall Street, Milton, Queensland 4064, Australia

John Wiley & Sons (Asia) Pte Ltd, 2 Clementi Loop #02-01, Jin Xing Distripark, Singapore 129809

John Wiley & Sons Canada Ltd, 6045 Freemont Blvd, Mississauga, Ontario, L5R 4J3, Canada

Wiley also publishes its books in a variety of electronic formats. Some content that appears in print may not be available in electronic books.

Anniversary Logo Design: Richard J. Pacifico

British Library Cataloguing in Publication Data

A catalogue record for this book is available from the British Library

ISBN-13: 978-0-470-01523-0

Typeset in 10/12 Times by Laserwords Private Limited, Chennai, India

Contents

Preface and acknowledgments

In the spring of 2001, Gaussian queues started to attract my attention. At that moment, I was working on a number of problems on networks of queues as well as queues operating under nonstandard scheduling disciplines–which turned out to be substantially harder to analyze than the classical single first-in first-out queue. What a mathematician does when he wishes to analyze this kind of hard problem is, to find a modelling framework that is 'clean' enough to enable (to some extent) explicit solutions, yet general enough to cover all interesting scenarios (in my case, all relevant arrival processes including the important class of long-range dependent inputs). At some point I came across a series of papers by Petteri Mannersalo and Ilkka Norros (VTT Research, Espoo, Finland) on Gaussian queues, and it turned out that the Gaussian input model they considered combined these attractive properties. I started to learn more about it, and it led to a very nice collaboration with Petteri and Ilkka (in the mean time, their annual spring-visit to Amsterdam has almost become a tradition...).

Coincidentally, more or less at the same time Krzysztof Dębicki took a leave of absence from the University of Wrocław, Poland, and became a colleague of mine at CWI in Amsterdam. He had been working already for a while on Gaussian queues, and I think it is fair to consider him as a real expert on their tail asymptotics. It was the start of a fruitful and very pleasant collaboration. Needless to say that this book would not have been the same without Petteri, Ilkka, and Krzys, and I would like to thank them for this.

The idea of writing a book on Gaussian queues came up in 2004. By then, I had been involved in a series of papers on the large deviations of Gaussian queues, and started to realize that they make up a nice and coherent body of theory. The Gaussian machinery proved to be a powerful tool, and I am convinced that it will be extremely useful for many other queuing problems as well. During the last years of research, my attention shifted somewhat from the asymptotics of Gaussian queues to their applications in communication networks; due to the generality and versatility of the Gaussian paradigm, it enabled the development of effective guidelines for several highly relevant networking problems.

To illustrate the strengths of the concept of Gaussian queues, I have laid much emphasis on applications. The book contains four application-related chapters, in which Quality-of-Service differentiation (through Generalized Processor Sharing), link dimensioning, and bandwidth trading are dealt with.

I have attempted to make the book as much self-contained as possible, so that it becomes accessible for a broader audience; unfortunately, it turned out that it is nearly impossible to write the book in such a way that it does not require any prior knowledge on standard probabilistic concepts.

The book could serve as a textbook for undergraduate students in mathematics (and 'mathematics-related' disciplines); as argued above, a rigorous background in probability is required. The book also targets graduate students in engineering, in particular, computer science and electrical engineering with interest in networking applications. It could also serve as a reference book for senior researchers both in academia and at telecommunication labs – particularly the application-part may be interesting to practice-oriented scientists.

My PhD students Abdelghafour Es-Saghouani, Pascal Lieshout, and Frank Roijers read earlier drafts of the manuscript, and provided me with many helpful suggestions. Much of the material presented in this book was drawn from my papers on Gaussian queues. These papers were not just my work – I would like to thank, besides the names I already mentioned above, the other co-authors: Hans van den Berg, Nam Kyoo Boots, Sem Borst, Ton Dieker, Krishnan Kumaran, Aiko Pras, Maya Ramakrishnan, Aleksandr Stolyar, Miranda van Uitert, Pieter Venemans, and Alan Weiss.

The first draft of this book was written while I had a 1-month leave at the University of Melbourne, Australia. This place proved to be a wonderful working environment, and I would like to thank Peter Taylor for hosting me, and Tony Krzesinski (University of Stellenbosch, South Africa) for being the ideal office mate during this period. I would also like to thank the publisher, Wiley Chichester, for its support – in particular Susan Barclay, Kelly Board, Wendy Hunter, Simon Lightfoot, and Kathryn Sharples. I should also mention Laserwords, Chennai, and in particular Gopika, who were responsible for the typesetting job (and I realize that I have probably been the most demanding author ever).

Work presented in this book has been partly carried out in the framework of the projects Equip (in the programme 'Netwerken', funded by the Netherlands Organization for Scientific Research, NWO), Logica (in the programme 'Open Competitie', also funded by NWO), Equanet (in the programme 'ICT Doorbraak', funded by Dutch Ministry of Economic Affairs through its agency Senter/Novem), and Bricks (in the programme 'Bsik', funded by several Dutch ministries).

I dedicate this book to my wife Miranda, for her continuous love and support, and our daughter Chloe.

Amsterdam, November 2007,
Michel Mandjes

Chapter 1

Introduction

Performance Analysis, Queuing Theory, Large Deviations. Performance analysis of communication networks is the branch of applied probability that deals with the evaluation of the level of efficiency the network achieves, and the level of (dis)satisfaction enjoyed by its users. Clearly, there is a broad variety of measures that characterize these two aspects. Focusing on the efficiency of the use of network resources, one could think of the throughput, i.e., the rate at which the network effectively works–in the case of a single network element, this could be the rate (in terms of, say, bits per second) at which traffic leaves. Another option is to use a relative measure, such as the utilization, commonly defined as the ratio of the throughput and the available service speed of the network element. Also the (dis)utility experienced by users can be expressed by a broad variety of measures. Realizing that at any network element traffic can be stored in a buffer when the input rate temporarily exceeds the available service rate, it seems justified to study performance indicators that describe the delay incurred when passing the network node. Buffers have a finite size, so there is the possibility of losing traffic, and as a result the fraction of traffic lost becomes a relevant metric.

Performance analysis is a probabilistic discipline, as the main underlying assumption is that user behavior is inherently *random*, and therefore described by a statistical model. This statistical model defines the probabilistic properties of the arrival process (or, input process) of traffic at the network. Traffic could arrive in a smooth way, but highly irregular patterns also occur; in the latter case, communication engineers call the arrival process *bursty*.

Justified by the above description of network elements as storage systems, we could model a communication network as a network of *queues*; at any node traffic arrives, is stored if it cannot be handled immediately, and is served. Performance analysis often relies heavily on results from the theory that describes the performance of these queues, i.e., *queuing theory*. A key element of performance analysis

Large deviations for Gaussian queues M. Mandjes

is the characterization of the impact of 'user parameters' on the performance offered by the network (how is the delay affected by the arrival rate? what is the impact of increased variability of the input traffic? etc.). On the other hand, one often studies the sensitivity of the performance in the system parameters (what is the impact of the buffer size on the loss probability? how does the service speed affect the mean delay? etc.)

A substantial part of the defined performance metrics relates to *rare events*. Often network engineers have the target to design the system such that the loss probability is below, say, 10^{-6}. Another common objective is that the probability that the delay is larger than some predefined excessive value is of the same order. This explains why we heavily rely on a subdomain of probability theory that exclusively focuses on the analysis of rare events: *large deviations theory*. This theory has a long history, but has been applied intensively for performance analysis purposes only during the last, say, two decades.

Traffic management, dimensioning. Once one is capable of evaluating the performance of a static situation (i.e., calculating performance metrics for a given arrival process and given network characteristics), the next step is often to choose the set of design parameters such that a certain condition is met, or such that some objective function is optimized. For instance, a requirement imposed upon the network element could be that just (on average) a fraction ϵ of the incoming traffic is lost. Evidently, when increasing the buffer size B, the loss probability decreases, and therefore it is legitimate to ask for which minimal B the loss probability is at most ϵ. Of course, there is often a cost incurred when increasing B. As a result one could imagine that one should maximize an objective function that consists of a 'utility part', minus a 'cost part', where both parts increase in B. Selecting an appropriate value for B is usually called *buffer dimensioning*; similarly the choice of a suitable service speed is referred to as *link rate dimensioning* (or, shortly, link dimensioning).

On the other hand, knowledge of the static situation enables the computation of conditions on the arrival process (both in terms of average input traffic rate and the variability of the arrival process) under which the network can offer some required performance level:

- In this way, one could develop mechanisms that decide what the maximum number of users is such that the mean delay stays within some predefined bound; such a mechanism is usually called *admission control*. To implement an admission control, one needs to be able to characterize the so-called admissible region, which is, in a situation of two classes of users, the combination of all numbers of users of both classes (n_1, n_2) for which for both classes the performance requirement is met.

- Also, insight into the static situation may tell us how to 'smooth' traffic (i.e., decrease the variability of the arrival process), such that the traffic stream becomes more 'benign', and the loss probability in some target queue can meet some set requirement (a technique known as *traffic shaping*). Traffic shaping is usually done by inserting an additional queue between the traffic sources and the target queue that is emptied at a service rate c' that is lower than the peak rate of the original stream (but higher than the service rate c of the target queue). Then the traffic stream arriving at the second queue is smoother than the original traffic stream, and therefore easier to handle, but this is at the expense of introducing additional delay.

This traffic shaping example explains the interest in *tandem queues*, i.e., systems of queues in series (in which the output of the first queue feeds into the second queue). In such a situation one would, for instance, like to dimension the shaping rate: given a buffer B and service rate c in the target queue, how should one choose the shaping rate c' to ensure that the loss probability in the target queue is below ϵ (where it assumed that the shaper queue has a relatively big buffer).

In the literature, the set of control measures that affect the network's efficiency or the user's (dis-)satisfaction is often called *traffic management*. Clearly, dimensioning is a traffic management action that relates to a relatively long timescale: one can choose a new value for the buffer size or the link rate only at a very infrequent rate; the process of updating the resource capacities is known as the *planning cycle*. Mechanisms like admission control serve to control fluctuations of the offered traffic at a relatively short timescale: admission control is done on the timescale that new users arrive (and hence the decision to accept or reject a new user has to be done essentially in real time).

Performance differentiation. We have described above the situation in which we wished to guarantee some performance requirement that is uniform across users; for instance, all users should be offered the same maximum loss probability. In practice, however, all applications have their own specific performance requirements. Think of a voice user, who tolerates a substantial amount of loss (up to the order of a few percents, if certain codecs are used) but whose delay is critical, versus a data user, who has very stringent requirements with respect to loss, but is less demanding with respect to delay. Of course one could treat all traffic in the same fashion, e.g., by using first-in-first-out (FIFO) queues; clearly, to meet the performance requirements of all users, the requirement of the most stringent users should be satisfied. Such an approach will, however, inevitably lead to a waste of resources, and therefore one has developed queuing disciplines that actively discriminate. An example of such a scheme is the (two-class) *priority queue*, in which one class has strict priority over another class. The high-priority class does not 'see' the low-priority class, so its performance can be evaluated as in the FIFO case. The low-priority class, however, sees a fluctuating service capacity, and therefore its performance is considerably harder to analyze.

Strict priority has the intrinsic drawback of 'starvation', i.e., the low-priority class can be excluded from service for relatively long periods of time (namely, the periods in which the high-priority class uses all the bandwidth). To avoid this starvation effect, one could guarantee the low-priority class at least some minimal service rate. This thought led to the idea of *generalized processor sharing* (GPS). In GPS, both classes have their own queue. Class i can always use a fraction ϕ_i of the total service rate C (where $\phi_1 + \phi_2 = 1$). If one of the classes does not use its full capacity, then the remaining capacity is allocated to the other class (thus making the service discipline work conserving). Note that the priority queue is a special case of GPS (choose $\phi_i = 1$ to give class i strict priority over the other class). One of the crucial engineering questions here is, for two user classes with given traffic arrival processes and performance targets, how should the weights be set?

Scope of this book. In view of the above, one could say that traffic management is all about the interrelationship between

(N) the network traffic offered (not only in terms of the average imposed load, but also in terms of its fluctuations, summarized in a certain arrival process);

(R) the amount of network resources available (link capacity, buffers, etc.);

(P) the performance level achieved.

With this interrelationship in mind, we conclude that there are three indispensable prerequisites for appropriate traffic management.

In the first place, we should have accurate traffic models at our disposal (i.e., N). Part A of this book is devoted to a class of models that has proven to be suitable in the context of communication networks: Gaussian traffic processes. An interesting feature of this class is that it is highly versatile, as it covers a broad class of correlation structures. We introduce this class and provide a number of generic properties. Then we explain why Gaussian models are likely to be an adequate statistical descriptor, and how this can be empirically verified. We also present a number of standard Gaussian models that are used throughout the book.

Secondly, we show in part B how to assess the performance of the network, for a given Gaussian traffic model, and for given amounts of available resources (i.e., (N,R) $\mapsto$ P). In other words, we analyze Gaussian queues, i.e., queues with Gaussian input. It turns out that only for a very limited subclass of inputs exact analysis is possible, and this explains why we resort to asymptotics. We present and explain several asymptotic results. Emphasis is on the so-called many-sources framework, which is an asymptotic regime in which the number of users grows large (where the traffic streams generated by these users have more or less similar statistical properties), and where the resources are scaled accordingly. Single queues are relatively easy to deal with in this framework, but we also focus on problems that are significantly harder, such as the analysis of a tandem queue, and a queue operating under GPS.

The final subject of the book is how these Gaussian queues can be used for traffic management purposes. Essentially, these problems all amount to questions of the type (N,P) $\mapsto$ R: given a traffic model and some performance target, how much resources are needed? Specific attention will be paid to link dimensioning in the single queue, the weight setting problem in generalized processor sharing, and bandwidth trading.

Bibliographical notes

This book focuses on large deviations for Gaussian queues, with applications to communication networking. There is a vast body of related literature, which we will cite at several occasions. Here, we briefly list a number of textbooks that can be used as background.

The literature on performance analysis is vast, and the key journals include *IEEE/ACM Transactions on Networking*, *Computer Networks*, and *Performance Evaluation*. A textbook that gives an excellent survey on performance evaluation techniques is by Roberts *et al.* [253], albeit with a focus on somewhat out-of-date technologies. We also recommend the book by Kurose and Ross [167], and the classical book by Bertsekas and Gallager [32].

There are several strong textbooks on queuing theory – without attempting to provide an exhaustive list, here we mention the books by Baccelli and Brémaud [17], Cohen [52], Prabhu [246], and Robert [250]. The beautiful survey by Asmussen [13] deserves some special attention, as it gives an excellent account of the state of the art on many topics in queuing theory. The leading journal in queuing is *Queueing Systems*, but there are many nice articles scattered over several other journals (including *Advances in Applied Probability*, *Journal of Applied Probability* and *Stochastic Models*).

During the last two decades a number of books on large deviations appeared with a focus on applications in performance and networking. In this context we mention the book by Bucklew [42] as a nice introduction to large deviations and the underlying intuition. The book by Shwartz and Weiss [267] is technically considerably more demanding, but the reader's efforts pay off when working through a beautiful series of appealing examples. Interestingly, Chang [46] connects deterministic network calculus methods with large deviations techniques. The book that is perhaps most related to the present book is Ganesh, O'Connell, and Wischik [109]. Also there the emphasis is on the application of large-deviations techniques in a queuing setting, albeit without focusing on Gaussian inputs, and without applying it (explicitly) in a communication networks context.

Apart from these books, there are a number of books on large deviations, but without a focus on queuing. Ellis [91] approaches large deviations from the angle of statistical mechanics, whereas in Dupuis and Ellis [87] control-theoretic elements appear. Perhaps the most complete, rigorous introductory book is by Dembo and Zeitouni [72]. Other useful textbooks include Deuschel and Stroock [75] and den Hollander [132]. Articles on large deviations appear in a broad variety of journals; besides the Applied Probability journals mentioned above, this also includes *Stochastic Processes and their Applications* and *Annals of Applied Probability*.

Part A: Gaussian traffic and large deviations

The first part of this book is of an introductory nature. It defines the basic concepts used throughout the book, by focusing on two topics. First, we introduce the notion of Gaussian traffic, and we argue why, under rather general circumstances, the Gaussian model offers an accurate description of network traffic. The second topic is large deviations: our main tool to probabilistically analyze rare events.

Large deviations for Gaussian queues M. Mandjes

Chapter 2

The Gaussian source model

In the introduction we argued that the traffic model is one of the cornerstones of performance evaluation of communication networks: it is a crucial building block in queuing models. This chapter introduces the class of traffic models that is studied in this book: the Gaussian traffic model. The main goal is to (qualitatively) argue why we feel that the Gaussian model is particularly suitable for modeling traffic streams in communication networks.

In Section 2.1, we identify a number of 'desirable properties' that a network traffic model should obey. After introducing some notation and some preliminaries on Gaussian random variables in Section 2.2, we define the Gaussian traffic model in Section 2.3. A few generic examples of Gaussian inputs are described in Section 2.4, viz. fractional Brownian motion and integrated Ornstein–Uhlenbeck; in this section also, a number useful concepts are introduced, most notably that of long-range dependence. Section 2.5 describes a number of other Gaussian source models that are used throughout the book; they are the Gaussian counterpart of 'classical' input models. Finally in Section 2.6 we return to the (qualitative) requirements stated in Section 2.1, and show that some Gaussian models are very well in line with these.

2.1 Modeling network traffic

There is a huge collection of traffic processes available from the literature, each having its own features. In general, an arrival process is an infinitely dimensional object $(A(t), t \in \mathbb{R})$, where $A(t)$ denotes the amount of traffic generated in time interval $[0, t)$, for $t > 0$; $(A(t), t \in \mathbb{R})$ is sometimes referred to as the *cumulative work process*. It is noted that $A(-t)$ is to be interpreted as the negative of the amount of traffic generated in $(-t, 0]$. Usually it is assumed that an arrival process

---

Large deviations for Gaussian queues M. Mandjes

is nondecreasing (as traffic cannot be negative); later we argue that this assumption is merely a technicality, and that it is, in many situations, not necessary. We also define, for $s < t$, the work that has arrived in time window $[s, t)$ as $A(s, t)$ (so that $A(s, t) = A(t) - A(s)$).

It is clear that some arrival processes fit better to 'real' network traffic than others. In this section, we list a number of properties that network traffic usually obeys. In the sequel, we use these properties as the motivation for our choice of Gaussian traffic models.

1. *Stationarity.* In quite general circumstances, it can be assumed that the traffic arrival process is stationary, at least over periods up to, say, one or more hours (over longer periods this usually does not hold–during a period of a day we often see the common diurnal pattern). In mathematical terms this stationarity means that the cumulative process $(A(t), t \in \mathbb{R})$ has *stationary increments*, i.e., in distribution,

$$A(s, t) = A(0, t - s) \quad \text{for all} \quad s < t.$$

In other words, the distribution of $A(s, t)$ is determined only by the *length* of the corresponding time interval (i.e., $t - s$), and not by the *position* of the interval.

2. *High aggregation level.* A common feature in modern communication networks is that the input stream of each node usually consists of the superposition of a large number of individual streams. It is clear that this is the case at the core links of networks, where the resources are shared by thousands of users, but even at the access of communication networks, the contributions of a substantial number of users is aggregated (at least in the order of tens).

Besides the fact that one can safely focus on relatively high aggregates, one can also argue that the behavior of a substantial part of the user population can be assumed homogeneous (i.e., obeying the same statistical law), as many users run the same applications over the network. In any case, one can usually subdivide the user population into a number of classes, within which the users are (nearly) homogeneous.

3. *Extreme irregularity of the traffic rate.* Measurements often indicate that the traffic rate exhibits extreme irregularity at a wide variety of timescales, including very small timescales. This 'burstiness' could be modeled by imposing the requirement that the instantaneous traffic rate process

$$R(t) := \lim_{s \uparrow t} \frac{A(s, t)}{t - s}$$

behaves irregularly (it could have nondifferentiable trajectories, for instance).

4. *Strong positive correlations on a broad range of timescales.* In the early 1990s, measurements, most notably those performed at Bellcore and at AT&T, gave the impression that network traffic exhibits significant positive correlation

on a broad range of timescales. A long series of measurement studies, performed at various networking environments, followed. These studies yielded convincing empirical evidence that network traffic is, under rather general circumstances, *long-range dependent*. We choose to postpone giving a precise definition of this notion, but on an intuitive level it means that the variance of $A(t)$ grows superlinearly in t. Traditional 'Markovian' traffic models had the implicit underlying assumption that this variance grows, at least for t large, linearly, and hence those models could not be used anymore.

2.2 Notation and preliminaries on Gaussian random variables

Let us give a few remarks on the notation used throughout this book, which we have tried to keep as light as possible. Here we introduce a number of general concepts.

- We denote the mean and variance of a random variable X by $\mathbb{E}X$ and $\mathbb{V}\text{ar}X$, respectively; recall that $\mathbb{V}\text{ar}X = \mathbb{E}(X^2) - (\mathbb{E}X)^2$. Also,

  $$\mathbb{C}\text{ov}(X, Y) := \mathbb{E}(XY) - (\mathbb{E}X)(\mathbb{E}Y)$$

  denotes the covariance between X and Y.

- $X =_{\rm d} Y$ indicates that the random variables X and Y are equally distributed.

- $X =_{\rm d} \mathcal{N}(\mu, \sigma^2)$ means that X is normally distributed with mean μ and variance σ^2. In other words, X has density

  $$f_{\mu,\sigma^2}(x) := \frac{1}{\sqrt{2\pi\sigma^2}} \exp\left(-\frac{1}{2}\left(\frac{(x-\mu)^2}{\sigma^2}\right)\right).$$

  We often use the notation $\Phi_{\mu,\sigma^2}(x)$ for the distribution function of a $\mathcal{N}(\mu, \sigma^2)$-distributed random variable:

  $$\Phi_{\mu,\sigma^2}(x) := \int_{-\infty}^{x} f_{\mu,\sigma^2}(y)\,{\rm d}y.$$

  This distribution function can be rewritten in terms of the distribution function of a standard normal random variable:

  $$\Phi_{\mu,\sigma^2}(x) = \mathbb{P}\left(\mathcal{N}_{\mu,\sigma^2} < x\right) = \mathbb{P}\left(\frac{\mathcal{N}_{\mu,\sigma^2} - \mu}{\sigma} < \frac{x-\mu}{\sigma}\right) = \Phi_{0,1}\left(\frac{x-\mu}{\sigma}\right).$$

  Throughout the book, we often abbreviate $\Phi_{0,1}(\cdot)$ as $\Phi(\cdot)$.

- We say that X has a multivariate normal distribution (of dimension $d \in \mathbb{N}$) with d-dimensional mean vector μ and (nonsingular) $d \times d$ covariance matrix Σ if X has density

$$f(x_1, \ldots, x_d) = \frac{1}{\sqrt{2\pi \det(\Sigma)}} \exp\left(-\frac{1}{2}(x - \mu)\mathrm{T}\Sigma^{-1}(x - \mu)\right),$$

where $\det(\Sigma)$ denotes the (non-zero) determinant of the matrix Σ.

- In the setting of multivariate normal random variables, conditional distributions can be given through elegant formulas. Consider for ease the case of (X, Y) being bivariate normal. The random variable $(X \mid Y = y)$, for some $y \in \mathbb{R}$ is then (univariate) normal with mean

$$\mathbb{E}(X \mid Y = y) = \mathbb{E}X + \frac{\mathbb{C}\mathrm{ov}(X, Y)}{\mathbb{V}\mathrm{ar}Y} \cdot (y - \mathbb{E}Y) \tag{2.1}$$

and variance

$$\mathbb{V}\mathrm{ar}(X \mid Y = y) = \mathbb{V}\mathrm{ar}X - \frac{(\mathbb{C}\mathrm{ov}(X, Y)^2}{\mathbb{V}\mathrm{ar}Y}; \tag{2.2}$$

notice that, interestingly, the conditional variance does not depend on the condition (i.e., the value y).

2.3 Gaussian sources

Now that we have, from Section 2.1, a list of properties that our traffic model should satisfy, this section introduces a versatile class of arrival processes: the so-called Gaussian sources. The aim of the remainder of this chapter is to show that certain types of Gaussian sources meet all the properties identified above.

For a Gaussian source, the entire probabilistic behavior of the cumulative work process can be expressed in terms of a mean traffic rate and a variance function. The mean traffic rate μ is such that $\mathbb{E}A(s,t) = \mu \cdot (t - s)$, i.e., the amount of traffic generated is proportional to the length of the interval. The variance function $v(\cdot)$ is such that $\mathbb{V}\mathrm{ar}A(s,t) = v(t - s)$; in particular $\mathbb{V}\mathrm{ar}A(t) = v(t)$.

Definition 2.3.1 Gaussian source. *$A(\cdot)$ is a Gaussian process with stationary increments, if for all $s < t$,*

$$A(s,t) =_{\mathrm{d}} \mathcal{N}(\mu \cdot (t - s), v(t - s)).$$

We say that $A(\cdot)$ is a Gaussian source. *We call a Gaussian source* centered *if, in addition, $\mu = 0$.*

The fact that the sources introduced in Definition 2.3.1 have stationary increments is an immediate consequence of the fact that the distribution of $A(s,t)$ just depends on the length of the time window (i.e., $t-s$), and not on its position.

The variance function $v(\cdot)$ fully determines the correlation structure of the Gaussian source. This can be seen as follows. First notice, assuming for ease $0<s<t$, that $\mathbb{C}\mathrm{ov}(A(s),A(t)) = \mathbb{V}\mathrm{ar}A(s) + \mathbb{C}\mathrm{ov}(A(0,s),A(s,t))$. Then, using the standard property that

$$\mathbb{V}\mathrm{ar}A(0,t) = \mathbb{V}\mathrm{ar}A(0,s) + 2\,\mathbb{C}\mathrm{ov}(A(0,s),A(s,t)) + \mathbb{V}\mathrm{ar}A(s,t),$$

we find the useful relation

$$\Gamma(s,t) := \mathbb{C}\mathrm{ov}(A(s),A(t)) = \frac{1}{2}(v(t)+v(s)-v(t-s)).$$

Indeed, knowing the variance function, we can compute all covariances. In particular $(A(s_1),\ldots,A(s_d))\mathrm{T}$ is distributed d-variate normal, with mean vector $(\mu s_1,\ldots,\mu s_d)\mathrm{T}$ and covariance matrix Σ, whose (i,j)th entry reads

$$\Sigma_{ij} = \Gamma(s_i,s_j), \quad i,j=1,\ldots,d.$$

The class of Gaussian sources with stationary increments is extremely rich, and this intrinsic richness is best illustrated by the multitude of possible choices for the variance function $v(\cdot)$. In fact, one could choose any function $v(\cdot)$ that gives rise to a positive semi-definite covariance function:

$$\sum_{s,t\in S} \alpha_s \mathbb{C}\mathrm{ov}(A(s),A(t))\alpha_t \geq 0,$$

for all $S\subseteq\mathbb{R}$, and $\alpha_s\in\mathbb{R}$ for all $s\in S$. The following example shows a way to find out whether a specific function can be a variance function.

Exercise 2.3.2 Consider increasing variance functions $v(\cdot)$ of the type $v(t)=t^{\alpha}$. Prove that necessarily $\alpha\in(0,2]$.

Solution. Notice that the fact that $v(\cdot)$ be increasing yields that $\alpha>0$. The fact that $v(\cdot)$ should correspond to a positive semi-definite covariance function rules out $\alpha>2$. Cauchy–Schwartz inequality implies that the correlation coefficient lies between -1 and 1; when applying this to $A(0,t)$ and $A(t,2t)$ we obtain

$$\frac{\mathbb{C}\mathrm{ov}(A(0,t),A(t,2t))}{\mathbb{V}\mathrm{ar}A(0,t)} \in [-1,1],$$

or, after simple algebraic manipulations, $v(2t)/2v(t)\in[0,2]$. Now inserting $v(t)=t^{\alpha}$ immediately leads to $\alpha\leq 2$.

In fact, the above reasoning indicates that $\alpha \in (1,2)$ corresponds to positive correlation, where the extreme situation $\alpha = 2$ can be regarded as 'perfect positive correlation' ($A(ft) = fA(t)$, for some $f \geq 0$); then the correlation coefficient is 1. On the other hand, $\alpha = 1$ relates to the complete lack of correlation (i.e., $\mathbb{C}\text{ov}(A(0,s), A(s,t)) = 0$, with $s < t$); in this case the correlation coefficient equals 0. Finally, if $\alpha \in (0,1)$, then there is negative correlation. ◇

2.4 Generic examples–long-range dependence and smoothness

This section highlights two basic classifications of Gaussian sources. These classifications can be illustrated by means of two generic types of Gaussian sources, which we also introduce in this section and which will be used throughout the book. The first directly relates to Exercise 2.3.2.

Definition 2.4.1 Fractional Brownian motion (or fBm). *A* fractional Brownian motion *source has variance function $v(\cdot)$ characterized by $v(t) = t^{2H}$, for an $H \in (0,1)$. We call H the* Hurst parameter.

The case with $H = 1/2$ is known as (ordinary) *Brownian motion.* In this case it is well known that the increments are independent, which is in line with the lack of correlation observed in Exercise 2.3.2.

Definition 2.4.2 Integrated Ornstein–Uhlenbeck (or iOU). *An* integrated Ornstein–Uhlenbeck *source has variance function $v(\cdot)$ characterized by $v(t) = t - 1 + e^{-t}$.*

Long-range dependence. The first way of classifying Gaussian sources relates to the correlation structure on long timescales: we are going to distinguish between srd sources and lrd sources.

To this end, we first introduce the notion of correlation on timescale t, for intervals of length ϵ. With $t \gg \epsilon > 0$, it is easily seen that

$$\mathbb{C}(t,\epsilon) := \mathbb{C}\text{ov}(A(0,\epsilon), A(t,t+\epsilon)) = \frac{1}{2}(v(t+\epsilon) - 2v(t) + v(t-\epsilon)).$$

For ϵ small, and $v(\cdot)$ twice differentiable, this looks like $\epsilon^2 v''(t)/2$. This argument shows that the 'intensity of the correlation' is expressed by the second derivative of $v(\cdot)$: 'the more convex (concave, respectively) $v(\cdot)$ at timescale t, the stronger the positive (negative) dependence between traffic sent 'around time 0' and traffic sent 'around time t'.

The above observations can be illustrated by using the generic processes fBm and iOU. As $v''(t) = (2H)(2H-1)t^{2H-2}$, we see that for fBm the correlation is positive when $H > \frac{1}{2}$ (the higher the H, the stronger this correlation; the larger the t, the weaker this correlation), and negative when $H < \frac{1}{2}$ (the lower the H, the stronger this correlation; the larger the t, the weaker this correlation), in line with the findings of Exercise 2.3.2. It is readily checked that for iOU $v''(t) = e^{-t}$. In other words: the correlation is positive, and decreasing in t.

Several processes could exhibit positive correlation, but the intensity of this correlation can vary dramatically; compare the (fast!) exponential decay of $v''(t)$ for iOU traffic with the (slow!) polynomial decay of $v''(t)$ for fBm traffic. The following definition gives a classification.

Definition 2.4.3 Long-range dependence. *We call a traffic source* long-range dependent *(lrd), when the covariances* $\mathbb{C}(k, 1)$ *are nonsummable:*

$$\sum_{k=1}^{\infty} \mathbb{C}(k, 1) = \infty,$$

and short-range dependent *(srd) when this sum is finite.*

Turning back to the case of fBm, with variance function given by $v(t) = t^{2H}$, it is easily checked that

$$\lim_{k\to\infty} \frac{\mathbb{C}(k, 1)}{k^{2H-2}} = \frac{1}{2} \cdot \lim_{k\to\infty} \frac{(1 + 1/k)^{2H} - 2 + (1 - 1/k)^{2H}}{1/k^2} = \frac{1}{2} \cdot v''(1).$$

This entails that we have to check whether k^{2H-2} is summable or not. We conclude that Gaussian sources with this variance function are lrd iff $2H > 1$, i.e., whenever they belong to the positively correlated case.

For iOU we have that

$$\mathbb{C}(k, 1) = \frac{1}{2}\left(e^{-k-1} - 2e^{-k} + e^{-k+1}\right),$$

which is summable. This implies that, according to Definition 2.4.3, iOU is srd.

Smoothness. A second criterion to classify Gaussian processes is based on the level of smoothness of the sample paths. We coin the following definition.

Definition 2.4.4 Smoothness. *We call a Gaussian source* smooth *if, for any* $t > 0$,

$$\lim_{\epsilon\downarrow 0} \frac{\mathbb{C}\mathrm{ov}(A(0, \epsilon), A(t, t + \epsilon))}{\sqrt{\mathbb{V}\mathrm{ar}(A(0, \epsilon))\mathbb{V}\mathrm{ar}(A(t, t + \epsilon))}} = \lim_{\epsilon\downarrow 0} \frac{\mathbb{C}(t, \epsilon)}{v(\epsilon)} \neq 0,$$

and nonsmooth *otherwise.*

An fBm source is nonsmooth, as is readily verified:

$$\lim_{\epsilon\downarrow 0} \frac{\mathbb{C}(t, \epsilon)}{v(\epsilon)} = \lim_{\epsilon\downarrow 0} \frac{1}{2}\epsilon^{2-2H} v''(t) = 0,$$

for any $t > 0$ and $H \in (0, 1)$. On the other hand, the iOU source is smooth, as, for any $t > 0$, applying that $2v(\epsilon)/\epsilon^2 \to 1$ as $\epsilon \downarrow 0$,

$$\lim_{\epsilon \downarrow 0} \frac{\mathbb{C}(t, \epsilon)}{v(\epsilon)} = v''(t) = e^{-t} > 0.$$

Generally speaking, one could say that Gaussian sources are smooth if there is a notion of a *traffic rate*. The following exercise gives more insight into this issue.

Exercise 2.4.5 Determine the distribution of the instantaneous traffic rate process

$$R(t) := \lim_{s \uparrow t} \frac{A(s, t)}{t - s},$$

for iOU; $t > 0$. And what happens in case of fBm?

Solution. For (centered) iOU, $R(t)$ has a normal distribution with mean 0 and variance

$$\lim_{s \uparrow t} \mathbb{V}\text{ar} \left(\frac{A(s, t)}{t - s} \right) = \lim_{s \uparrow t} \frac{\mathbb{V}\text{ar} A(t - s)}{(t - s)^2} = \frac{1}{2}.$$

It can also be verified that $\mathbb{C}\text{ov}(R(0), R(t)) = e^{-t}$. In other words, the rate process is again Gaussian.

In case of fBm, $\mathbb{V}\text{ar} R(t) = \infty$ for all $t > 0$. ◇

The above findings can be interpreted in the following, more concrete way. One could say that for fBm there is full independence of the direction in which the process is moving (in line with the nondifferentiable nature of the sample-paths for fBm). For iOU there is some positive dependence between the rates (but this dependence vanishes fast, viz. at an exponential rate).

2.5 Other useful Gaussian source models

The example of fBm in the previous section may have left the impression that nonsmoothness is a necessary condition for long-range dependence. This is by no means true. In this section we present a number of other useful Gaussian processes, including one that is smooth and lrd at the same time. The common feature of these processes is that they are the so-called *Gaussian counterpart* of well-known, classical (non-Gaussian) arrival processes. In this context, we say that an arrival process $(A(t), t \in \mathbb{R})$ (with stationary increments) has the Gaussian counterpart $(\overline{A}(t), t \in \mathbb{R})$ if $\overline{A}(\cdot)$ is Gaussian and, in addition, $\mathbb{E}A(t) = \mathbb{E}\overline{A}(t)$ and $\mathbb{V}\text{ar}A(t) = \mathbb{V}\text{ar}\overline{A}(t)$ for all t. Generally speaking, $\overline{A}(\cdot)$ inherits the correlation structure of $A(\cdot)$.

A. The Gaussian counterpart of the Poisson stream. Let jobs of size 1 arrive according to a Poisson process with rate $\lambda > 0$. Then it is well known that $A(t)$, denoting the amount of traffic generated in an interval of length t, has a Poisson distribution with mean λt (and hence also variance λt). As a consequence, the Gaussian counterpart of this model is (a scaled version of) Brownian motion.

B. The Gaussian counterpart of the M/G/∞ input model. In the M/G/∞ input model, jobs arrive according to a Poisson process of rate λ, stay in the system during some random time D (where the durations of the individual jobs constitute a sequence of i.i.d. random variables), and transmit traffic at rate, say, 1 while in the system. Clearly, this model does not correspond to a Gaussian source, but of course we could approximate it by a Gaussian source with the same mean and correlation structure.

It follows immediately that the mean rate μ of this Gaussian counterpart should be chosen as $\mu = \lambda \mathbb{E}D$, assuming that D has a finite mean, say δ. The variance is somewhat harder to compute. With $A(t)$ denoting the amount of traffic generated by the M/G/∞ input model in an interval of length t, we have that $A(t)$ can be decomposed into the contribution of the M jobs that were already present at the start of the interval (say, time 0), and the contribution of the N_t jobs arriving in $(0, t)$:

$$A(t) =_{\rm d} \sum_{i=0}^{M} X_i(t) + \sum_{i=0}^{N_t} Y_i(t); \tag{2.3}$$

here $X_i(t)$ is the amount of traffic generated by an arbitrary job that was present at time 0, whereas $Y_i(t)$ denotes the amount of traffic of an arbitrary job arriving in $(0, t)$. Observe that these two sums in the right-hand side of Equation (2.3) are independent. In more detail,

- standard theory on M/G/∞ queues says that M has a Poisson distribution with mean $\lambda\delta$, and the $X_i(t)$ are i.i.d. (as a random variable $X(t)$), independently of M;
- on the other hand, N_t is Poisson with mean λt, and, again, the $Y_i(t)$ are i.i.d. (as some random variable $Y(t)$), independently of N_t.

Recalling the property that, with Z_i i.i.d. (as a random variable Z), and N Poisson with mean ν (independent of the Z_i), we have that

$$\mathbb{V}\mathrm{ar}\left(\sum_{i=1}^{N} Z_i\right) = \nu\, \mathbb{V}\mathrm{ar} Z,$$

we find that we are left with finding $\mathbb{V}\mathrm{ar} X(t)$ and $\mathbb{V}\mathrm{ar} Y(t)$.

To this end, we wonder how $X(t)$ and $Y(t)$ are distributed. Let $f_D(\cdot)$ $(F_D(\cdot))$ be the density (distribution function) of D. In the sequence we need the notion of the so-called *integrated tail* (or residual lifetime) of D, which is a random variable that we denote by D^r. The density of this D^r is given by [13, Ch. 5]

$$f_{D^r}(t) = \frac{1 - F_D(t)}{\delta};$$

the fact that $\mathbb{E}D \equiv \delta = \int_0^\infty (1 - F_D(t))\,\mathrm{d}t$ implies that this is indeed a density. $F_{D^r}(\cdot)$ is the distribution function corresponding to $f_{D^r}(\cdot)$.

First consider $X(t)$. It is standard result that the remaining time the job stays in the system for a time that has density $f_{D^r}(\cdot)$. If this remaining time is $s < t$, then $X(t) = s$' otherwise it is t. In other words, the kth moment of $X(t)$ equals

$$\mathbb{E}(X(t))^k = \int_0^t s^k f_{D^r}(s)\,\mathrm{d}s + t^k(1 - F_{D^r}(t)).$$

This immediately yields a formula for $\mathbb{V}\mathrm{ar}X(t)$.

Then focus on $Y(t)$. It is well known that the epoch the job arrives is uniformly distributed on $(0, t)$. With a similar reasoning as above, where the variable u corresponds to the epoch the job arrives, we find that

$$\mathbb{E}(Y(t))^k = \int_0^t \frac{1}{t}\left((t-u)^k(1 - F_D(t-u)) + \int_0^{t-u} s^k f_D(s)\,\mathrm{d}s\right)\mathrm{d}u.$$

Again, this enables computation of $\mathbb{V}\mathrm{ar}Y(t)$.

Taking all terms together yields

$$\begin{aligned}\mathbb{V}\mathrm{ar}A(t) = \lambda\delta\left(\int_0^t s^2 f_{D^r}(s)\,\mathrm{d}x + t^2(1 - F_{D^r}(t))\right)\\ + \lambda\left(\int_0^t (t-u)^2(1 - F_D(t-u))\,\mathrm{d}u\right.\\ \left. + \int_0^t\int_u^t (s-u)^2 f_D(s-u)\,\mathrm{d}s\,\mathrm{d}u\right);\end{aligned} \tag{2.4}$$

see also [197].

Exercise 2.5.1 (i) Let D be exponentially distributed with mean δ. Show that

$$\mathbb{V}\mathrm{ar}A(t) = 2\lambda\delta^3\left(\frac{t}{\delta} - 1 + \exp\left(-\frac{t}{\delta}\right)\right). \tag{2.5}$$

Compare this with the variance function of iOU, and conclude that iOU is (up to a scaling) the Gaussian counterpart of the M/G/∞ input model with exponential jobs.

(ii) Let D have a Pareto distribution, i.e., $F_D(t) = 1 - (1/(1+t))^\alpha$. To have that $\delta < \infty$, we assume $\alpha > 1$. Verify that

$$v(t) = \frac{2\lambda}{(3-\alpha)(2-\alpha)(1-\alpha)}\left(1 - (t+1)^{3-\alpha} + (3-\alpha)t\right), \tag{2.6}$$

with $\alpha = (1+\delta)/\delta$, excluding $\delta = 1$ or $\frac{1}{2}$.

Solution. The claims follow after straightforward calculus. ◇

Exercise 2.5.2 (i) Show that the Gaussian counterpart of the M/G/∞ input model is smooth.

(ii) Assuming D exponential, is the Gaussian counterpart lrd? Assuming D as Pareto, is the Gaussian counterpart lrd?

Solution. (i) Using standard rules for differentiation of integrals,

$$v'(t) = \lambda\delta \cdot 2t\,(1 - F_{D^r}(t)) + \lambda \int_0^t 2(t-u)(1 - F_D(t-u))\,\mathrm{d}u$$

and hence $v'(0) = 0$. Similarly,

$$\begin{aligned} v''(t) &= 2\lambda\delta\,(1 - F_{D^r}(t) - t f_{D^r}(t)) + 2\lambda \int_0^t (1 - F_D(t-u) - (t-u) f_D(t-u))\,\mathrm{d}u \\ &= 2\lambda \int_t^\infty (1 - F_D(s))\,\mathrm{d}s. \end{aligned}$$

Hence, $v''(0) = 2\lambda\delta < \infty$. As a result,

$$\lim_{\epsilon \downarrow 0} \frac{\mathbb{C}(t,\epsilon)}{v(\epsilon)} = \lim_{\epsilon \downarrow 0} \frac{\epsilon^2 v''(t)/2}{\epsilon^2 v''(0)} > 0.$$

(ii) When D is exponential, the Gaussian counterpart is srd (recall that we know from Exercise 2.5.1 that it has the same correlation structure as iOU). From Equation (2.6) it is readily seen that for Pareto D the Gaussian counterpart is lrd iff $\alpha \in (1,2)$. For these parameters, $\mathbb{V}\mathrm{ar}A(t)$ grows superlinearly in t, whereas for $\alpha \geq 2$, it grows essentially linearly. ◇

C. The Gaussian counterpart of the purely periodic stream. Many networking applications inject information packets at a (more or less) constant rate. Perhaps the most prominent example is *voice* traffic: sound is put into equally sized packets, which are fed into the network after constant time intervals (say, every D units of time). In fact, the only random element in this traffic model is the *phasing*. More precisely, when considering the first interval of length D, say $[0, D)$, it is obvious that exactly one packet will arrive; the epoch that this packet arrives, however, is uniformly distributed over $[0, D)$. From then on, the process is deterministic; if the packet arrived at $t \in [0, D)$, then packets also arrive at epoch $t + iD$, with integer i.

Now consider the Gaussian counterpart of this arrival process. It is not hard to check that (if packets have size 1) $\mu = 1/D$. Observe that $A(t)$ equals either $\lfloor t/D \rfloor$ or $\lceil t/D \rceil$. Note that for $t \in [0, D)$, $A(t)$ is 1 (0) with probability t/D $(1 - t/D$, respectively). This entails that

$$\mathbb{V}\text{ar}A(t) = (t \bmod D)(1 - t \bmod D),$$

where $t \bmod D$ denotes the fractional part of t/D, i.e., $t \bmod D = t/D - \lfloor t/D \rfloor$. We conclude that the Gaussian counterpart of the purely period stream has a variance function that equals (up to some scaling) $t(1 - t)$, repeated periodically.

With $B(t)$ denoting a Brownian motion, the process (defined for $t \in [0, 1]$)

$$\overline{B}(t) = (B(t) \mid B(1) = 0)$$

is usually referred to as a *Brownian bridge*. Using the formula (2.2) for conditional variances given in Section 2.2, we find that

$$\mathbb{V}\text{ar}\overline{B}(t) = \mathbb{V}\text{ar}B(t) - \frac{(\mathbb{C}\text{ov}(B(t), B(1))^2}{\mathbb{V}\text{ar}B(1)} = t - t^2 = t(1 - t);$$

Here, that $\mathbb{C}\text{ov}(B(t), B(1)) = t$ is also used. Observe that the shape of $\mathbb{V}\text{ar}\overline{B}(t)$ makes sense: (i) for small t it looks like the variance of $B(t)$ (i.e., t), as the effect of the condition $B(1) = 0$ will still be minimal; (ii) for t close to 1 it goes to 0.

We conclude from the above that the Gaussian counterpart of the periodic stream is a scaled Brownian bridge, repeated periodically.

D. Two-timescale models. In many applications in communication networks, a purely periodic traffic model is too crude. Often sources alternate between an on-mode and an off-mode, where during on-times packets are transmitted (as before) periodically. The typical example where such a model can be used is *voice with silence suppression*: a codec recognizes whether there is a voice signal to transmit or not, and if there is, then packets are generated in a periodic fashion. This type of models is called *two-timescale models*, as it combines a model that relates to the very short timescale (i.e., the timescale of the transmission of packets, which is in the order of, say, tens of milliseconds) with the timescale of a user alternating between an active and silent mode (in the order of seconds).

As said, within the on-time, packets (of fixed size, say, 1) are sent periodically; for ease, we normalize time such that the corresponding packet interarrival time is 1. The on-times (or *bursts*) are random variables with length T_{on}, which is a natural number. The off-times (or *silence* periods) are random variables with length T_{off}, which is also integer valued. We emphasize that–although T_{on} and T_{off} only attain values in $\mathbb{N}$–the arrival process is *not* a discrete-time process, as the 'phase' of the source is uniformly distributed on $[0, 1)$.

Whereas in the examples A, B, and C we computed the variance function of $A(t)$ in a direct way, we now follow an indirect approach. We compute the so-called *moment generating function* (mgf) of $A(t)$, i.e.,

$$\alpha_t(\theta) := \mathbb{E}\exp(\theta A(t)),$$

for $t > 0$ and $\theta \in \mathbb{R}$. From this mgf, the mean $\mathbb{E}A(t)$ (which equals μt) can be found through differentiation with respect to θ and letting θ go to 0:

$$\mathbb{E}A(t) = \left(\frac{\partial}{\partial\theta}\alpha_t(\theta)\right)\Big|_{\theta=0}.$$

Similarly, the second moment of $A(t)$ can be found by taking the second derivative, so that we find for the variance

$$\mathbb{V}\mathrm{ar}A(t) = \left(\frac{\partial^2}{\partial\theta^2}\alpha_t(\theta)\right)\Big|_{\theta=0} - \left(\left(\frac{\partial}{\partial\theta}\alpha_t(\theta)\right)\Big|_{\theta=0}\right)^2.$$

For ease we assume that $T_{\rm on}$ is geometric with probability p, i.e.,

$$\mathbb{P}(T_{\rm on} = k) = (1-p)^{k-1}p$$

(and even then the *residual* on-time has this distribution, due to the memoryless property of the geometric distribution). The off-time $T_{\rm off}$ (and the residual off-time) is assumed to be geometrically distributed with probability q.

Define, for $t > 0$, $t_m := t \bmod 1$. It is clear that $A(t_m) = 0$ when the source is off, whereas $A(t_m) = 1$ (0) with probability t_m ($1 - t_m$, respectively) when the source is on. In other words,

$$\mathbb{E}_1 e^{\theta A(t_m)} = t_m e^\theta + (1 - t_m), \qquad \mathbb{E}_0 e^{\theta A(t_m)} = 1,$$

with $\mathbb{E}_i(\cdot)$, for $i = 0, 1$, defined in a self-evident way. This leaves us with the task of computing $\mathbb{E}_i e^{\theta A(T)}$ for integer T. A straightforward argument immediately gives us the system of coupled difference equations

$$\mathbb{E}_1 e^{\theta A(T)} = (1-p)e^\theta \mathbb{E}_1 e^{\theta A(T-1)} + p\mathbb{E}_0 e^{\theta A(T-1)},$$

$$\mathbb{E}_0 e^{\theta A(T)} = qe^\theta \mathbb{E}_1 e^{\theta A(T-1)} + (1-q)\mathbb{E}_0 e^{\theta A(T-1)}.$$

This system can be solved in a standard way. Let $\lambda_+(\theta)$ and $\lambda_-(\theta)$ be the eigenvalues of the matrix

$$\begin{pmatrix} (1-p)e^\theta & p \\ qe^\theta & 1-q \end{pmatrix},$$

i.e., the conjugate square roots

$$\lambda_{+,-}(\theta) = \frac{1 - q + (1-p)e^\theta \pm \sqrt{(1 - q + (1-p)e^\theta)^2 + 4(p+q-1)e^\theta}}{2}.$$

Then, for $i = 0, 1$, we have $\mathbb{E}_i e^{\theta A(T)} = f_i(\theta)(\lambda_+(\theta))^T + (1 - f_i(\theta))(\lambda_-(\theta))^T$, where the $f_i(\theta)$ are determined by the initial conditions. They turn out to be

$$f_1(\theta) := \frac{(1-p)e^\theta + p - \lambda_-(\theta)}{\lambda_+(\theta) - \lambda_-(\theta)} \text{ and } f_0(\theta) := \frac{qe^\theta + 1 - q - \lambda_-(\theta)}{\lambda_+(\theta) - \lambda_-(\theta)}.$$

It is easily seen that in case of $p + q < 1$ there are positive correlations. Then both eigenvalues are positive. For $p + q = 1$ there is no correlation: every unit of time a packet is generated with probability $q(p+q)^{-1}$. In case of $p + q > 1$, there is a negative correlation structure; $\lambda_-(\theta)$ is negative. As said, the variance function can be found by differentiating $\mathbb{E}\exp(\theta A(t))$; this is done in Section 6.2, and it clearly shows the impact of the two timescales involved.

2.6 Applicability of Gaussian source models for network traffic

In the first section of this chapter we identified four requirements that we argued needed to be fulfilled by a 'good' model for network traffic. We now come back to these requirements, showing that particular Gaussian models (e.g., fBm) seem to be logical candidates.

1. *Stationarity.* The Gaussian source model, as introduced in Definition 2.3.1, has stationary increments, as desired. Indeed, the distribution of $A(s, t)$ depends on s and t only through their difference $t - s$.

2. *High aggregation level.* If the input $(A(t), t \in \mathbb{R})$ of a network element is the superposition of a large number n (more or less) i.i.d. arrival processes $(A_i(t), t \in \mathbb{R})$ with stationary increments, then one expects on the basis of central-limit type of arguments that, for $t \in \mathbb{R}$,

$$A(t) \approx_{\rm d} n\,\mathbb{E}A_i(1)\,t + \sqrt{n\,\mathbb{V}{\rm ar}A_i(t)}\,\mathcal{N}(0,1),$$

where $\mathcal{N}(0, 1)$ denotes a standard normal random variable. Denoting $\mathbb{E}A_i(1)$ by μ, and $\mathbb{V}{\rm ar}A_i(t)$ by $v(t)$, we conclude that $A(\cdot)$ is 'close to' a Gaussian process, where $A(t)$ has mean $n\mu t$ and variance $nv(t)$. In other words, the fact that the aggregation level is high, makes the Gaussian model well applicable.

The above has an interesting consequence. If $A(\cdot)$ is indeed Gaussian, then we can do 'as if' the individual sources $A_i(\cdot)$ are also Gaussian with mean μ and variance $v(t)$.

3. *Extreme irregularity of the traffic rate.* As we have seen above, the class of Gaussian source models nicely meets requirements I and II. Requirement III, however, is not fulfilled by all Gaussian sources. In fact, this requirement asks for traffic models that are nonsmooth. This leaves fBm as a candidate model; iOU gives differentiable paths, and is therefore less appropriate.

4. *Strong positive correlations on a broad range of timescales.* This property can be translated into requiring that the arrival process be lrd. Again, iOU does not fulfill this condition, but fBm (with $H > \frac{1}{2}$) does.

Negative traffic. One problem, at least on a conceptual level, of Gaussian source models is that they allow negative traffic, or, in other words, $A(t)$ is not necessarily nondecreasing. In fact, the probability that $A(t)$ is negative is

$$\mathbb{P}(A(t) < 0) = \mathbb{P}\left(\mathcal{N}(0, 1) < -\frac{\mu t}{\sqrt{v(t)}}\right).$$

This shows that the probability of negative traffic on timescale t is small when $\mu t/\sqrt{v(t)}$ is large. Particularly for μ and t large, this is usually the case – for instance, in case of fBm input $\mu t/\sqrt{v(t)}$ equals μt^{1-H}. In practical situations, we will typically have that, indeed, already for relatively small t, μt is orders of magnitude larger than $\sqrt{v(t)}$.

The fact that we allow negative traffic may seem unattractive at first sight, but we remark that similar problems appear in many other situations. Consider, for instance, a sequence of n i.i.d. coin flips $Z_1, \ldots, Z_n$, and identify '1' with head and '0' with tail. If the probability of head is p, it is fully acceptable to approximate $Z := \sum_{i=1}^{n} Z_i$ by

$$\overline{Z} := np + \sqrt{np(1-p)}\,\mathcal{N}(0, 1)$$

for n large (use the central limit theorem). It is clear however that Z attains only values in $\{0, \ldots, n\}$, whereas $\overline{Z}$ attains values in on $\mathbb{R}$. We come back to the issue negative traffic in Section 5.2.

The above, qualitative observations suggest that fBm (or perhaps another nonsmooth and lrd Gaussian process) could be a good candidate to describe, under quite general circumstances, network traffic. Of course, such a statement asks for thorough statistical verification, before it is used as input in performance analysis studies. One of the main goals of the next chapter is to illustrate techniques to do this verification.

The other issue that will be addressed in the next chapter relates to the central-limit type of arguments mentioned above: The conditions under which the superposition of i.i.d. on-off processes indeed converges to a Gaussian process is shown. We then explain a possible reason why it is so often observed that fBm gives a nice fit: if in addition it is assumed that the on- or off-times are heavy-tailed, and an appropriate rescaling of time is applied, then the limiting Gaussian process is fBm.

Bibliographical notes

The most complete introduction on Gaussian processes is by Adler [6], with emphasis on sample-path properties, such as boundedness and continuity. The setup is fairly abstract, and relies on concepts such as entropy and majorizing measures. Also, considerable attention is paid to the distribution of extreme values attained by Gaussian processes; we will use some of these results later on.

Some of the examples of Gaussian source models are taken from Addie, Mannersalo, and Norros [5] and Mandjes [185]. A textbook introduction to long-range dependence can be found in, for instance, Beran [25].

Chapter 3

Gaussian sources: validation, justification

The previous chapter introduced the concept of Gaussian sources. It also showed, on the basis of qualitative arguments, that the Gaussian source model could be a 'good' model to accurately describe network traffic. The next question is, of course, whether the use of these Gaussian sources, in the context of communication networking, can also be quantitatively justified.

The answer to this question is, fortunately, 'yes', and this chapter presents two types of support for this claim. In the first place, we describe how one can validate Gaussianity for real traffic traces. We present a collection of visualization techniques and statistical tests, and demonstrate how these can be performed. They have been applied to a large set of traffic traces, see, e.g., the recent measurement studies [102, 156, 211]. A common conclusion is that, under fairly general circumstances, it is reasonable to assume Gaussianity. We also present a brief summary of the findings of [102, 156, 211].

A few caveats are in place here.

- Of course there are many situations in which Gaussianity does not apply. Studies such as [102, 156, 211] however, provide explicit conditions under which the Gaussian model can serve, at least, as an accurate approximation. These conditions are typically in terms of a minimum and maximum timescale, and a minimum aggregation level required.
- It is also noted that $A(\cdot)$ is, in principle, an infinitely dimensional object. A test on Gaussianity should therefore formally correspond to 'any timescale'. The tests described in Section 3.1, however, focus on a single timescale (and can be extended to a finite set of timescales).

Large deviations for Gaussian queues M. Mandjes

The second type of support for Gaussianity, presented in this chapter, is based on central-limit type of arguments. In many situations, the traffic stream that is offered to a network element will be the aggregate of a considerable number of flows, and, by virtue of the central limit theorem, one could expect that this aggregate is approximately Gaussian. In Section 3.2, we present a series of illustrative results regarding this convergence:

- The first result is a central limit theorem for on-off arrival processes: the scaled superposition of on-off sources converges to a Gaussian process with some appropriate correlation structure.
- Then we describe a famous result that focuses on the special case of *heavy-tailed* on-off sources; interestingly, under a particular rescaling of time the limiting Gaussian process is fBm.

  This result can be seen as some sort of 'explanation' why fBm is a good traffic model. In practice, the aggregate process is the superposition of many user streams, where each user stream is approximately on-off. One could say that the on-times correspond to the retrieval of documents, while the off-times could be considered, for instance, as 'read times'. Measurement studies [58] have indicated that document sizes (and hence also the on-times) are heavy-tailed. This reasoning explains why one could expect that the aggregate stream resembles fBm.

We remark here that in Section 3.2 no formal proofs are included; this stochastic-process limit theory is a subject on its own.

3.1 Validation

As argued above, Gaussianity is expected on the basis of central-limit type of arguments: the sum of a large number of 'small' independent (or weakly dependent), statistically more or less identical, random variables (users) has an approximately normal (i.e., Gaussian) distribution. Thus, one can expect that an aggregated traffic stream consisting of contributions of many individual users may be modeled by a Gaussian process.

However, the above argumentation is rather imprecise. In the first place, it can be expected that the central limit reasoning does not apply to any timescale: on the timescale of transmission of (minimum size) packets, the traffic stream is always of the on-off type (either there is transmission at link speed, or silence) – which is obviously not Gaussian. Thus, apart from the number of users (referred to as 'vertical aggregation'), there should also be sufficient aggregation in time ('horizontal aggregation'). The necessity for some aggregation in both directions for traffic to be Gaussian, was pointed out in [156].

In this section we present a number of basic techniques that can be used to assess Gaussianity. Then, these can also be used to find the limitations of the Gaussian model, i.e., they may give the 'minimal aggregation level' (both horizontally and vertically) needed to safely assume Gaussianity.

QQ-plot. We first describe a visualization technique that is known as the QQ-plot method. This nonparametric method is used to assess whether two samples are taken from the same distribution. In our situation, we could use this method to verify whether it is reasonable to assume that, for some given timescale t, the observations, say $a_i := A(i\tau, i\tau + t)$ could stem from a normal distribution; to make the intervals $i\tau, i\tau + t$ nonoverlapping, assume that τ is chosen such that $t \leq \tau$. Suppose there are n observations.

First observe that (unbiased) estimates $\hat{\mu}$ and $\hat{v(t)}$ of the average and (sample) variance of the traffic rates in our traces can be determined in a straightforward fashion:

$$\hat{\mu} = \frac{1}{nt}\sum_{i=1}^{n} a_i, \quad \text{and} \quad \hat{v(t)} = \frac{1}{n-1}\sum_{i=1}^{n}(a_i - \hat{\mu})^2. \tag{3.1}$$

Note that the convergence of the estimator of the sample variance could be rather slow when traffic is lrd, which can be expected for real network traffic [25].

Now let $a_{(i)}$ be the ith order statistic of $(a_j)_{j=1}^n$, i.e., $a_{(i)} = a_j$ if there are $i-1$ values smaller than a_j, and $n-i$ larger than a_j (for ease assuming that there are no ties). Recall that $\Phi_{0,1}(\cdot)$ denotes the distribution function of a standard normal random variable, i.e., $\Phi_{0,1}(x) = \int_{-\infty}^{x}(1/\sqrt{2\pi})\exp(-y^2/2)\,dy$.

Exercise 3.1.1 Prove that the inverse distribution function of a $\mathcal{N}(\mu, \sigma^2)$ random variable is a linear transformation of the inverse distribution function of a standard normal random variable.

Solution. In self-evident notation,

$$\Phi_{\mu,\sigma^2}^{-1}(y) = \{x : \mathbb{P}(\mathcal{N}(\mu,\sigma^2) \leq x) = y\} = \left\{x : \mathbb{P}\left(\mathcal{N}(0,1) \leq \frac{x-\mu}{\sigma}\right) = y\right\}$$

$$= \{\mu + \sigma z : \mathbb{P}(\mathcal{N}(0,1) \leq z) = y\} = \mu + \sigma\Phi_{0,1}^{-1}(y).$$

Hence, the particular linear transformation has intercept μ and slope σ. ◇

Intuitively, the more the points

$$(q_i, a_{(i)}) := \left(\Phi_{\hat{\mu},\hat{v(t)}}^{-1}\left(\frac{i}{n+1}\right), a_{(i)}\right)$$

look like the line $x = y$, the better the fit between the sample and $\mathcal{N}(\hat{\mu}, \hat{v(t)})$; then the quantiles of the sample match with what could be expected if the distribution were indeed $\mathcal{N}(\hat{\mu}, \hat{v(t)})$. Exercise 3.1.1, however, implies that a simplification can be done: also the points

$$\left(\Phi_{0,1}^{-1}\left(\frac{i}{n+1}\right), a_{(i)}\right) \tag{3.2}$$

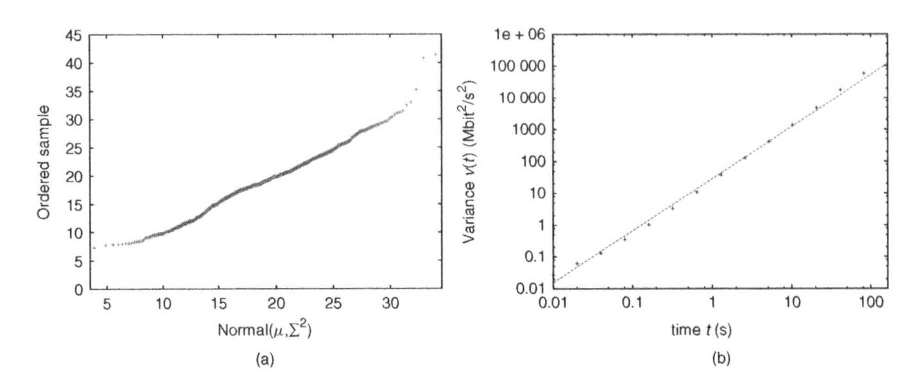

Figure 3.1: (a) is a QQ-plot consisting of 1800 data points. (b) depicts $\log \hat{v(t)}$ against $\log t$.

need to be on a straight line; the corresponding μ and $v(t)$ can be estimated from the QQ-plot.

The purpose of this section is to provide some basic tools for assessing Gaussianity–it is not our goal to claim that network traffic is Gaussian, nor to present explicit conditions under which Gaussianity holds (in terms of horizontal and vertical aggregation). Nevertheless, to illustrate the applicability of Gaussian models we include the following example. Figure 3.1(a) is a QQ-plot consisting of 1800 data points; these data points are taken from a 30-min trace, and each point corresponds to 1 s. The points are plotted against the quantiles of a Normal distribution, and hence the (almost) straight line indicates that the 1-s measurements follow approximately a normal distribution. For many other data sets, and other time-windows (i.e., different from 1 s), we usually see a similar QQ-plot; only for very small and very long time windows, the Gaussian source model does not seem to apply. The fact that the upper tail (and in our graph, the lower tail also) does not have a good match, seems to be a phenomenon that occurs more frequently, see [156].

Goodness-of-fit test. The QQ-plot being essentially a visual tool, there is need for more quantitative tests. An often-used tool is the so-called linear correlation coefficient:

$$\gamma_n = \frac{\sum_{i=1}^{n}(a_{(i)} - \hat{\mu})(q_i - \overline{q})}{\sqrt{\sum_{i=1}^{n}(a_{(i)} - \hat{\mu})^2 \sum_{i=1}^{n}(q_i - \overline{q})^2}};$$

here the q_i are the quantile of our model distribution (in our case, those corresponding to $\mathcal{N}(\hat{\mu}, \hat{v(t)})$), and $\overline{q} := n^{-1} \cdot \sum_{i=1}^{n} q_i$. From the Cauchy-Schwartz inequality, we know that γ_n is a number in the interval $[-1, 1]$; it has value 1 (-1) in case of perfect positive (negative) linear correlation, whereas γ_n close to 0 corresponds to independence (or 'weak dependence'). Unfortunately, it is hard to state general claims on the statistical meaning of a certain value of γ_n. Clearly an extremely high value of γ_n (say, 0.99) seems convincing support for Gaussianity, but does, for instance, $\gamma_n = 90\%$ suffice? Stated differently, it remains unclear what the critical value is, see also Section IV.C of [156].

Other tests. The *Kolmogorov-Smirnov test* is a well-accepted tool for testing whether a sample stems from a certain distribution. This test is often abused, as its quite restrictive conditions are frequently ignored. In principle, the test can be used only to test against a *given* distribution–in other words: it is not allowed to first estimate μ and $v(t)$, and to test then against $\mathcal{N}(\hat{\mu}, \widehat{v(t)})$. Also, the test assumes that the observations are independent, which clearly does not hold in our situation. Due to the lack of alternatives, and due to the fact that its p-values are nicely tabulated, still this test is often used. In [211], it is empirically verified that a high linear correlation coefficient indeed corresponds to a low Kolmogorov-Smirnov statistic (which means that Gaussianity is not rejected).

Estimation of H. So far, we concentrated on addressing the Gaussianity issue on a single timescale t. Of course, to make the arrival process Gaussian, the test should be performed on 'any' timescale. Experiments performed in a variety of networking environments indicate [102, 156, 210, 211] that there is usually Gaussianity on a substantial range of timescales, but not those below, say, 100 μs, nor those above, say, 1 h. The reason for these limitations is that, as mentioned before, on very short timescales there is either a transmission of a packet or not, which is clearly not Gaussian; on longer timescales stationarity does not apply anymore.

On the timescales where Gaussianity applies, it is often insightful to make a so-called variance-time plot [102]. Figure 3.1(b) plots $\log \widehat{v(t)}$ as a function of $\log t$. The fact that the resulting curve is (approximately) linear, expresses that $v(t)$ is of the form $\alpha\, t^{2H}$ for some positive α, H, i.e., is a (scaled) fBm. It can be concluded that in this case we have that the Hurst parameter H is about 0.82.

The above method to estimate H is all but statistically sound. The estimation of H is an art on its own. The survey [273] is an excellent source in this respect; see also [25].

Interestingly, the data in [102] suggests that often $\log v(t)$ is not linear, but *piecewise* linear (with two pieces). To shorter timescales, another H applies than to longer timescales: there is a weak dependence on short timescales, whereas a stronger dependence takes over on longer timescales. This could be modeled by the variance curve

$$v(t) = \max\{\alpha_1 t^{2H_1}, \alpha_2 t^{2H_2}\}; \tag{3.3}$$

remark that indeed $\log v(t)$ is piecewise linear. If one assumes, without loss of generality, that $H_1 < H_2$, then for small timescales (i.e., those below some critical timescale t_{crit}) H_1 is dominant: $v(t) = \alpha_1 t^{2H_1}$. On the other hand, for long timescales H_2 dominates: $v(t) = \alpha_2 t^{2H_2}$. It is readily checked that the critical timescale equals

$$t_{\text{crit}} = \left(\frac{\alpha_1}{\alpha_2}\right)^{1/(2H_2-2H_1)}.$$

Having verified long-range dependence (at least, in an empirical way), the next question is: what phenomenon causes network traffic to have this extremely bursty correlation structure? The next section shows that one of the potential causes of lrd is the aggregation of many users sending heavy-tailed documents over the network.

Bibliographical notes

Key articles on Gaussianity of network traffic are [102, 156, 211]. Interestingly, Fraleigh, Tobagi, and Diot [102] give an explicit indication of the aggregation level (viz. 50 Mbps) above which Gaussianity can be safely assumed. Kilpi and Norros [156] were the first to decouple horizontal and vertical aggregation, and gave a set of testing procedures. The limitations of Gaussianity were further explored in Van de Meent, Mandjes, and Pras [211], for an extensive set of traces corresponding to highly heterogeneous networking environments.

The breakthrough paper on statistical analysis of network traffic was doubtlessly Leland, Taqqu, Willinger, and Wilson [172], where the selfsimilar nature of Ethernet LAN traffic was discovered. There are dozens of other relevant traffic modeling papers; we mention here Juva, Susitaival, Peuhkuri, and Aalto [143], Ben Azzouna, Clérot, Fricker, and Guillemin [24], and Barakat, Thiran, Iannaccone, Diot, and Owezarski [21]. Interesting background reading on traffic modeling is Erramilli, Veitch, Willinger *et al.* [4, 97, 98].

3.2 Convergence of on-off traffic to a Gaussian process

Where the previous section presented statistical techniques to validate the Gaussianity assumption as well as (some) empirical evidence in favor of Gaussian source models, this section focuses on analytical support. The setting considered is the superposition of a large number of i.i.d. on-off sources (with general on- and off-times). It is explained that this aggregate converges, after a certain normalization, to a Gaussian source; this result can be regarded as a central limit theorem on the process level. In addition, when the on- or off-times are heavy-tailed, we show that the limiting Gaussian process is fBm, if time is rescaled in an appropriate way.

We consider n i.i.d. fluid sources. The traffic rate of each source alternates between on and off; during the on-times traffic is generated continuously at a (normalized) peak rate of 1. The activity periods constitute an i.i.d. sequence of random variables, each of them distributed as a random variable T_{on} with values in $\mathbb{R}_+$. The silence periods are also an i.i.d. sequence, distributed as a random variable T_{off} with values in $\mathbb{R}_+$. In addition, both sequences are mutually independent. Let T_{on} and T_{off} have distribution functions $F_{\text{on}}(\cdot)$ and $F_{\text{off}}(\cdot)$, respectively, and complementary distribution functions $\overline{F}_{\text{on}}(\cdot)$ and $\overline{F}_{\text{off}}(\cdot)$. Assuming that $\mu_{\text{on}} := \mathbb{E}(T_{\text{on}}) < \infty$ and $\mu_{\text{off}} := \mathbb{E}(T_{\text{off}}) < \infty$, we have that the mean input rate equals $n\mu$, with $\mu := \mu_{\text{on}}/(\mu_{\text{on}} + \mu_{\text{off}})$. Following [67], we impose the following regularity properties.

Assumption 3.2.1 *$F_{\text{on}}(\cdot)$ and $F_{\text{off}}(\cdot)$ are absolutely continuous with densities $f_{\text{on}}(\cdot)$ and $f_{\text{off}}(\cdot)$, respectively, such that both*

$$\lim_{t\downarrow 0^+} f_{\text{on}}(t) < \infty \text{ and } \lim_{t\downarrow 0^+} f_{\text{off}}(t) < \infty.$$

Let $\{\xi_i(t), t \in \mathbb{R}\}$, for $i = 1, \ldots, n$, denote the i.i.d. sequence of the stationary traffic rate processes, where $\xi_i(t) = 1$ (0) if source i is on (off) at time t. This stationary process has the property that is on at time 0 with probability μ; the time until it switches to the off-state T^r_{on} has density $\overline{F}_{\text{on}}(\cdot)/\mu_{\text{on}}$; then an off-time with density $f_{\text{off}}(\cdot)$ follows, and then an on-time with density $f_{\text{on}}(\cdot)$, etc. The situation that the source is off at time 0 is dealt with similarly. Let $R(\cdot)$ denote the covariance function of a generic source, i.e., $R(s) := \mathbb{C}\text{ov}(\xi_1(s+t), \xi_1(t))$.

Now introduce the aggregate traffic rate at time Ts of the n multiplexed sources, and the traffic generated in the interval $[Ts, Tt]$, with $s \le t$, $s, t \in \mathbb{R}$:

$$Z_{T,n}(s) := \sum_{i=1}^{n} \xi_i(Ts), \quad A_{T,n}(s,t) := \int_s^t Z_{T,n}(u)\,\mathrm{d}u.$$

Define $\{Z(t), t \in \mathbb{R}\}$ as a stationary centered *Gaussian* process having the same covariance structure as a single on-off source, i.e., $\mathbb{C}\text{ov}(Z(s+t), Z(t)) = R(s)$. We denote by $\{Z_T(s), s \in \mathbb{R}\}$ a stationary centered Gaussian process such that $\{Z_T(s), s \in \mathbb{R}\} =_{\text{d}} \{Z(Ts), s \in \mathbb{R}\}$. Observe that

$$R_T(s) := \mathbb{C}\text{ov}(Z_T(s+t), Z_T(t)) = R(Ts).$$

Furthermore, we write $A_T(s,t) := \int_s^t Z_T(u)\,\mathrm{d}u$.

Exercise 3.2.2 (i) Show that

$$\sigma(t) := \sqrt{\mathbb{V}\text{ar}(A_{1,1}(0,t))} = \sqrt{2\int_0^t \int_0^s R(u)\,\mathrm{d}u\mathrm{d}s}.$$

(ii) Verify that $A_T(0, t)$ is a centered Gaussian process with variance function

$$2\int_0^t \int_0^{Ts} \frac{R(u)}{T} \,du\,ds.$$

Solution. (i) This is shown as follows:

$$\begin{aligned}\mathbb{V}\mathrm{ar} A_{1,1}(0,t) &= \mathbb{V}\mathrm{ar}\left(\int_0^t Z_{1,1}(u)\mathrm{d}u\right) = \int_0^t \int_0^t \mathbb{C}\mathrm{ov}(Z_{1,1}(u), Z_{1,1}(s))\,\mathrm{d}u\,\mathrm{d}s \\ &= 2\int_0^t \int_0^s \mathbb{C}\mathrm{ov}(Z_{1,1}(u), Z_{1,1}(s))\,\mathrm{d}u\,\mathrm{d}s \\ &= 2\int_0^t \int_0^s R(s-u)\,\mathrm{d}u\,\mathrm{d}s = 2\int_0^t \int_0^s R(u)\mathrm{d}u\,\mathrm{d}s;\end{aligned}$$

in the second equality the 'continuous counterpart' of

$$\mathbb{V}\mathrm{ar}\left(\sum_{i=1}^n X_i\right) = \sum_{i=1}^n \sum_{j=1}^n \mathbb{C}\mathrm{ov}(X_i, X_j)$$

is used.
(ii) Follows similarly. ◇

We write

$$\lim_{n\to\infty} \{Y_n(t), t \in \mathcal{T}\} =_{\mathrm{d}} \{Y(t), t \in \mathcal{T}\}$$

in order to denote that sequence of stochastic processes $\{Y_n(t), t \in \mathcal{T}\}$ weakly converges in $C(\mathcal{T})$ to $\{Y(t), t \in \mathcal{T}\}$ as $n \to \infty$; see [286] for more background on weak convergence, and precise definitions. Moreover, by $\lim_{n\to\infty} Y_n =_{\mathrm{d}} Y$, we mean that sequence of random variables Y_n converges in distribution to the random variable Y as $n \to \infty$.

The next proposition follows directly from the results in [67], in particular Corollary 4.1. It states that what we expected is indeed true: the traffic rate due to the superposition of a large number of i.i.d. on-off sources converges to a Gaussian process.

Proposition 3.2.3 *If Assumption 3.2.1 is satisfied, then, for any $T > 0$,*

$$\lim_{n\to\infty} \left\{\frac{Z_{T,n}(s) - \mu n}{\sqrt{n}}, s \geq 0\right\} =_{\mathrm{d}} \{Z_T(s), s \geq 0\};$$

Also,

$$\lim_{n\to\infty} \left\{\frac{A_{T,n}(0,t) - \mu n t}{\sqrt{n}}, t \geq 0\right\} =_{\mathrm{d}} \{A_T(0,t), t \geq 0\}.$$

The proofs of weak convergence results on the process level are usually rather technical. In general, a three-step procedure is followed to show that $X_n(\cdot) \to X(\cdot)$: (i) the convergence of $X_n(t)$ to $X(t)$ for any t, (ii) the convergence of the finite-dimensional distributions, (iii) the convergence $X_n(\cdot) \to X(\cdot)$ using a tightness argumentation. For the general theory, we refer to, for instance, [286]; in [64, 67] it is demonstrated for specific situations how the three-step procedure works.

Regularly varying on- or off-times. Now that we know that the superposition of n i.i.d. on-off sources, appropriately scaled, tends to a Gaussian process, the next step is to consider the special case that the on- and off-times have an extremely heavy tail. More specifically: we assume that the complementary distribution functions $\overline{F}_{\rm on}(\cdot)$ and $\overline{F}_{\rm off}(\cdot)$ have some sort of a polynomial tail, viz. a *regularly varying* tail. Here a function $\phi(\cdot)$ is said to be regularly varying at ∞, of index α, if $\phi(x)$ can be written as $x^\alpha L(x)$, with $L(\cdot)$ *slowly varying* at ∞ (i.e., $L(tx)/L(x) \to 1$ for $x \to \infty$, for all $t > 0$).

It is clear that if a complementary distribution function is regularly varying of index $-\alpha$, then the kth moment is finite for all $k < \alpha$. In the sequel, we assume that $\overline{F}_{\rm on}(\cdot)$ and $\overline{F}_{\rm off}(\cdot)$ are regularly varying, both of an index that implies that the variances $\mathbb{V}{\rm ar}\, T_{\rm on}$ and $\mathbb{V}{\rm ar}\, T_{\rm off}$ are infinite. The main goal of the remainder of this section is to explain that, under a rescaling of time, the limiting process turns out to be fBm.

In our analysis, we impose the following assumption on the on- and off-times. For definitions and backgound on slowly and regularly varying functions, see [34].

Assumption 3.2.4 *The complementary distribution functions are regularly varying, and the on- and off-times have infinite variance, cf.* [130, 216]:

1. $\overline{F}_{\rm on}(x) = x^{-\alpha_{\rm on}} L_{\rm on}(x)$, *with* $\alpha_{\rm on} \in (1, 2)$;
2. $\overline{F}_{\rm off}(x) = x^{-\alpha_{\rm off}} L_{\rm off}(x)$, *with* $\alpha_{\rm off} \in (1, 2)$;
3. $\overline{F}_{\rm on}(x) = o(\overline{F}_{\rm off}(x))$ *as* $x \to \infty$ *or* $\overline{F}_{\rm off}(x) = o(\overline{F}_{\rm on}(x))$ *as* $x \to \infty$,

where $L_{\rm on}(\cdot)$ *and* $L_{\rm off}(\cdot)$ *are slowly varying at* ∞.

In the sequel, we use the following identification:

$$H \equiv \frac{3 - \min(\alpha_{\rm on}, \alpha_{\rm off})}{2}.$$

Above we already considered the limit $n \to \infty$; now we let, in addition, also T grow large. Below $B_H(\cdot)$ is standard fBm, i.e., centered with $\mathbb{V}{\rm ar} B_H(t) = t^{2H}$. The following result applies.

Proposition 3.2.5 *If Assumptions 3.2.1 and 3.2.4 hold, then*

$$\lim_{T\to\infty} \lim_{N\to\infty} \left\{ \frac{T}{\sigma(T)} \cdot \frac{A_{T,n}(0,t) - \mu n t}{\sqrt{n}}, t \geq 0 \right\} =_{\rm d} \{B_H(t), t \geq 0\}.$$

This is essentially [274, Theorem 1], but see [286, Theorem 7.2.5] for the tightness arguments to prove weak convergence. It would be beyond the scope of this book to give the proof of Proposition 3.2.5. Instead, we give a more intuitive explanation.

From Proposition 3.2.3 it is already clear that the limiting process is Gaussian. We now give a heuristic argumentation why this Gaussian process is fBm. This is done by showing that

$$\mathbb{V}\text{ar}\left(\frac{T}{\sigma(T)\sqrt{n}}A_{T,n}(0,t)\right) \tag{3.4}$$

indeed converges (after taking the appropriate limits) to t^{2H}.

- First observe that

$$\begin{aligned}\mathbb{V}\text{ar}A_{T,n}(0,t) &= \mathbb{V}\text{ar}\left(\int_0^t \sum_{i=1}^n \xi_i(Tu)\,\mathrm{d}u\right)\\ &= n\mathbb{V}\text{ar}\left(\int_0^t \xi_i(Tu)\,\mathrm{d}u\right) = \frac{n}{T^2}\mathbb{V}\text{ar}\left(\int_0^{tT} \xi_i(v)\,\mathrm{d}v\right).\end{aligned}$$

  It immediately follows that

$$\begin{aligned}&\lim_{T\to\infty}\lim_{n\to\infty}\mathbb{V}\text{ar}\left(\frac{T}{\sigma(T)\sqrt{n}}A_{T,n}(0,t)\right)\\ &= \lim_{T\to\infty}\lim_{n\to\infty}\left(\frac{T^2}{\sigma^2(T)n}\right)\mathbb{V}\text{ar}A_{T,n}(0,t) = \lim_{T\to\infty}\frac{\mathbb{V}\text{ar}A_{1,1}(0,Tt)}{\mathbb{V}\text{ar}A_{1,1}(0,T)}.\end{aligned} \tag{3.5}$$

- Now consider the kth moment of $A_{1,1}(0,T)$ for large T. Take for the moment (without loss of generality) $\alpha_{\text{on}} < \alpha_{\text{off}}$; the opposite case can be dealt with similarly. Because the on-times are heavy-tailed, there is a substantial probability that the source was on at time 0, and has remained on during the entire interval $[0,T]$. Notice that the tail of the distribution of a residual on-time, denoted by T_{on}^r, is one degree heavier than that of a normal on-time; use Karamata's theorem, see [34, Proposition 1.5.11(i)]. In other words, where T_{on} was regularly varying of degree $-\alpha_{\text{on}}$, we have that T_{on}^r is regularly varying of degree $-\alpha_{\text{on}}+1$.

  We thus find, modulo the slowly varying function,

$$\mathbb{P}(T_{\text{on}}^r > x) = \frac{1}{\mu_{\text{on}}}\frac{1}{\alpha_{\text{on}}-1}x^{-\alpha_{\text{on}}+1}.$$

  In fact, one can show that, for T large, the contribution of this event will dominate all other contributions, in a way that is made precise in [188].

- Recalling that $\mu \in (0, 1)$ is the probability that the source is on at an arbitrary time instant, this entails for the kth moment that

$$\begin{aligned}\mathbb{E}A_{1,1}^k(0,T) &\approx T^k \cdot \mu \cdot \mathbb{P}(T_{\rm on}^r > x)\\ &\approx T^k \cdot \mu \cdot \frac{1}{\mu_{\rm on}} \frac{1}{\alpha_{\rm on}-1} T^{-\alpha_{\rm on}+1}\\ &= \frac{\mu}{\mu_{\rm on}} \frac{1}{\alpha_{\rm on}-1} T^{-\alpha_{\rm on}+k+1}.\end{aligned}$$

  It follows that, for T large, up to a slowly varying function, $\mathbb{V}{\rm ar}A_{1,1}(0,T)$ looks like $\mathbb{E}A_{1,1}^2(0,T)$, or, more precisely, $\mathbb{V}{\rm ar}A_{1,1}(0,T)$ is regularly varying of degree $-\alpha_{\rm on}+3$. Substituting $T^{-\alpha_{\rm on}+3}$ for $\mathbb{V}{\rm ar}A_{1,1}(0,T)$ in the right-hand side of Equation (3.5), we find $t^{-\alpha_{\rm on}+3} = t^{2H}$, as desired.

Proposition 3.2.5 provides an appealing possible explanation for the appearance of fBm in communication networks. Measurements have indicated that document sizes, e.g., those transmitted over the Internet, have a heavy-tailed distribution. Approximating a user's traffic stream by an on-off process (a user is alternating between transmitting at a constant rate while retrieving documents, and being silent), and superimposing the streams generated by the individual users, we are in the setting of Proposition 3.2.5.

The limit result of Proposition 3.2.5 should be treated with care, and a caveat is in place. In this respect we mention that when taking the limits $T \to \infty$ and $n \to \infty$ in reverse order, the limiting process is *not* fBm, but rather α-stable Lévy motion, which is an infinite variance process whose increments are stationary and independent.

Bibliographical notes

There are several papers on the convergence of on-off sources to their Gaussian counterpart. The convergence result we quote above, Proposition 3.2.3, is due to [67]; its proof heavily relies on Szczotka [271]. A similar result for the M/G/∞ input model is described in [103]. For an overview of this type of results, we refer to Chapter 7 of Whitt [286] and references therein; [286] and [139] are standard references on stochastic-process limits.

Crovella and Bestavros [58] were among the first researchers to systematically study the distribution of Web document sizes. It was concluded that these were indeed heavy-tailed. The fact that a small part of the web documents is extremely large while the vast majority consists of just a few packets is often referred to as the *mice-elephants effect*, a term coined by Guo and Matta [125].

Taqqu, Willinger, and Sherman [274] is generally considered as the seminal paper on the convergence of on-off sources with heavy-tailed on- and off-times to fBm. Also the convergence to α-stable Lévy motion, with the limits taken in reverse order, is established in that paper; see also [288]. A recent article by Mikosch, Resnick, Rootzén, and Stegeman [216] explores how these two limits relate; in fact it is proven that under *slow growth* (connection rates are modest relative to the connection length distribution tails) the limit is an α-stable Lévy motion, whereas under *fast growth* (the opposite situation) there is an fBm limit. In this respect, also the recent work (on the so-called 'telecom process') by Kaj and Taqqu [144] should be mentioned.

Chapter 4

Large deviations for Gaussian processes

Large deviations is a theory that is concerned with rare event probabilities. It considers a sequence of probabilities p_n that decay to 0 when the scaling parameter n goes to ∞. Then, conditions are stated under which these p_n decay exponentially, and the corresponding decay rate, say γ, is identified:

$$\gamma := -\lim_{n\to\infty} \frac{1}{n} \log p_n.$$

For obvious reasons, this theory has appeared to be extremely useful in situations in which the perceived performance is strongly affected by outliers and extreme values.

This chapter describes large deviations results for Gaussian processes. As Gaussian processes are infinitely dimensional objects and therefore relatively hard to handle, we start with a number of more concrete results relating to the finite-dimensional setting. Section 4.1 reviews the main large deviations results for univariate and multivariate random variables. Particular attention is paid to the large deviations for sample means: we present the classical results by Cramér, Chernoff, and Bahadur-Rao.

Large deviations for Gaussian processes are dealt with in Section 4.2. The main theorem is (the generalized version of) Schilder's theorem. This powerful result is stated, and illustrated by means of a number of examples.

4.1 Cramér's theorem

Consider a sequence $X_1, \ldots, X_n$ of i.i.d. random variables, distributed like a generic random variable X that has mean $\mu := \mathbb{E}X < \infty$. Laws of large numbers

Large deviations for Gaussian queues M. Mandjes

state that, under appropriate conditions, the sample mean $n^{-1}\sum_{i=1}^{n} X_i$ converges to μ almost surely as $n \to \infty$. The next question is, what is the probability that, despite the fact that $n \to \infty$, there is still a serious deviation from the mean? In other words: we are looking for ways to characterize

$$f(n) := \mathbb{P}\left(\frac{1}{n}\sum_{i=1}^{n} X_i > a\right), \tag{4.1}$$

for $a > \mu$ (of course, we could equivalently consider the probability of the sample mean being below a for $a < \mu$). Perhaps the most well-known result in this respect is Cramér's theorem. However, to get some intuition, we first derive a strong but elementary bound: the Chernoff bound.

Chernoff bound. Recall that for any nonnegative random variable X we have that $\mathbb{P}(X \geq a)$ is bounded from above by $\mathbb{E}X/a$:

$$\mathbb{E}X = \int_0^\infty x \, d\mathbb{P}(X \leq x) \geq \int_a^\infty x \, d\mathbb{P}(X \leq x) \geq \int_a^\infty a \, d\mathbb{P}(X \leq x) = a\mathbb{P}(X \geq a);$$

this is the *Markov inequality*. Define the moment generating function (mgf) by $M(\theta) := \mathbb{E}\exp(\theta X)$; assume that this mgf is finite in a neighborhood of 0 (which implies that all moments of X are finite!). Now apply the Markov inequality to the random variable $\exp(\theta \sum_{i=1}^{n} X_i)$: for any nonnegative θ,

$$\begin{aligned} f(n) = \mathbb{P}\left(\frac{1}{n}\sum_{i=1}^{n} X_i > a\right) &= \mathbb{P}\left(\exp\left(\theta\sum_{i=1}^{n} X_i\right) > e^{n\theta a}\right) \\ &\leq e^{-n\theta a}\mathbb{E}\left(\exp\left(\theta\sum_{i=1}^{n} X_i\right)\right) = e^{-n\theta a}(M(\theta))^n. \end{aligned}$$

But, because this holds for any $\theta \geq 0$, we may as well take the tightest among these upper bounds:

$$f(n) \leq \inf_{\theta\geq 0} e^{-n\theta a}(M(\theta))^n = \exp\left(-n\sup_{\theta\geq 0}(\theta a - \log M(\theta))\right).$$

We thus have found, through an elementary argument, an explicit upper bound on the probability of the sample mean attaining a large deviation. Evidently, the larger the number of random variables, the smaller $f(n)$ becomes (as we know $f(n) \to 0$ as n grows large), but the above result also shows that the corresponding decay is of an exponential nature, and that the decay rate is

$$I(a) := \sup_{\theta\geq 0}(\theta a - \log M(\theta)).$$

The function $I(a)$ is usually referred to as the *Fenchel–Legendre transform* or *convex conjugate*. It has a number of nice properties. The following exercise shows that $I(a)$ essentially measures some sort of 'distance' to the mean μ.

Exercise 4.1.1 (i) Show that $I(a) > 0$ for $a > \mu$. (ii) Show that $I(\mu) = 0$. (iii) Prove that $I(\cdot)$ is convex.

Solution. (i) It can be proven that $\log M(\theta)$ is convex by applying Hölder's inequality. Clearly $\log M(\theta)$ goes through the origin, where it has slope μ. As a consequence, $a > \mu$ implies that for small positive θ we have that $\theta a - \log M(\theta)$ is positive, yielding $I(a) > 0$.

(ii) Now the convexity of $\log M(\theta)$ entails that $\log M(\theta) \geq \theta\mu$ for all $\theta \geq 0$. As both $\log M(\theta)$ and $\theta\mu$ go through the origin, it follows that $I(\mu) = 0$.

(iii) $I(\cdot)$ is the maximum of the straight lines (as a function of a!) $\theta a - \log M(\theta)$, and is consequently convex. ◇

For later reference we state the Chernoff bound as a proposition. We also add the 'classical' version, that relates to individual random variables rather than sample means, and which can be proven analogously [48].

Proposition 4.1.2 Chernoff bound. *Let $X \in \mathbb{R}$ be a random variable with mean μ and mgf $M(\theta) := \mathbb{E}e^{\theta X}$ finite in a neighborhood of 0. Then, for $a > \mu$,*

$$\mathbb{P}(X > a) \leq e^{-I(a)}.$$

Let $X_i \in \mathbb{R}$ be i.i.d. random variables, distributed as a random variable X with mean μ and mgf $M(\theta) := \mathbb{E}e^{\theta X}$ finite in a neighborhood of 0. Then, for $a > \mu$,

$$\mathbb{P}\left(\frac{1}{n}\sum_{i=1}^{n} X_i > a\right) \leq e^{-nI(a)}.$$

The following exercise shows that, under the (demanding) criterion of a finite mgf around 0, the weak and strong law of large numbers can be proven in an elementary way.

Exercise 4.1.3 Let $X_i \in \mathbb{R}$ be i.i.d. random variables, distributed as a random variable X with mean μ and mgf $M(\theta) := \mathbb{E}e^{\theta X}$ which is finite in a neighborhood of 0. (i) Prove that $n^{-1}\sum_{i=1}^{n} X_i \to \mu$ in probability as $n \to \infty$. (ii) Prove that $n^{-1}\sum_{i=1}^{n} X_i \to \mu$ almost surely as $n \to \infty$.

Solution. (i) follows directly from the Chernoff bound, whereas (ii) is implied by the Chernoff bound in conjunction with the Borel–Cantelli lemma. ◇

Cramér's theorem and Bahadur-Rao asymptotics. One of the interesting features of the Chernoff bound is that this bound is, on a logarithmic scale, *tight*. This is expressed by the Cramér's theorem [57], but before stating this, we first need the following definition, cf. [91, 72, 109].

Definition 4.1.4 Large-deviations principle. *A sequence $Y_1, Y_2, \ldots$ obeys the* large-deviations principle *(ldp) with rate function $J(\cdot)$ if*

(a) For any closed set F,

$$\limsup_{n\to\infty} \frac{1}{n} \log \mathbb{P}\left(\frac{1}{n}\sum_{i=1}^{n} Y_i \in F\right) \leq -\inf_{a\in F} J(a);$$

(b) For any open set G,

$$\liminf_{n\to\infty} \frac{1}{n} \log \mathbb{P}\left(\frac{1}{n}\sum_{i=1}^{n} Y_i \in G\right) \geq -\inf_{a\in G} J(a).$$

Theorem 4.1.5 Cramér. *Let $X_i \in \mathbb{R}$ be i.i.d. random variables, distributed as a random variable X with mean μ and mgf $M(\theta) := \mathbb{E}e^{\theta X}$ that is finite in a neighborhood of 0. Then $X_1, X_2, \ldots$ obeys the ldp with rate function $I(\cdot)$.*

Now specialize to our probabilities $f(n)$, as introduced in Equation (4.1), of the sample mean exceeding some $a > \mu$. It is readily checked that Cramér's theorem (in conjunction with the convexity derived in Exercise 4.1.1) says that

$$\lim_{n\to\infty} \frac{1}{n} \log f(n) = -I(a). \tag{4.2}$$

Compared to the Chernoff bound this statement is perhaps somewhat imprecise, as it gives information on the *logarithm* of the probability of our interest, rather than the probability itself. It would be tempting to conclude from Equation (4.2) that for large n it is accurate to approximate $f(n) \approx e^{-nI(a)}$, but this is not necessarily true. Instead, from Equation (4.2) we can, in fact, only conclude that $f(n) = \phi(n)e^{-nI(a)}$ for some 'subexponential' function $\phi(\cdot)$; here $\phi(\cdot)$ is called *subexponential* (at ∞) if $\log\phi(n) = o(n)$, or, equivalently, $n^{-1}\log\phi(n) \to 0$ as $n \to \infty$.

In other words, if $\phi(\cdot)$ is subexponential, it still can have a broad variety of shapes. It can be a slowly varying function, such as $\log n$, and then the error made by approximating $f(n) \approx e^{-nI(a)}$ is relatively small. On the other hand, $\phi(\cdot)$ can be any polynomial function n^α (where α can be both positive and negative), or even functions of the type $\exp(n^{1-\epsilon})$, with ϵ a small positive number. We conclude that the approximation $e^{-nI(a)}$ may be very inaccurate.

Asymptotics of the type Equation (4.2) are often referred to as *logarithmic asymptotics*. As they suffer from the 'impreciseness problems' described above, one could also consider a more accurate type of asymptotics, namely the so-called *exact asymptotics*. In the latter type of asymptotics, the goal is to find an explicit function $g(\cdot)$ such that $\lim_{n\to\infty} f(n)/g(n) = 1$.

Exact asymptotics for $f(n)$ were found by Bahadur and Rao [18]. Their result shows that $f(n)$ decays, in fact, faster than exponential: the Chernoff bound is off by a factor proportional to $n^{-1/2}$.

Proposition 4.1.6 Bahadur-Rao. *Define* $\theta^\star \equiv \theta^\star(a) = \arg\sup_{\theta \geq 0}(\theta a - \log M(\theta))$, *and*

$$\eta^\star := \left. \frac{d^2}{d\theta^2} \log M(\theta) \right|_{\theta=\theta^\star}.$$

Then $\lim_{n\to\infty} f(n)/g(n) = 1$, *where*

$$g(n) := \frac{1}{\theta^\star \sqrt{2\pi n \eta^\star}} e^{-nI(a)}.$$

Exercise 4.1.7 Consider the special case in which $X_1, X_2, \ldots$ are i.i.d. $\mathcal{N}(\mu, \sigma^2)$ random variables. Show that the Bahadur-Rao asymptotics follow directly from the standard inequality for $y > 0$

$$\frac{1}{y + y^{-1}} e^{-y^2/2} \leq \int_y^\infty e^{-t^2/2}\,\mathrm{d}t \leq \frac{1}{y} e^{-y^2/2}. \tag{4.3}$$

Solution. Let, as before, $M(\theta) := \mathbb{E}e^{\theta X}$. First notice that $M(\theta) = e^{\theta\mu} N(\theta\sigma)$, where $N(\cdot)$ is the mgf of a $\mathcal{N}(0, 1)$ random variable. Furthermore

$$N(\theta) = \int_{-\infty}^\infty e^{\theta x} \frac{1}{\sqrt{2\pi}} e^{-x^2/2}\,\mathrm{d}x = e^{\theta^2/2} \int_{-\infty}^\infty \frac{1}{\sqrt{2\pi}} e^{-(x-\theta)^2/2}\,\mathrm{d}x = e^{\theta^2/2}, \tag{4.4}$$

so that $M(\theta) = \exp(\theta\mu + \sigma^2\theta^2/2)$. With $h(a) = (a - \mu)/\sigma$, it is readily verified that $I(a) = (h(a))^2/2$. Now, with $a > \mu$,

$$f(n) = \int_{h(a)\sqrt{n}}^\infty \frac{1}{\sqrt{2\pi}} e^{-t^2/2}\,\mathrm{d}t.$$

It can be checked that the inequalities in Equation (4.3) imply that $f(n)/g(n) \to 1$ as $n \to \infty$, with

$$g(n) := \frac{1}{\sqrt{2\pi}} \frac{1}{\sqrt{n}h(a)} \exp\left(-\frac{1}{2} n (h(a))^2\right).$$

It is then a matter of straightforward calculus to show that this agrees with the Bahadur-Rao asymptotics, as $\theta^\star = h(a)/\sigma$, and $\eta^\star = \sigma^2$. ◇

Multivariate Cramér. There exists also a multivariate (i.e., d-dimensional, with $d \in \mathbb{N}$) version of Cramér's theorem. It deals with the probability that the sample mean of d-dimensional vectors $X_1, X_2, \ldots$ attains a remote value. Let $\langle a, b\rangle$ denote the usual (d-dimensional) inner product $\sum_{j=1}^d a_j b_j$.

Theorem 4.1.8 Multivariate Cramér. *Let $X_i \in \mathbb{R}^d$ be i.i.d. d-dimensional random vectors, distributed as a random vector X with moment-generating function $M(\theta) := \mathbb{E}e^{\langle\theta,X\rangle}$ that is finite in a neighborhood of 0. Then $X_1, X_2, \ldots$ obeys the ldp with rate function $I_d(\cdot)$, defined through*

$$I_d(a) := \sup_{\theta\in\mathbb{R}^d} \left(\langle\theta, a\rangle - \log M(\theta)\right). \tag{4.5}$$

In the context of this book, the special case that X has a multivariate normal distribution is of particular importance; see the definition in Section 2.2. Suppose X has mean vector μ and $(d \times d)$ nonsingular covariance matrix Σ. Analogously to Equation (4.4), it can be shown that $\log M(\theta) = \log \mathbb{E}e^{\langle\theta,X\rangle} = \langle\theta,\mu\rangle + \frac{1}{2}\theta^{\mathrm{T}}\Sigma\theta$. Then it is not hard to derive that, with $(a-\mu)^{\mathrm{T}} \equiv (a_1-\mu_1, \ldots, a_d-\mu_d)$,

$$\theta^\star = \Sigma^{-1}(a-\mu) \quad \text{and} \quad I_d(a) = \frac{1}{2}(a-\mu)^{\mathrm{T}}\Sigma^{-1}(a-\mu),$$

where $\theta^\star$ optimizes Equation (4.5); as in dimension 1, it can be proven that $I_d(\cdot)$ is convex.

The result proved in the following exercise will be used several times later on.

Exercise 4.1.9 Consider a sequence of i.i.d. bivariate random variables (X_1, Y_1), $(X_2, Y_2), \ldots$, with mean vector $(\mu_X, \mu_Y)^{\mathrm{T}}$, and two-dimensional covariance matrix

$$\Sigma = \begin{pmatrix} \sigma_X^2 & \varrho(X,Y) \\ \varrho(X,Y) & \sigma_Y^2 \end{pmatrix}.$$

Fix $a > \mu_X$ and $b > \mu_Y$. Determine

$$\lim_{n\to\infty} \frac{1}{n} \log \mathbb{P}\left(\frac{1}{n}\sum_{i=1}^n X_i \geq a, \frac{1}{n}\sum_{i=1}^n Y_i \geq b\right).$$

Solution. Due to 'multivariate Cramér', we have to minimize the 'quadratic form'

$$I_2(x,y) = \frac{1}{2}(x-\mu_X, y-\mu_Y)^{\mathrm{T}}\Sigma^{-1}\begin{pmatrix} x-\mu_X \\ y-\mu_Y \end{pmatrix}$$

over all $x \geq a$ and $y \geq b$. Three cases may occur.

- In the first place, it could be that the contour that touches the line $x = a$ has a y-value, say $y^\star$, larger than b; see the left graph in Figure 4.1. Then the optimum is clearly attained in $(a, y^\star)$. It is not hard to prove that

$$y^\star = \mu_Y + \frac{\varrho(X,Y)}{\sigma_X^2}(a-\mu_X);$$

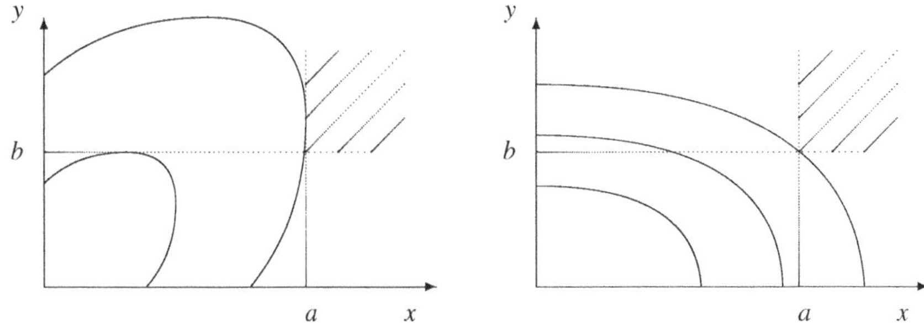

Figure 4.1: Contour lines of the (two-dimensional) rate function; the objective function is to be minimized over the shaded region.

this expression for $y^\star$ has the appealing interpretation $\mathbb{E}(Y \mid X = a)$; see the explicit formula (2.1) for conditional expectations. The resulting value for the decay rate is

$$\frac{1}{2}\frac{(a-\mu_X)^2}{\sigma_X^2}; \tag{4.6}$$

notice that in this expression neither the characteristics of the Y_i nor b plays a role. In fact, given that $\sum_{i=1}^n X_i \geq na$, then also $\sum_{i=1}^n Y_i \geq nb$ with overwhelming probability, making the latter requirement redundant.

- Similarly, if $x^\star := \mathbb{E}(X \mid Y = b) > a$, then the optimum is attained in $(x^\star, b)$, see the right graph in Figure 4.1. The decay rate is analogous to Equation (4.6); replace a by b, and (μ_X, σ_X^2) by (μ_Y, σ_Y^2).

- If both $x^\star < a$ and $y^\star < b$, then the optimum of $I_2(\cdot)$ is attained at (a, b), leading to the decay rate

$$\frac{1}{2}(a-\mu_X, b-\mu_Y)^{\mathrm{T}}\Sigma^{-1}\begin{pmatrix} a-\mu_X \\ b-\mu_Y \end{pmatrix};$$

intuitively, one could say that in this case both constraints are tight. ◇

4.2 Schilder's theorem

This section describes the extension of Cramér's theorem to an infinitely dimensional framework. The key result in this context is Schilder's theorem, or, to be more precise, *the generalized version of* Schilder's theorem (as formally 'Schilder' focuses only on the special case of Brownian motion, rather than general Gaussian processes). Whereas 'Cramér' can be used to describe the likelihood of a sample mean of normal random variables (or random vectors) attaining an uncommon value, 'Schilder' considers the large deviations of the sample mean of Gaussian processes (i.e., infinitely dimensional objects).

More precisely, the setting is as follows. Let $A_1(\cdot), A_2(\cdot), \ldots$ be a sequence of i.i.d. Gaussian processes. Then consider the 'sample mean path' $n^{-1}\sum_{i=1}^n A_i(\cdot)$. For n large, it is clear that $n^{-1}\sum_{i=1}^n A_i(t) \to \mu t$, if μ is the mean rate of the Gaussian processes. 'Schilder' describes the probability of deviations from this 'mean path': it characterizes the exponential decay rate of the sample mean path $n^{-1}\sum_{i=1}^n A_i(\cdot)$ being in a remote set.

Schilder's theorem is a rather abstract result, and we therefore start this section by giving a heuristic description. Then, as an example, we address the special case of Brownian motion, in which the resulting expressions are relatively transparent. After this, we are ready to provide the formal setup, and to state the result. We conclude this section with a couple of examples, and a computation that shows that 'multivariate Cramér' is nothing else than a special (finite-dimensional) case of 'Schilder'.

Heuristics – most likely path. Informally, one could rewrite 'Cramér' as

$$\mathbb{P}\left(\frac{1}{n}\sum_{i=1}^n X_i \approx a\right) \approx e^{-nI(a)},$$

as we saw in the previous section. In words: the probability of the sample mean of n i.i.d. random variables being approximately equal to a decays (roughly) exponentially in n, and the corresponding decay rate is $I(a)$. One may wonder whether we could do something similar for the sample mean of Gaussian processes. It would read

$$\mathbb{P}\left(\frac{1}{n}\sum_{i=1}^n A_i(t) \approx f(t),\ t \in \mathbb{R}\right) \approx e^{-n\mathbb{I}(f)}. \tag{4.7}$$

Here $f : \mathbb{R} \to \mathbb{R}$ is a given function (or 'path'; it is a function of time); we are wondering what the probability is of $n^{-1}\sum_{i=1}^n A_i(\cdot)$ remaining 'close to' $f(\cdot)$. The decay rate is $\mathbb{I}(f)$, where $\mathbb{I}(\cdot)$ is a functional that assigns a nonnegative value to any path. This $\mathbb{I}(\cdot)$ should be such that $\mathbb{I}(f_\mu) = 0$, where $f_\mu(t) = \mu t$. Schilder's theorem says that, in a logarithmic sense, Equation (4.7) is indeed true.

Now return to the setting of 'Cramér'. Considering the probability that the sample mean of n i.i.d. random variable is in a set A (rather than close to some specific value a), it yields the approximation

$$\mathbb{P}\left(\frac{1}{n}\sum_{i=1}^n X_i \in A\right) \approx \exp\left(-n \inf_{a\in A} I(a)\right);$$

in words, the decay rate is dominated by the 'most likely point in A', i.e., the point $a^\star := \arg\inf_{a\in A} I(a)$. Evidently, if $\mu := \mathbb{E}X \in A$, then $\inf_{a\in A} I(a) = 0$.

Likewise, 'Schilder' gives an expression for the probability of the sample mean of n i.i.d. Gaussian processes (recall that this sample mean is now a *path*) being in some set $\mathcal{S}$:

$$p_n[\mathcal{S}] := \mathbb{P}\left(\frac{1}{n}\sum_{i=1}^{n} A_i(\cdot) \in \mathcal{S}\right) \approx \exp\left(-n \inf_{f\in\mathcal{S}} \mathbb{I}(f)\right).$$

Here, the set $\mathcal{S}$ represents a collection of paths; later in this section, we give a number of examples of such sets. Informally, Schilder's theorem says that $p_n[\mathcal{S}]$ decays exponentially in n, and the corresponding exponential decay rate of $p_n[\mathcal{S}]$ equals the minimum of $\mathbb{I}(f)$ over all $f \in \mathcal{S}$.

An intrinsic difficulty of 'Schilder' is that the functional $\mathbb{I}(\cdot)$ cannot be given explicitly. As we will see, only if f corresponds to a linear combination of covariance functions, then a closed-form expression for $\mathbb{I}(f)$ can be given. We remark that the case of Brownian arrival processes is an exception; in this case, due to the independence of the increments, we are able to give for this special case an explicit formula for $\mathbb{I}(f)$, which is done below.

From the above it is concluded that, in general, finding the minimum of $\mathbb{I}(f)$ over all $f \in \mathcal{S}$ is a hard variational problem: the optimization should be done over all paths in $\mathcal{S}$ (which are infinitely dimensional objects), and the objective function $\mathbb{I}(f)$ is only explicitly given if f is a mixture of covariance functions. However, if we succeed in finding such a minimizing $f^\star(\cdot)$ in $\mathcal{S}$, then this path has an appealing interpretation. Conditional on the sample-mean path being in the set $\mathcal{S}$, with overwhelming probability this happens via a path that is 'close to' $f^\star(\cdot)$. We call $f^\star$ the *most likely path* in the set $\mathcal{S}$. Stated differently, the decay rate of $p_n[\mathcal{S}]$ is fully dominated by the likelihood of the most likely element in $\mathcal{S}$: as $n \to \infty$, we have that $n^{-1}\log p_n[\mathcal{S}] \to -\mathbb{I}(f^\star)$. Knowledge of the most likely path often gives insight into the dynamics of the problem; several insightful examples are given in Section 6.1.

Brownian motion. As argued above, one of the main problems with 'Schilder' is the implicitness of the rate function. In general, as seen above, only for paths that are linear combinations of covariance functions, the 'rate functional' $\mathbb{I}(\cdot)$ can be evaluated explicitly. In the Brownian case, however, the fact that increments are independent allows an explicit calculation of the rate function. We give here an intuitive derivation of the rate function; for more details and a rigorous treatment, see [72, Theorem 5.2.3].

Consider the superposition of n i.i.d. standard Brownian motions; each individual Brownian motion $A_1(\cdot), \ldots, A_n(\cdot)$ is a centered Gaussian process with variance function $v(t) = t$. Our goal is to analyze the likelihood of the event that the sample mean of the processes follows the function $f(\cdot)$ on the interval $[0, T]$, where it is assumed that $f(\cdot)$ is differentiable on this interval. Discretizing, we require that for $k = 0, \ldots, T/\Delta t$,

$$\frac{1}{n}\sum_{i=1}^{n} A_i(k\Delta t) \approx f(k\Delta t),$$

or equivalently, for $k = 1, \ldots, T/\Delta t$,

$$\frac{1}{n}\frac{\sum_{i=1}^{n} A_i(k\Delta t) - A_i((k-1)\Delta t)}{\Delta t} \approx \frac{f(k\Delta t) - f((k-1)\Delta t)}{\Delta t};$$

notice that, because of the independent increments in Brownian motion, these $T/\Delta t$ events are independent. Now consider the kth of these events. It is readily verified that the random variable

$$Y_i := \frac{A_i(k\Delta t) - A_i((k-1)\Delta t)}{\Delta t}$$

has a normal distribution with mean 0 and variance $\Delta t/(\Delta t)^2 = 1/\Delta t$. The decay rate of the sample mean of n i.i.d. replicas of this random variable attaining value a is, according to 'Cramér', equal to

$$\frac{(a - \mathbb{E}Y)^2}{2\mathbb{V}\mathrm{ar}Y} = \frac{1}{2}a^2\Delta t.$$

Hence, for the decay rate of all the $T/\Delta t$ events we get

$$\sum_{k=1}^{T/\Delta t} \frac{1}{2}\left(\frac{f(k\Delta t) - f((k-1)\Delta t)}{\Delta t}\right)^2 \Delta t.$$

Now let $\Delta t \downarrow 0$, and use the differentiability of $f(\cdot)$. We thus find the limit $\int_0^T \frac{1}{2}(f'(t))^2\,\mathrm{d}t$. Summarizing, for Brownian motions we have an explicit characterization for the 'rate functional' of a path $f(\cdot)$ (which is required to be continuous):

$$\mathbb{I}(f) = \int_{-\infty}^{\infty} \frac{1}{2}(f'(t))^2\,\mathrm{d}t.$$

Informally, this expression for $\mathbb{I}(f)$ says that, in the Brownian case, the 'cost' of a path f is exclusively determined by the derivative along the path. This is further illustrated in the following exercise.

Exercise 4.2.1 (i) Consider the path $f(\cdot)$ that is equal to 0 for $t < 0$, equal to κt for $t \in [0, T]$, and equal to κT for $t > T$. What is the 'Brownian norm' of this path?

(ii) Consider the set $\mathcal{S}$ defined by, for $a > 0$ and $\nu \geq 0$,

$$\mathcal{S}_t := \{f : f(t) \geq a + \nu t\}, \quad t \geq 0;$$

$$\mathcal{S} := \bigcup_{t\geq 0} \mathcal{S}_t = \{f : \exists t \geq 0 : f(t) \geq a + \nu t\}.$$

Assume that $\mathcal{S}$ is most likely reached through a straight line from the origin (which is reasonable, as the increments are independent). Find the most likely path.

Solution. (i) $f'(t) = \kappa$ for $t \in (0, T)$ and 0 otherwise. Hence

$$\mathbb{I}(f) = \int_0^T \frac{1}{2}\kappa^2 \, \mathrm{d}t = \frac{1}{2}\kappa^2 T.$$

(ii) A straight line with slope κ reaches $a + \nu t$ after $a/(\kappa - \nu)$ time (notice that this requires that κ is larger than ν). In order to find the most likely slope κ to reach $\mathcal{S}$, we consider the following optimization problem

$$\inf_{\kappa>\nu} \frac{1}{2}\kappa^2 \frac{a}{\kappa - \nu}.$$

Notice that this formula nicely shows the trade-off involved. Taking a slope just larger than ν is 'cheap' per unit time, but the time it takes to reach $a + \nu t$ is long; if one chooses a large slope the 'cost' per unit time is large, but the time needed is short. In fact the variational problem identifies the optimal slope, that turns out to be $\kappa^\star = 2\nu$. The line $a + \nu t$ is reached after a/ν time units. ◇

Framework – reproducing kernel Hilbert space. After having described Schilder's theorem in a heuristic manner above (including the explicit formulae for the case of Brownian inputs), we now proceed with a formal treatment of the result. It requires the introduction of a number of concepts: (i) a *path space* Ω, (ii) a *reproducing kernel Hilbert space* R, (iii) an *inner product* $\langle\cdot,\cdot\rangle_R$, and (iv) finally a *norm* $||\cdot||_R$. This norm turns out to be intimately related to the 'rate functional' $\mathbb{I}(\cdot)$. Having defined these notions, we are able to state Schilder's theorem, which we do in Theorem 4.2.3.

The framework of Schilder's theorem is formulated as follows. Consider a sequence of i.i.d. processes $A_1(\cdot), A_2(\cdot), \ldots,$ distributed as a Gaussian process with variance function $v(\cdot)$. We assume for the moment that the processes are centered, but it is clear that the results for centered processes can be translated immediately into results for noncentered processes; we return to this issue in more detail in Remark 5.2.1. Recall that, taking without loss of generality $t \geq s$,

$$\Gamma(s,t) := \mathbb{C}\mathrm{ov}(A(s), A(t)) = \frac{1}{2}(v(t) + v(s) - v(t-s)).$$

Define the path space Ω as

$$\Omega := \left\{\omega : \mathbb{R} \to \mathbb{R},\ \text{continuous},\ \omega(0) = 0,\ \lim_{t\to\infty} \frac{\omega(t)}{1+|t|} = \lim_{t\to-\infty} \frac{\omega(t)}{1+|t|} = 0\right\},$$

which is a separable Banach space by imposing the norm

$$||\omega||_\Omega := \sup_{t\in\mathbb{R}} \frac{|\omega(t)|}{1+|t|}.$$

In [5] it is pointed out that $A_i(\cdot)$ can be realized on Ω under the following assumption, which is supposed to be in force throughout the remainder of this monograph. We already found that $v(\cdot)$ cannot increase faster than quadratically, see Exercise 2.3.2; Assumption 4.2.2 requires that $v(\cdot)$ increases slower than quadratically.

Assumption 4.2.2 *There is an* $\alpha < 2$ *such that*

$$\lim_{t\to\infty} \frac{v(t)}{t^\alpha} = 0.$$

Next we introduce and define the *reproducing kernel Hilbert space* $R \subseteq \Omega$ – see [6] for a more detailed account – with the property that its elements are roughly as smooth as the covariance function $\Gamma(s,\cdot)$. We start from a 'smaller' space $R^\star$, defined by linear combinations of covariance functions:

$$R^\star := \left\{\omega : \mathbb{R} \to \mathbb{R},\ \omega(\cdot) = \sum_{i=1}^n a_i \Gamma(s_i,\cdot),\ \ a_i, s_i \in \mathbb{R}, n \in \mathbb{N}\right\}.$$

The inner product on this space $R^\star$ is, for $\omega_a, \omega_b \in R^\star$, defined as

$$\langle \omega_a, \omega_b\rangle_R := \left\langle \sum_{i=1}^n a_i\Gamma(s_i,\cdot), \sum_{j=1}^n b_j\Gamma(s_j,\cdot)\right\rangle_R = \sum_{i=1}^n\sum_{j=1}^n a_i b_j \Gamma(s_i,s_j); \tag{4.8}$$

notice that this implies $\langle \Gamma(s,\cdot), \Gamma(\cdot,t)\rangle_R = \Gamma(s,t)$. This inner product has the following useful property, which we refer to as the *reproducing kernel* property,

$$\omega(t) = \sum_{i=1}^n a_i\Gamma(s_i,t) = \left\langle \sum_{i=1}^n a_i\Gamma(s_i,\cdot), \Gamma(t,\cdot)\right\rangle_R = \langle \omega(\cdot), \Gamma(t,\cdot)\rangle_R. \tag{4.9}$$

From this we introduce the norm $||\omega||_R := \sqrt{\langle \omega,\omega\rangle_R}$. The closure of $R^\star$ under this norm is defined as the space R.

Schilder's ldp. Having introduced the norm $||\cdot||_R$, we can now define the rate function that will apply in Schilder's ldp:

$$\mathbb{I}(\omega) := \begin{cases} \frac{1}{2}||\omega||_R^2 & \text{if } \omega \in R; \\ \infty & \text{otherwise.}\end{cases} \tag{4.10}$$

Remark that for f that can be written as a linear combination of covariance functions (i.e., $f \in R^\star$), Equations (4.8) and (4.10) yield an explicit expression for $\mathbb{I}(f)$.

Now we have the following ldp.

Theorem 4.2.3 (Generalized) Schilder. *Let $A_i(\cdot) \in \Omega$ be i.i.d. centered Gaussian processes, with variance function $v(\cdot)$. Then $A_1(\cdot), A_2(\cdot), \ldots$ obeys the ldp with rate function $\mathbb{I}(\cdot)$.*

Recall that this theorem, informally, states that

$$p_n[\mathcal{S}] := \mathbb{P}\left(\frac{1}{n}\sum_{i=1}^{n} A_i(\cdot) \in \mathcal{S}\right)$$
$$\approx \exp\left(-n \inf_{f\in\mathcal{S}} \mathbb{I}(f)\right) = \exp\left(-\frac{n}{2}\inf_{f\in\mathcal{S}} ||f||_R^2\right).$$

In other words, if we can write the probability of our interest as $p_n[\mathcal{S}]$ for some set of paths $\mathcal{S}$, then 'Schilder' provides us (at least in principle) with the corresponding decay rate.

The following exercise can be regarded as an example how Theorem 4.2.3 can be applied. We define a set $\mathcal{S}$, determine $\inf_{f\in\mathcal{S}} \mathbb{I}(f)$, and interpret $f^\star := \arg\inf_{f\in\mathcal{S}} \mathbb{I}(f)$.

Exercise 4.2.4 Let $A_i(\cdot) \in \Omega$ be i.i.d. centered Gaussian processes, with variance function $v(\cdot)$. Define $\mathcal{S}_t$ and $\mathcal{S}$ as in Exercise 4.2.1.

(i) Show that

$$\lim_{n\to\infty} \frac{1}{n} \log p_n[\mathcal{S}_t] = -\gamma(t), \quad \text{where} \quad \gamma(t) := \frac{(a+\nu t)^2}{2v(t)}.$$

(ii) Determine the most likely path in $\mathcal{S}_t$.

(iii) Determine $\lim_{n\to\infty} n^{-1} \log p_n[\mathcal{S}]$. Assume that $v(t) \to \infty$ as $t \to \infty$.

Solution. (i) Realize that

$$p_n[\mathcal{S}_t] = \mathbb{P}\left(\frac{1}{n}\sum_{i=1}^{n} A_i(t) \geq a + \nu t\right).$$

Then the statement follows directly from applying 'Cramér'.

(ii) Consider the path $f(\cdot)$ given by

$$f(r) := \mathbb{E}(A(r) \mid A(t) = a + \nu t) = \frac{\Gamma(r,t)}{v(t)}(a + \nu t);$$

recall the formula for conditional means Equation (2.1). Then, it follows that this path is the most likely path in $\mathcal{S}_t$, as we indeed have that its rate function has the value that we found in (i):

$$\frac{1}{2}||f||_R^2 = \frac{(a+\nu t)^2}{2v^2(t)}\Gamma(t,t) = \gamma(t).$$

We conclude that the most likely path has the appealing interpretation of a conditional mean: given that $n^{-1}\sum_{i=1}^{n} A_i(\cdot)$ reaches $a+\nu t$ at time t, its most likely value at time r can be expressed through a conditional mean $f(r)$.

(iii) Using that $\mathcal{S}=\cup_{t\geq 0}\mathcal{S}_t$, application of 'Schilder' yields that

$$\lim_{n\to\infty}\frac{1}{n}\log p_n[\mathcal{S}] = -\inf_{f\in\mathcal{S}}\mathbb{I}(f) = -\inf_{t\geq 0}\left(\inf_{f\in\mathcal{S}_t}\mathbb{I}(f)\right).$$

We know from (i) that the infimum over $\mathcal{S}_t$ equals $\gamma(t)$, and hence the decay rate of $p_n[\mathcal{S}]$ equals $\inf_{t\geq 0}\gamma(t)$.

If $\nu=0$, then $v(t)\to\infty$ as $t\to\infty$ implies that $\gamma(t)\to 0$ as $t\to\infty$; hence, in this case, the decay rate is 0. ◇

'Multivariate Cramér' as a special case of 'Schilder'. One of the conceptual difficulties of working with 'Schilder' is its abstractness. To get more insight in what the theorem does, and to get a feel of what the reproducing kernel property, norm, etc. mean, we consider the following finite-dimensional setting. The computations below show that 'Schilder' is essentially nothing else than the infinite-dimensional version of 'multivariate Cramér'.

More specifically, if the set $\mathcal{S}$ is such that it imposes conditions on just a *finite* set of time points (rather than a continuum), the decay rate can be found by applying 'multivariate Cramér'. Then we 'guess' the accompanying path, and show by using 'Schilder' that this path gives the desired decay rate; we conclude that the path is therefore the most likely path.

Let the event under consideration consist of all realizations of $n^{-1}\sum_{i=1}^{n} A_i(\cdot)$ that go through a finite number of points:

$$\mathcal{S} := \{f\in\Omega : f(t_k)\approx\phi_k, k=1,\ldots,K\}.$$

for predefined $t_1,\ldots,t_K$ and $\phi_1,\ldots,\phi_K$. Using 'multivariate Cramér' it is seen directly that

$$\lim_{n\to\infty}\frac{1}{n}\log p_n[\mathcal{S}] = -\frac{1}{2}\phi^{\mathrm{T}}\Sigma^{-1}\phi,$$

where Σ is the (symmetric) $K\times K$-matrix in which the (j,k)th entry is the covariance $\Gamma(t_j,t_k)$, and $\phi\equiv(\phi_1,\ldots,\phi_K)^{\mathrm{T}}$.

We can also try to compute this decay rate, but now relying on the sample-path ldp of 'Schilder'. Let us first see whether we can find a linear combination of covariance functions, say f, such that $f(t_k)=\phi_k$. We immediately see that, in general, we need (at least) K of such functions:

$$f(r)=\sum_{k=1}^{K} a_k\Gamma(r,t_k).$$

As we require that $f(t_k)=\phi_k$, we have that the a_k should be chosen such that $a=\Sigma^{-1}\phi$. As this path is a linear combination of covariance functions, we can compute the corresponding decay rate explicitly. We obtain

$$\begin{aligned}\mathbb{I}(f) &= \frac{1}{2}||f||_R^2 = \frac{1}{2}\sum_{j=1}^{K}\sum_{k=1}^{K} a_j a_k \Gamma(t_j, t_k) \\ &= \frac{1}{2}a^{\mathrm{T}}\Sigma a = \frac{1}{2}(\Sigma^{-1}\phi)^{\mathrm{T}}\Sigma(\Sigma^{-1}\phi) \\ &= \frac{1}{2}\phi^{\mathrm{T}}\Sigma^{-1}\phi.\end{aligned}$$

We conclude that $f(\cdot)$, as defined above is indeed the most likely path in $\mathcal{S}$, as it gives the right decay rate.

Interestingly, a nice interpretation can be given for the shape of the paths $f(\cdot)$. Extending the result on conditional distributions of Section 2.2, i.e., Equation (2.1), to *multiple* conditions, it can be verified that

$$\begin{aligned}\mathbb{E}(A(r) \mid A(t_k) = \phi_k, k = 1, \ldots, K) &= \sum_{j=1}^{K}\sum_{k=1}^{K} \Gamma(r, t_j) \left(\Sigma^{-1}\right)_{jk} \phi_k \\ &= \sum_{k=1}^{K} a_k \Gamma(r, t_k).\end{aligned}$$

In other words, we find that $f(r) = \mathbb{E}(A(r) \mid A(t_k) = \phi_k, k = 1, \ldots, K)$. This could perhaps be expected: one makes sure that the process goes through the points (t_k, ϕ_k) by choosing the coefficients a_j appropriately, and on the other points in time one gives the path its most likely value, namely its (conditional) expectation.

Bibliographical notes

There are dozens of good introductory texts on the fundamentals of large deviations that address the most prominent results, such as the Chernoff bound and Cramér's theorem. For references, see Chapter 1.

The sample-path ldp for Gaussian processes, i.e., (generalized) Schilder's theorem, was established in Azencott [16] and Bahadur and Zabell [19], see also Deuschel and Stroock [75]. In our approach we have adopted the notation and setup of Addie, Mannersalo and Norros [5, 206].

Part B: Large deviations for Gaussian queues

In the second part of this book, large-deviations results (in particular Cramér's theorem and Schilder's theorem) are used to analyze overflow behavior of queues with Gaussian inputs. To this end we first give, in Chapter 5, a short introduction on queues fed by arrival processes with stationary increments, and then specialize to the case of queues with Gaussian input traffic. We present a few key results on the tail asymptotics under the so-called large-buffer scaling.

Our emphasis in this book is on another asymptotic regime: the many-sources scaling. Chapter 6 analyzes the single queue under this scaling; special emphasis is laid on insights into the impact of the correlation structure on the overflow probability.

The larger part of Part B, however, is devoted to queuing systems that are intrinsically more complex than the single queue, viz. tandem networks, priority queues, and queues operating under generalized processor sharing. For these systems, relying mainly on Schilder's theorem, we explicitly identify (bounds on) the many-sources asymptotics. For the special case of Gaussian inputs with a weak dependence structure (short-range dependent inputs), and Gaussian inputs with even independent increments (Brownian motion), closed-form expressions are derived.

Large deviations for Gaussian queues M. Mandjes

Chapter 5

Gaussian queues: an introduction

In part A of this monograph, we have introduced Gaussian sources. In a communication networking context, these sources typically feed into *queues*. Part B is devoted to the analysis of the *Gaussian queues*, with emphasis on large deviations results. This chapter first recalls, in Section 5.1, a number of properties of the basic single-server queuing model. Then we specialize in Section 5.2 to queues fed by Gaussian sources. It turns out that just in a few special cases the buffer-content distribution can be explicitly solved, see Section 5.1. We then sketch in Section 5.4 a framework that facilitates approximations for the overflow probability. Section 5.5 presents several types of asymptotics, which we make precise in the regime of large buffers in Section 5.6.

5.1 Lindley's recursion, the steady-state buffer content

In the previous chapter, we have introduced theorems that describe the likelihood that the arrival process attains some exceptional value (a 'large deviation'). One of the purposes of this book, however, is to analyze the probability that a high value is attained by the *buffer content* of the queue the arrival process feeds into. To enable statements of the latter kind, we clearly need a way to express the steady-state buffer content (i.e., a random variable) in terms of the arrival process $A(\cdot)$.

The goal of this section is to show that the steady-state buffer content can be written as some explicit functional of the arrival process $A(\cdot)$. In this section, we

Large deviations for Gaussian queues M. Mandjes

only assume that the arrival process has stationary increments; the assumption of Gaussianity is imposed from Section 5.2 onward.

To make the exposition more transparent, we first focus on the case of slotted time, i.e., a system in which traffic only enters at integer epochs. Where we wrote $A(0,t)$ to denote, in continuous time, the amount of traffic offered to the system in $[0,t)$, we now write $A\{1,n\}$ for the amount of traffic arrived in slots 1 up to n in our discrete-time setting. Furthermore, $A\{1,n\} = X_1 + \cdots + X_n$, with X_i the amount of traffic entering the queue in slot i. In the queuing systems of our interest, there is usually a constant service speed; therefore we suppose that the traffic is fed into a queue that is served at a deterministic service rate c.

Let μ be the arrival rate at the queue. Due to the stationary increments we have that, if a stationary version of the arrival process is used,

$$\mu = \mathbb{E}X_1 = \lim_{n\to\infty} \frac{\mathbb{E}A\{1,n\}}{n}.$$

It is intuitively clear that a sufficient condition for stability of the queuing system is that the mean input rate μ is smaller than the service rate: $\mu < \mathrm{c}$.

Now consider the buffer content of the queue at time 0, say Q_0. Assume that the processes have already been active for a while, so that at time 0 the queue is in *steady state*. It is clear that in this slotted system Q_0 is closely related to the buffer content one slot earlier, Q_{-1}. In fact, Q_0 equals Q_{-1}, increased by the amount of traffic arrived, and decreased by the amount that was served. As queues obviously cannot become negative, this yields

$$Q_0 = \max\{Q_{-1} + X_{-1} - \mathrm{c}, 0\}.$$

This relation is called *Lindley's recursion*; it says that the dynamics of the buffer content of a queue are essentially described by an additive recursion that is reflected at 0. Of course we can further iterate; after 2 steps we find

$$\begin{aligned} Q_0 &= \max\{\max\{Q_{-2} + X_{-2} - \mathrm{c}, 0\} + X_{-1} - \mathrm{c}, 0\} \\ &= \max\{Q_{-2} + A\{-2,-1\} - 2\mathrm{c}, A\{-1,-1\} - \mathrm{c}, 0\}, \end{aligned}$$

and after i steps

$$\begin{aligned} Q_0 = \max\{&Q_{-i} + A\{-i,-1\} - i\mathrm{c}, A\{-i+1,-1\} - (i-1)\mathrm{c}, \\ &\cdots, A\{-1,-1\} - \mathrm{c}, 0\}. \end{aligned}$$

Now notice that if there is an i such that $Q_{-i} = 0$ (which is true if the queue is stable, i.e., $\mu < \mathrm{c}$), then we find that the steady-state buffer content is distributionally equivalent to the supremum of a 'free process' with negative drift:

$$Q_0 =_{\mathrm{d}} \sup_{n\in\mathbb{N}_0} A\{-n,-1\} - n\mathrm{c}.$$

This is in fact quite a remarkable property: apparently, we have that in distribution there is equality of

1. The steady-state distribution of the queuing process. The queuing process is a reflected additive recursion (which consequently lives on $[0, \infty)$).

2. The supremum of a free (i.e., nonreflected) process with negative drift. This free process lives on $\mathbb{R}$; realize though that the supremum is nonnegative.

It turns out that this framework can be directly extended to continuous time. Then we obtain the following relation between the steady-state queue length Q_0 and the arrival process $A(\cdot)$, which plays a central role in our book. This fundamental distributional identity is often attributed to Reich [249].

Theorem 5.1.1 Reich. *In a queuing system, fed by an arrival process with stationary increments, that is stable, the following distributional identity holds:*

$$Q_0 =_{\rm d} \sup_{t \geq 0} A(-t, 0) - {\rm c}t,$$

where Q_0 is the steady-state buffer content. If the arrival process is time-reversible, we have in addition

$$Q_0 =_{\rm d} \sup_{t \geq 0} A(t) - {\rm c}t.$$

Our goal was to relate the steady-state buffer content to the arrival process. Reich's theorem gives this relation, in that it provides us explicitly with the functional that maps the arrival process $A(\cdot)$ on Q_0. Through an exercise, we illustrate the equivalence of the supremum of the free process with the steady-state buffer content.

Exercise 5.1.2 Consider the classical random walk Y on $\mathbb{Z}$:

$$\mathbb{P}(Y_{i+1} = m + 1 \mid Y_i = m) = 1 - \mathbb{P}(Y_{i+1} = m - 1 \mid Y_i = m) =: p,$$

$m \in \mathbb{Z}$. We also introduce Z, the reflected version of Y, which lives on $\mathbb{N}_0$. It has the same transition probabilities as Y, except that $\mathbb{P}(Z_{i+1} = 0 \mid Z_i = 0) = 1 - p$. Show that in this specific situation Reich's identity holds, assuming that $p < 1/2$.

Solution. First realize that the reflected process is a queue in slotted time with service rate ${\rm c} = 1$, and, with X_i again the amount of traffic arriving in slot i, the X_i being i.i.d. with value 2 with probability p and value 0 with probability $1 - p$. This queue is stable, as $\mathbb{E}X_1 = 2p < 1 = {\rm c}$. In other words, we have to show that

$$Q_0 =_{\rm d} M := \sup_{n \in \mathbb{N}_0} A\{0, n\} - n,$$

where it is used that the arrival process is time-reversible.

First consider the distribution π of Q_0, being the steady-state distribution of the queue described above. Basic Markov chain analysis yields that

$$\pi_m = p\pi_{m-1} 1\{m > 0\} + (1-p)\pi_{m+1},$$

for $m \in \mathbb{N}_0$. The solution of these equations is $\pi_n = \kappa(p/(1-p))^n$, where κ is a normalizing constant. Because the probabilities should add up to 1, we have that

$$\pi_n = \left(\frac{p}{1-p}\right)^n \frac{1-2p}{1-p}. \tag{5.1}$$

Now consider the distribution of M. Define $\varrho_n := \mathbb{P}(M \geq n)$. It is seen directly that $\varrho_n = r^n$, with $r := \varrho_1$. Now $r = \varrho_1$ can be computed as follows. Starting in 0, one can reach level 1 in one step, which happens with probability p. On the other hand, with probability $1-p$, the process moves to -1, and then two steps upwards are needed. Thus,

$$r = p + (1-p)r^2.$$

From this equation one can solve r, and it turns out that $r = p/(1-p)$ (use that $p < 1/2$ and $r \in (0, 1)$). Now using that $\mathbb{P}(M = n) = \varrho_n - \varrho_{n+1}$, we conclude that M has distribution (5.1) as well.

In other words, the steady-state distribution of the reflected process coincides with the distribution of the maximum of the free process. We have thus checked Reich's theorem for this model. ◇

5.2 Gaussian queues

When the process $A(\cdot)$ is a (superposition of) Gaussian source(s), then

$$Q_0 := \sup_{t \geq 0} A(-t, 0) - ct$$

is called a *Gaussian queue*. To ensure stability, the assumption needs to be imposed that the mean rate of the Gaussian input process, say μ, is smaller than the service rate c. Notice that Gaussian sources are time-reversible by nature, and consequently we may as well replace $A(-t, 0)$ by $A(0, t) = A(t)$, see Reich's theorem (Theorem 5.1.1).

As noticed in Chapter 2, Gaussian sources have the conceptual problem that the possibility of *negative traffic* is not ruled out–in contrast with 'classical' input processes. It is clear however, that, irrespective of whether $A(\cdot)$ can correspond to negative traffic or not, we can still evaluate the functional $\sup_{t \geq 0}(A(-t, 0) - ct)$. Hence, it does make sense to define the steady-state queue length for Gaussian queues in this way.

Gaussian queues as limit of queues with on-off input. In Section 3.2 we mentioned that the superposition of on-off sources satisfies a central limit theorem, as it converges after some scaling to a Gaussian process. Moreover, it turned out that in case of heavy-tailed on- or off-times the limiting process is fBm (where, in addition, a certain time-rescaling is applied). Importantly, these convergence results carry over to the corresponding queues. This is not a trivial result: it is true that Q is distributed as $\sup_{t\geq 0} A(t) - \mathrm{c}t$, but the supremum operator is not continuous in $C([0,\infty))$. As a result, the continuous mapping theorem cannot be applied. In other words, although the superposition of on-off sources converges to some Gaussian process, it is not at all obvious that the queue fed by the on-off sources converges to the corresponding Gaussian queue; there is really something to prove.

Whereas many papers deal with the convergence of arrival processes to some limiting process, considerably less attention is paid to the relevant question whether this convergence carries over to the buffer content distribution. A few important contributions, however, have been made. In this respect we mention [161]: under a specific parameterization the queue fed by n exponential on-off sources (the so-called Anick–Mitra–Sondhi model) converges to a queue with Gaussian input (with an iOU-type variance structure), as n grows large. This result was generalized in [67] to general on-off sources. In [64] the case of heavy-tailed on- or off-times is considered: after the rescaling of time introduced in Section 3.2, it is proven that in a certain heavy-traffic regime the corresponding queue indeed converges to a Gaussian queue with fBm input, as could be expected from Proposition 3.2.5.

This type of convergence results for queues provides a further justification for the use of Gaussian queues, despite the somewhat unsatisfactory feature of negative traffic.

Remark 5.2.1 Let $A(\cdot)$ be a Gaussian process with mean rate μ and variance function $v(\cdot)$. Consider the 'centered version' $\overline{A}(\cdot)$ of $A(\cdot)$, i.e., the Gaussian process with mean rate 0 and variance function $v(\cdot)$. With the stability $\mu < \mathrm{c}$ in force, it is trivial that

$$\mathbb{P}\left(\sup_{t\geq 0} A(t) - \mathrm{c}t \geq \mathrm{B}\right) = \mathbb{P}\left(\sup_{t\geq 0} \overline{A}(t) - (\mathrm{c}-\mu)t \geq \mathrm{B}\right).$$

As a consequence, when we have reduced the service rate c by the mean rate μ of the input process, we can restrict ourselves, without loss of generality, to considering just the centered sources. $\diamond$

We emphasize that Gaussian queues are notoriously hard to analyze. In fact, only the cases of the Brownian motion and Brownian bridge have been solved explicitly, in the sense that a clean expression for the steady-state buffer content distribution has been given. We describe these results in Section 5.3. For general variance functions, we have to resort to approximations, asymptotic results, and simulations.

Fortunately, a powerful, and intuitively well understood, heuristic exists, that is based on first principles. It turns out that for the special cases of Brownian motion and Brownian bridge the resulting approximation is exact (see Section 5.4). Then we argue in Section 5.5 that there are several asymptotic regimes that are relevant in the context of Gaussian queues, most notably the large-buffer asymptotics and the many-sources asymptotics. Asymptotics of the former kind are further treated in Section 5.6, whereas those of the second kind are addressed in detail in the next chapters.

5.3 Special cases: Brownian motion and Brownian bridge

As mentioned above, for two types of Gaussian queues an explicit solution for the buffer content distribution has been found: the Brownian motion and the Brownian Bridge.

Brownian motion. As argued in Section 2.5, (scaled) Brownian motion can be regarded as the Gaussian counterpart of the Poisson process, and as a result the queue with Brownian input may serve as an approximation to the M/D/1 queue.

More precisely, consider an arrival process in which at a Poisson rate λ unit-sized jobs are generated. Suppose these feed into a queue that is drained at a constant rate 1 (where $\lambda < 1$). Then the arrival stream could be approximated by $\sqrt{\lambda}\, B(t) + \lambda t$, with $B(\cdot)$ standard Brownian motion. As a consequence, we could use

$$\mathbb{P}\left(\sup_{t\geq 0} \sqrt{\lambda} B(t) - (1-\lambda)t \geq \mathrm{B}\right)$$

as an approximation for the probability of the steady-state buffer content in the M/D/1 queue exceeding B.

Let us therefore focus on the distribution of $Q = \sup_{t\geq 0} B(t) - \mathrm{c}t$. It is not hard to see that Q has an exponential structure, due to the independent increments and continuous paths of Brownian motion: as a consequence, we have that $\alpha(\mathrm{B}, \mathrm{C}) := \mathbb{P}(Q > \mathrm{B}) = \exp(-\delta \mathrm{B})$. It remains to determine the positive constant δ. To this end, we first observe that the process $\exp(2\mathrm{c}B(t) - 2\mathrm{c}^2 t)$ is a martingale. Introduce the stopping time

$$T := \inf\{t \geq 0 : B(t) = \mathrm{B} + \mathrm{c}t\}.$$

Then, by virtue of the optional stopping theorem, we conclude that

$$1 = \mathbb{E}e^{2\mathrm{c}B(T) - 2\mathrm{c}^2 T}\ \alpha(\mathrm{B}, \mathrm{C}).$$

As $B(T) = \mathrm{B} + \mathrm{c}T$ almost surely, we have found that $\alpha(\mathrm{B}, \mathrm{C}) = \exp(-2\mathrm{BC})$. Put differently, the steady-state buffer content distribution Q of reflected Brownian motion has an exponential distribution with mean $1/2\mathrm{c}$.

Brownian Bridge. Consider the $N{\cdot}$D/D/1 queue (with N i.i.d. periodic input streams, transmitting a packet every time unit, feeding into a queue that is emptied at rate $N\textsc{c}$, with $\textsc{c} \geq 1$). This queuing system received substantial attention in the context of constant bit rate applications in ATM in the early 1990s [137, 254, 228]. Its exact solution, however, turned out to have been around (in an entirely different branch of the literature) for several decades [74, 247, 272].

As this exact solution is rather implicit, there has been interest in finding accurate approximations. We have seen in Section 2.5 that a periodic stream has the same variance function as the Brownian Bridge. This gives rise to the idea of approximating the $N{\cdot}$D/D/1 queue by

$$\mathbb{P}\left(\sup_{t\geq 0} B(t) - (\textsc{c}-1)t > \frac{\textsc{b}}{N} \mid B(1) = 0\right).$$

To compute this probability, it is sufficient to know $\beta(\textsc{b}, \textsc{c}, 0)$, with

$$\beta(\textsc{b}, \textsc{c}, a) := \mathbb{P}\left(\sup_{t\geq 0} B(t) - \textsc{c}t \geq \textsc{b} \mid B(1) = a\right).$$

We remark that it is straightforward to check (use the formulas for conditional means and variances of bivariate Normal random variables) that $\beta(\textsc{b}, \textsc{c}, a) = \beta(\textsc{b}, \textsc{c} - a, 0)$.

It holds that $\beta(\textsc{b}, \textsc{c}, 0)$ equals $\exp(-2\textsc{b}(\textsc{b} + \textsc{c}))$. We do not give a constructive derivation of this explicit formula, but rather show that this solution must be right through an indirect argument; we choose to follow this approach, as a similar construction will be used again in Chapter 12.

We have derived for reflected Brownian motion that $\alpha(\textsc{b}, \textsc{c}) = \exp(-2\textsc{bc})$. Suppose that we want to verify this formula for $\alpha(\textsc{b}, \textsc{c})$ by using that $\beta(\textsc{b}, \textsc{c}, 0) = \exp(-2\textsc{b}(\textsc{b} + \textsc{c}))$, or, equivalently, that $\beta(\textsc{b}, \textsc{c}, a) = \exp(-2\textsc{b}(\textsc{b} + \textsc{c} - a))$. This can be done in the following manner. It is clear that

$$1 - \alpha(\textsc{b}, \textsc{c}) = \mathbb{P}(\forall t \geq 0 : B(t) < \textsc{b} + \textsc{c}t).$$

Now let us try to find an alternative expression for the right-hand side of the previous display by conditioning on the value of $B(1)$. $B(1)$ has a $\mathcal{N}(0, 1)$ distribution, and to make sure that $B(t) < \textsc{b} + \textsc{c}t$ for all $t \geq 0$, we should have that $B(1)$ lies somewhere in the range $(-\infty, B + C)$. This leads to the following identity:

$$1 - \alpha(\textsc{b}, \textsc{c}) = \int_{-\infty}^{B+C} \frac{1}{\sqrt{2\pi}} e^{-a^2/2} (1 - \beta(\textsc{b}, \textsc{c}, a))(1 - \alpha(\textsc{b} + \textsc{c} - a, \textsc{c}))\, \mathrm{d}a,$$

where (i) the term $1 - \beta(\textsc{b}, \textsc{c}, a)$ accounts for the requirement that $B(t) < \textsc{b} + \textsc{c}t$ for all $t \in (0, 1)$, conditional on $B(1) = a$, and (ii) the term $1 - \alpha(\textsc{b} + \textsc{c} - a, \textsc{c})$ for the requirement that $B(t) < \text{B} + \text{C}t$ for all $t > 1$, conditional on $B(1) = a$ (which is equivalent to requiring that $B(t) < \text{B} + \text{C} - a + \text{C}t$ for all $t > 0$).

The integral of the previous display can be expanded into four terms, and we can insert $\alpha(\text{B}, \text{C}) = \exp(-2\text{BC})$ and $\beta(\text{B}, \text{C}, a) = \exp(-2\text{B}(\text{B} + \text{C} - a))$. After tedious calculations, the four terms reduce to

$$\Phi(\text{B} + \text{C}) - \Phi(-\text{B} + \text{C})e^{-2\text{BC}} - \Phi(\text{B} - \text{C})e^{-2\text{BC}} + \Phi(-\text{B} - \text{C}),$$

with $\Phi(\cdot)$, as before, the standard normal distribution function. Using that $\Phi(x) + \Phi(-x) = 1$, we indeed find, adding the terms up, $1 - \exp(-2\text{BC})$, as desired. We conclude that apparently the formula we used for $\beta(\text{B}, \text{C}, a)$ was correct, in that it is consistent with the result for reflected Brownian motion derived earlier.

5.4 A powerful approximation

Unfortunately, the only queues that allow an explicit solution are those studied in the previous section, i.e., those associated with Brownian motion and Brownian Bridge. This has led researchers to try to find approximations for the situation of a general correlation structure. In this section, we present an elegant and powerful approximation, that turns out to be exact for Brownian motion and Brownian bridge; the next sections show that the approximation appears to work well in various asymptotic regimes. For ease, we assume that the Gaussian source is centered (see Remark 5.2.1).

The approximation approach is based on two principles: (1) the principle of the largest term, and (2) the Chernoff bound. Let us first rewrite our overflow probability as

$$\begin{aligned}\mathbb{P}(Q \geq \text{B}) &= \mathbb{P}\left(\sup_{t\geq 0} A(t) - \text{C}t \geq \text{B}\right) = \mathbb{P}\left(\exists t \geq 0 : A(t) \geq \text{B} + \text{C}t\right) \\ &= \mathbb{P}\left(\bigcup_{t\geq 0}\{A(t) \geq \text{B} + \text{C}t\}\right).\end{aligned}$$

In other words, the probability of our interest is the probability of a union of events (where it is noted that the number of events is uncountably infinite). Clearly, the probability of a union of events is larger than the probability of each individual event, and hence also larger than the largest of these. This reasoning yields the lower bound

$$\mathbb{P}\left(\bigcup_{t\geq 0}\{A(t) \geq \text{B} + \text{C}t\}\right) \geq \sup_{t\geq 0} \mathbb{P}\left(A(t) \geq \text{B} + \text{C}t\right). \tag{5.2}$$

The next question is, of course, how big a gap there is between the probability of our interest and the lower bound in Equation (5.2). To this end, recall that in

Chapter 4 we have seen that the probability of a rare event is usually strongly dominated by the likelihood of the most probable realization. This explains why it can be expected that the lower bound in Equation (5.2) is reasonably tight, so that we may write

$$\mathbb{P}\left(\bigcup_{t\geq 0}\{A(t)\geq \mathrm{B}+\mathrm{C}t\}\right)\approx \sup_{t\geq 0}\mathbb{P}\left(A(t)\geq \mathrm{B}+\mathrm{C}t\right)=\mathbb{P}\left(A(t^\star)\geq \mathrm{B}+\mathrm{C}t^\star\right), \tag{5.3}$$

where $t^\star$ is the maximizer in the right-hand side of Equation (5.2). This procedure is usually referred to as the *principle of the largest term.*

To further evaluate Equation (5.3), we now focus, for fixed t, on the probability $\mathbb{P}(A(t)\geq \mathrm{B}+\mathrm{C}t)$. We can use the Chernoff bound, i.e., Proposition 4.1.2, to bound this probability from above:

$$\begin{aligned}\mathbb{P}\left(A(t)\geq \mathrm{B}+\mathrm{C}t\right)&\leq \exp\left(-\sup_{\theta\geq 0}\left(\theta(\mathrm{B}+\mathrm{C}t)-\log\mathbb{E}e^{\theta A(t)}\right)\right)\\&=\exp\left(-\frac{(\mathrm{B}+\mathrm{C}t)^2}{2v(t)}\right).\end{aligned}$$

Again this is just a bound (but now an upper bound, rather than a lower bound...). However, also in this case it can be argued that one could expect it to be rather tight; in the context of sample means of i.i.d. random variables, for instance, the Bahadur-Rao estimate indicates that the Chernoff bound is reasonably accurate, in the sense that the exponent that appeared in the bound was 'correct'.

The resulting approximation is the following:

Approximation 5.4.1 *For a centered Gaussian source with variance function $v(\cdot)$,*

$$\mathbb{P}(Q\geq \mathrm{B})\approx \sup_{t\geq 0}\exp\left(-\frac{(\mathrm{B}+\mathrm{C}t)^2}{2v(t)}\right)=\exp\left(-\inf_{t\geq 0}\frac{(\mathrm{B}+\mathrm{C}t)^2}{2v(t)}\right). \tag{5.4}$$

In line with the interpretation of $f^\star := \arg\inf_{f\in\mathcal{S}}\mathbb{I}(f)$ as the most likely path in set $\mathcal{S}$, we can regard the optimizing t in Equation (5.4) as the most likely epoch for $A(t)$ to exceed $\mathrm{B}+\mathrm{C}t$. Conditional on $A(t)$ being larger than $\mathrm{B}+\mathrm{C}t$ for some t, this most likely happens at $t=t^\star$. In queuing language, given that we see the buffer content exceeding level B at time 0, then the busy period preceding this exceedance started at time $-t^\star$. We return to this *most likely timescale of overflow* several times later in this book.

Interestingly, for the cases that could be analyzed explicitly (see Section 5.1) Approximation 5.4.1 turns out to be exact.

Exercise 5.4.2 Show that Approximation 5.4.1 is exact for the Brownian motion and the Brownian bridge.

Solution. For Brownian motion we have to choose $v(t) = t$. It turns out that $t^\star = \text{B}/\text{C}$, and the approximation yields $\exp(-2\text{BC})$, as desired.

For Brownian bridge we insert $\mu = 0$ and $v(t) = t(1-t)$. Now

$$t^\star = \frac{\text{B}}{\text{C} + 2\text{B}},$$

which yields the approximation $\exp(-2\text{B}(\text{B}+\text{C}))$, as desired. ◇

Exercise 5.4.3 Determine Approximation 5.4.1 for fBm.

Solution. Take $v(t) = t^{2H}$, with $H \in (0,1)$. Direct calculations show that

$$t^\star = \frac{\text{B}}{\text{C}} \frac{H}{1-H}.$$

It yields the approximation

$$\mathbb{P}(Q \geq \text{B}) \approx \exp\left(-\frac{1}{2}\left(\frac{\text{B}}{1-H}\right)^{2-2H}\left(\frac{\text{C}}{H}\right)^{2H}\right).$$

Interestingly, as a function of B, $\mathbb{P}(Q \geq \text{B})$ decays in a 'Weibullian' way, i.e., roughly as $\exp(-\text{B}^{2-2H})$. If $H > 1/2$ this is slower than exponential (which could be expected on the basis of the long-range dependence behavior of the input process); for $H < 1/2$ the decay is faster than exponential (the negative correlations help to prevent long queues). ◇

Exercise 5.4.4 Determine Approximation 5.4.1 for iOU, and B large.

Solution. We sketch a derivation; we come back to this later, in Sections 6.1 and 6.2. Large values of B correspond to large value of the optimizing t. In that region, $v(t) \approx t - 1$. So we have to find

$$\inf_{t \geq 0} \frac{(\text{B} + \text{C}t)^2}{t-1}.$$

It turns out that $t^\star = \text{B}/\text{C} + 2$. This leads to $\mathbb{P}(Q \geq \text{B}) \approx \exp(-2\text{C}(\text{B}+\text{C}))$. In other words, this probability decays (roughly) exponentially in B, approximately as $\exp(-2\text{BC})$. We observe that this resembles the approximation for the case of Brownian input. This is in line with the srd nature of iOU. ◇

5.5 Asymptotics

As said above, only in a few special cases, the distribution of $\sup_{t \geq 0} A(t) - \text{C}t$ can be analyzed explicitly. We proposed an approximation, which was exact for those special cases, but its accuracy remained unclear in general. One way to assess the performance of the approximation could be to study asymptotic regimes in which it is, according to some metric, exact.

In queuing literature, many asymptotic scalings are used. In the context of this book, two regimes are of particular interest, viz. the *large-buffer regime* and the *many-sources* regime.

- In the large-buffer regime, the objective is to find asymptotic expansions of the probability $\mathbb{P}(Q \geq \text{B})$ for $\text{B} \to \infty$. Then again there are two types of results. In the first place there are what could be called 'logarithmic asymptotics': find an explicit function $f(\cdot)$ such that

  $$\frac{\log \mathbb{P}(Q \geq \text{B})}{f(\text{B})} \to 1, \quad \text{as } \text{B} \to \infty.$$

  Secondly there are 'exact asymptotics': find an explicit function $g(\cdot)$ such that

  $$\frac{\mathbb{P}(Q \geq \text{B})}{g(\text{B})} \to 1, \quad \text{as } \text{B} \to \infty.$$

  It is clear that exact asymptotics are 'stronger', in that the exact asymptotics immediately yield the logarithmic asymptotics (but not vice versa). Exact asymptotics are often considerably harder to prove, though.

  In formulae, logarithmic asymptotics provide us with an explicit function $f(\cdot)$ such that $\mathbb{P}(Q \geq \text{B}) = \phi(\text{B}) \exp(f(\text{B}))$, for an (unknown) function $\phi(\cdot)$ such that

  $$\frac{\log \phi(\text{B})}{f(\text{B})} \to 0$$

  as $\text{B} \to \infty$; exact asymptotics on the contrary characterize this $\phi(\text{B})$ for B large.

  (Notice that we have seen the distinction between logarithmic and exact asymptotics earlier in this monograph: 'Cramér' provides the logarithmic asymptotics of the large deviations of the sample mean, whereas 'Bahadur-Rao' gives the corresponding exact asymptotics.)

  Section 5.6 reviews the most important large-buffer results.

- The focus in this book, however, is on the many-sources setting. In the many-sources setting it is, not surprisingly, assumed that the number of sources n grows large. At the same time, however, the other resources are scaled: the buffer threshold B is replaced by nb, whereas the service capacity C is replaced by nc. Despite the fact that the load imposed on the system remains constant, it is clear that the probability of exceeding nb decays to 0.

  More concretely, our goal is to analyze buffer content under the many-sources scaling, defined as Q_n, exceeds nb; we have added the subscript n here, as the buffer content is obviously a function of n (there are n i.i.d. sources feeding into a buffer emptied at rate nc). Then we focus on the analysis of

$$p_n(b,c) := \mathbb{P}(Q_n \geq nb) = \mathbb{P}\left(\sup_{t\geq 0}\sum_{i=1}^{n} A_i(-t,0) - nct \geq nb\right).$$

Again, the results can be divided into logarithmic asymptotics and exact asymptotics.

5.6 Large-buffer asymptotics

In this section, we comment on the accuracy of Approximation 5.4.1 in the regime of large buffers.

Logarithmic asymptotics. In [60] the following result was proven. It shows that Approximation 5.4.1 is what could be called 'logarithmically correct'.

Proposition 5.6.1 *For any* $\mathrm{C} > 0$,

$$\lim_{\mathrm{B}\to\infty} \log \mathbb{P}(Q \geq \mathrm{B}) \Bigg/ \left(\inf_{t\geq 0} \frac{(\mathrm{B}+\mathrm{C}t)^2}{2v(t)}\right) = -1.$$

The upper bound of the proof in [60] heavily relies on one of the cornerstones in the analysis of extreme values of Gaussian processes: Borell's inequality; we will use this strong result later, in Section 7.4. The lower bound is essentially the 'principle of the largest term', as explained in Section 5.4.

In the special case of fBm input, we obtain

$$\lim_{\mathrm{B}\to\infty} \log \mathbb{P}(Q \geq \mathrm{B}) \Bigg/ \left(\frac{1}{2}\left(\frac{\mathrm{B}}{1-H}\right)^{2-2H}\left(\frac{\mathrm{C}}{H}\right)^{2H}\right) = -1. \tag{5.5}$$

Exact asymptotics. It turned out that exact asymptotics were substantially harder to obtain than logarithmic asymptotics. By the end of the 1990s, the exact asymptotics corresponding to the case of fBm input were found. In the resulting asymptotic expansions, a somewhat mysterious constant played a crucial role: the so-called *Pickands constant.* It is defined through the following limit:

$$\mathcal{H}_\alpha := \lim_{T\to\infty} \frac{1}{T} \cdot \mathbb{E}\exp\left(\sup_{t\in[0,T]}\left(\sqrt{2}B_{\alpha/2}(t) - t^\alpha\right)\right),$$

where $\alpha \in (0, 2]$, and B_H is fBm with Hurst parameter H. Only for special values of α this expression can be explicitly evaluated (viz. $\mathcal{H}_1 = 1$ and $\mathcal{H}_2 = 1/\sqrt{\pi}$). It remained long unclear whether this function was continuous or not. Recently this continuity was established by [61]. Remarkably enough, it has turned out to be extremely hard, if not unfeasible, to obtain numerical values for $\mathcal{H}_\alpha$ through simulation.

The exact asymptotics were found in [138], but see also [220]. They indicate that the proposed approximation is, up to a hyperbolic 'prefunction', asymptotically exact. In [68, Theorem 4.3] the result is stated as follows. Define:

$$\alpha(H) := \frac{\mathcal{H}_{2H}\sqrt{\pi}}{2^{(1-H)/2H}\sqrt{H}}\left(\frac{H}{c(1-H)}\right)^{H-1}\left(\frac{1}{1-H}\right)^{(2-H)/H};$$

$$\beta(H) := \left(\frac{c(1-H)}{H}\right)^{H}\frac{1}{1-H}.$$

Then, it turns out that the asymptotics of $\mathbb{P}(Q \geq \mathrm{B})$ factorize into a Weibullian term (which was dominant in the logarithmic asymptotics, see Proposition 5.6.1) and a polynomial term.

Proposition 5.6.2 *For the queue with fBm input,* $\mathrm{C} > 0$*,*

$$\lim_{\mathrm{B}\to\infty} \mathbb{P}(Q \geq \mathrm{B}) \Big/ \mathrm{B}^{2H-3+1/H} \exp\left(-\frac{1}{2}\left(\frac{\mathrm{B}}{1-H}\right)^{2-2H}\left(\frac{\mathrm{C}}{H}\right)^{2H}\right) = \frac{\alpha(H)}{\sqrt{2\pi}\,\beta(H)}.$$

Interestingly,

$$2H - 3 + \frac{1}{H} = (1-H)\cdot\left(\frac{1-2H}{H}\right) < 0 \text{ iff } H > 1/2.$$

In other words, the hyperbolic prefunction is increasing for $H > 1/2$, and decreasing for $H < 1/2$.

After the establishment of Proposition 5.6.2, several extensions followed. Short-range dependent processes were studied in [61]; this was done by considering so-called Gaussian integrated processes. The most complete framework was however presented in a recent contribution [78], where the case of regularly varying (at ∞) variance function is studied in detail. These studied rely on a substantial amount of rather technical machinery. Most notably, the so-called *double sum method* was intensively applied. The results enabled the derivation of interesting reduced-load relations [79], cf. also [300]; 'reduced-load equivalence' means that in the queuing asymptotics (some of) the light-tailed sources can be replaced by their mean [7].

The following exercise shows an interesting connection between large-buffer asymptotics and many-sources asymptotics, for the case of fBm input.

Exercise 5.6.3 Let $Q := \sup_{t\geq 0} A(t) - \mathrm{c}t$ and

$$Q_n := \sup_{t\geq 0} \sum_{i=1}^{n} A_i(t) - n\mathrm{c}t,$$

where $A(\cdot)$ and the $A_i(\cdot)$ are centered fBm s with variance function t^{2H}. Prove that

$$\lim_{\mathrm{B}\to\infty} \frac{1}{\mathrm{B}^{2-2H}} \log \mathbb{P}(Q \geq \mathrm{B}) = \lim_{n\to\infty} \frac{1}{n} \log \mathbb{P}(Q_n \geq n).$$

(Notice that we know the right-hand side from Equation (5.5). In other words, this knowledge of the logarithmic large-buffer asymptotics also immediately yields the logarithmic many-sources asymptotics.)

Solution. This is done as follows. We first replace B^{2-2H} by n, and then rescale time by a factor $n^{1/(2-2H)}$:

$$\begin{aligned}
&\lim_{\mathrm{B}\to\infty} \frac{1}{\mathrm{B}^{2-2H}} \log \mathbb{P}\left(\exists t \geq 0 : A(t) - \mathrm{c}t \geq \mathrm{B}\right) \\
&\quad = \lim_{n\to\infty} \frac{1}{n} \log \mathbb{P}\left(\exists t \geq 0 : A(t) - \mathrm{c}t \geq n^{1/(2-2H)}\right) \\
&\quad = \lim_{n\to\infty} \frac{1}{n} \log \mathbb{P}\left(\exists t \geq 0 : A(tn^{1/(2-2H)}) - \mathrm{c}tn^{1/(2-2H)} \geq n^{1/(2-2H)}\right).
\end{aligned}$$

Now use the selfsimilarity of fBm: $A(\alpha t)$ is distributed as $\alpha^H A(t)$. Hence the previous display can be rewritten as:

$$\lim_{n\to\infty} \frac{1}{n} \log \mathbb{P}\left(\exists t \geq 0 : n^{H/(2-2H)} A(t) - \mathrm{c}tn^{1/(2-2H)} \geq n^{1/(2-2H)}\right).$$

By multiplying both sides of the inequality by $n^{-1/(2-2H)}$, we obtain

$$\lim_{n\to\infty} \frac{1}{n} \log \mathbb{P}\left(\exists t \geq 0 : n^{(H-1)/(2-2H)} A(t) - \mathrm{c}t \geq 1\right).$$

As $n^{(H-1)/(2-2H)} = 1/\sqrt{n}$, we arrive at

$$\lim_{n\to\infty} \frac{1}{n} \log \mathbb{P}\left(\exists t \geq 0 : \sqrt{n} A(t) - n\mathrm{c}t \geq n\right).$$

Now recall $\sqrt{n}A(t)$ is distributed as $\sum_{i=1}^{n} A_i(t)$ (it is easily verified that both are normally distributed, with variance nt^{2H}), so that we obtain

$$\lim_{n\to\infty}\frac{1}{n}\log\mathbb{P}\left(\exists t \geq 0 : \sum_{i=1}^{n} A_i(t) - n\mathrm{c}t \geq n\right),$$

as desired. 

Bibliographical notes

The first to consider a queue with 'nontrivial' Gaussian input was Norros [221], who found a lower bound for $\mathbb{P}(Q \geq \mathrm{B})$ for the case of fBm input (in fact, it corresponds to the bound obtained by applying the 'principle of the largest term'); see also [222]. In the next years, the queue fed by fBm remained an extremely challenging object of study; Duffield and O'Connell [84] found logarithmic asymptotics for fBm (and other input processes). These were extended by Dębicki to general Gaussian sources [60]. The approximation of Section 5.4 appeared in several papers; see also, for instance, Section 10.3 of [109].

In the area of exact asymptotics, we mention the contributions on fBm by Hüsler and Piterbarg [138], Narayan [220], and Massoulié and Simonian [208]. A key result in this area was derived in a pioneering study by Pickands [240]; he characterized the tail distribution of certain centered Gaussian processes over a finite interval (in terms of $\mathcal{H}_{2H}$). Dębicki [61] extended the results to different classes of Gaussian sources, most notably the so-called Integrated Gaussian sources, which give rise to srd behavior; this led to the introduction of so-called *generalized Pickands constants*, see also [157]. We refer to the nice survey by Dębicki and Rolski [68].

The body of literature devoted to large-buffer asymptotics of Gaussian processes is vast. We mention interesting work by Dębicki and Rolski [69] and Duncan, Yan, and Yan [86] on maxima of Gaussian processes over *finite* time intervals, and, for the case of fBm input, work by Zeevi and Glynn [295] and Piterbarg [244] on the maximum of the stationary buffer-content process.

As mentioned above, the most general setting treated so far is presumably by Dieker [78], which relies on a delicate extension of the double-sum method. He finds exact asymptotics, under the assumption of a regularly varying (at ∞) variance function. The double-sum technique was originally developed by Pickands [240]; see also [242, 243].

Here, we also mention a series of articles by Choe and Shroff [49, 50, 51] who focus on logarithmic large-buffer asymptotics. In [49] the srd case is dealt with, and sharp estimates of the type $\mathbb{P}(Q \geq \mathrm{B}) \leq \alpha e^{-\beta\mathrm{B}}$ are given. One of the main results of [51] is that, under a set of mild conditions imposed on the variance function $v(\cdot)$ (which allow the input to be lrd),

$$\log \mathbb{P}(Q \geq \mathrm{B}) + \left(\inf_{t \geq 0} \frac{(\mathrm{B} + \mathrm{C}t)^2}{2v(t)} \right) \in O(\log \mathrm{B}),$$

as $\mathrm{B} \to \infty$. Notice that this type of asymptotics are stronger than those of Proposition 5.6.1, but weaker than the exact asymptotics. For instance for the case of fBm, the result reads

$$\log \mathbb{P}(Q \geq \mathrm{B}) + \frac{1}{2} \left(\frac{\mathrm{B}}{1 - H} \right)^{2-2H} \left(\frac{\mathrm{C}}{H} \right)^{2H} \in O(\log \mathrm{B}),$$

while from Proposition (5.6.2) we know that this $O(\log \mathrm{B})$ term roughly behaves as $(2H - 3 + 1/H) \log \mathrm{B}$ (and in fact the exact asymptotics even give, besides this hyperbolic term, the multiplicative constant $\alpha(H)/\sqrt{2\pi}\,\beta(H)$).

Chapter 6

Logarithmic many-sources asymptotics

In the previous chapter, we presented the fundamental approximation for the buffer-content distribution

$$\mathbb{P}(Q \geq \text{B}) \approx \exp\left(-\inf_{t\geq 0} \frac{(\text{B} + \text{C}t)^2}{2v(t)} \right) \tag{6.1}$$

(Approximation 5.4.1) as well as a number of results on large-buffer asymptotics. The main conclusion was that the Approximation (6.1) was backed up by the large-buffer asymptotics. In this chapter (and the next chapter), we study the powerful many-sources regime, and we will see that even in this regime important additional support for Approximation (6.1) can be found.

This chapter is structured as follows. In Section 6.1 we first use Schilder's theorem to find the logarithmic asymptotics. These give clean, explicit formulas that provide much additional insight into the loss statistics. In the first place, knowledge of the logarithmic asymptotics yields an interesting duality relation between the decay rate (as a function of the buffer level b), and the variance function (on the most likely timescale of exceeding b); this duality is explored in Section 6.2. In the second place, we show in Section 6.3 that, for a fixed decay rate δ, the buffer size b and link capacity c trade off in a convex manner.

6.1 Many-sources asymptotics: the loss curve

The goal of this chapter is to analyze the logarithmic asymptotics of the probability that the buffer content under the many-sources scaling, defined as Q_n, exceeds nb:

$$p_n(b, c) := \mathbb{P}(Q_n \geq nb)$$

Large deviations for Gaussian queues M. Mandjes

$$= \mathbb{P}\left(\sup_{t\geq 0}\sum_{i=1}^{n} A_i(-t,0) - nct \geq nb\right)$$

$$= \mathbb{P}\left(\sup_{t\geq 0}\frac{1}{n}\sum_{i=1}^{n} A_i(-t,0) - ct \geq b\right).$$

To use 'Schilder', we have to define a set of 'overflow paths':

$$\mathcal{S}^{(f)} := \{f \in \Omega : \exists t \geq 0 : -f(-t) \geq b + ct\}.$$

Here we use the superscript (f) as a mnemonic for FIFO, since we consider single work-conserving queues here, of which the FIFO queue is the most prominent example; we turn to larger networks and other scheduling disciplines in later chapters.

Clearly, the observation that $A(-t,0) \equiv -A(-t)$ shows that indeed $p_n(b,c) = p_n[\mathcal{S}^{(f)}]$. This entails that we can apply Schilder's theorem to obtain

$$\lim_{n\to\infty}\frac{1}{n}\log p_n(b,c) = -\inf_{f\in\mathcal{S}^{(f)}}\mathbb{I}(f).$$

The logarithmic asymptotics of $p_n(b,c)$ now immediately follow from Exercise 4.2.4. We obtain the following result. With Remark 5.2.1 in mind, we restrict ourselves without loss of generality to centered sources.

Theorem 6.1.1 Logarithmic asymptotics. *For any $b, c > 0$,*

$$I_c^{(f)}(b) := -\lim_{n\to\infty}\frac{1}{n}\log p_n(b,c) = \inf_{t\geq 0}\frac{(b+ct)^2}{2v(t)}. \tag{6.2}$$

We call the decay rate $I_c^{(f)}(b)$, seen as a function of the buffer size b, and with c held fixed, the *loss curve*. In this chapter, the impact of b on the optimizing t in Equation (6.2) plays a crucial role; we therefore use the notation $t(b)$. The path $f^\star \in \mathcal{S}^{(f)}$ that optimizes $\mathbb{I}(f)$ is (see again Exercise 4.2.4)

$$f^\star(r) := \mathbb{E}(A(r) \mid A(-t(b),0) = b + ct(b))$$

$$= \frac{\Gamma(r,-t(b))}{v(t(b))}(b + ct(b)); \tag{6.3}$$

we call this the *most likely path to overflow*. Here $-t(b)$ can be interpreted as the most likely time at which the buffer starts to build up in order to exceed level nb at time 0; we therefore call $t(b)$ the *most likely timescale of overflow*.

The following exercises investigate the shape of the loss curve, and interpret the shapes of the most likely path (6.3). As Gaussian processes are reversible in time, it is equivalent to optimize over

$$\mathcal{S}_{+}^{(\mathrm{f})} := \{f \in \Omega : \exists t \geq 0 : f(t) \geq b + ct\}.$$

As it is conceptually and notationally easier to look forward in time, the rest of this section focuses on the most likely path in $\mathcal{S}_{+}^{(\mathrm{f})}$, i.e.,

$$f^{\star}(r) := \mathbb{E}(A(r) \mid A(0, t(b)) = b + ct(b)) = \frac{\Gamma(r, t(b))}{v(t(b))}(b + ct(b))$$

that reaches level $b + ct(b)$ at time $t(b)$, starting empty at time 0.

Exercise 6.1.2 (i) Determine the most likely path to overflow for the case of Brownian motion input. (ii) Determine the most likely path to overflow for the (periodically repeated) Brownian bridge; i.e., the variance curve should be taken $v(t) = t_m(1 - t_m)$, where $t_m := t \bmod 1$.

Solution. (i) Take $v(t) = t$. Then $t(b) = b/c$. In the interval $(0, t(b))$ traffic is generated at rate $(b + ct(b))/t(b) = 2c$, so that the queue fills at rate c. After $t(b)$ the queue drains at rate $-c$, such the queue is empty again at time $2b/c$. It follows directly from Equation (6.2) that the loss curve is linear: $I_c^{(\mathrm{f})}(b) = 2bc$.

(ii) In this case $t(b) = b/(c + 2b)$ (which is smaller than $1/2$). In the interval $(0, t(b))$ traffic is generated at rate $2(b + c)$, so that the queue fills at rate $2b + c$. After $t(b)$ the queue drains at rate $-2b$, such that the queue is empty again at time $2t(b)$ (which is smaller than 1). Then the queue starts to build up again at time 1, etc. The loss curve can be computed easily, and turns out to be convex: $I_c^{(\mathrm{f})}(b) = 2b(b + c)$.

For a graphical illustration of both (i) and (ii), see Figure 6.1. ◇

In the case of fBm input, the shape of the most likely path and the loss curve critically depend on the value of H.

Exercise 6.1.3 Consider a Gaussian queue with fBm input. Show that the loss curve is concave for $H > \frac{1}{2}$, and convex for $H < \frac{1}{2}$.

Solution. Take $v(t) = t^{2H}$, for some $H \in (0, 1)$. If we perform the optimization in the right-hand side of Equation (6.2), we obtain, for $b > 0$,

$$t(b) = \frac{b}{c}\frac{H}{1 - H}, \quad I_c^{(\mathrm{f})}(b) = \frac{1}{2}\left(\frac{b}{1 - H}\right)^{2-2H}\left(\frac{c}{H}\right)^{2H}. \tag{6.4}$$

We see that the loss curve $I_c^{(\mathrm{f})}(\cdot)$ is convex (concave) when the Hurst parameter is smaller (larger) than $\frac{1}{2}$.

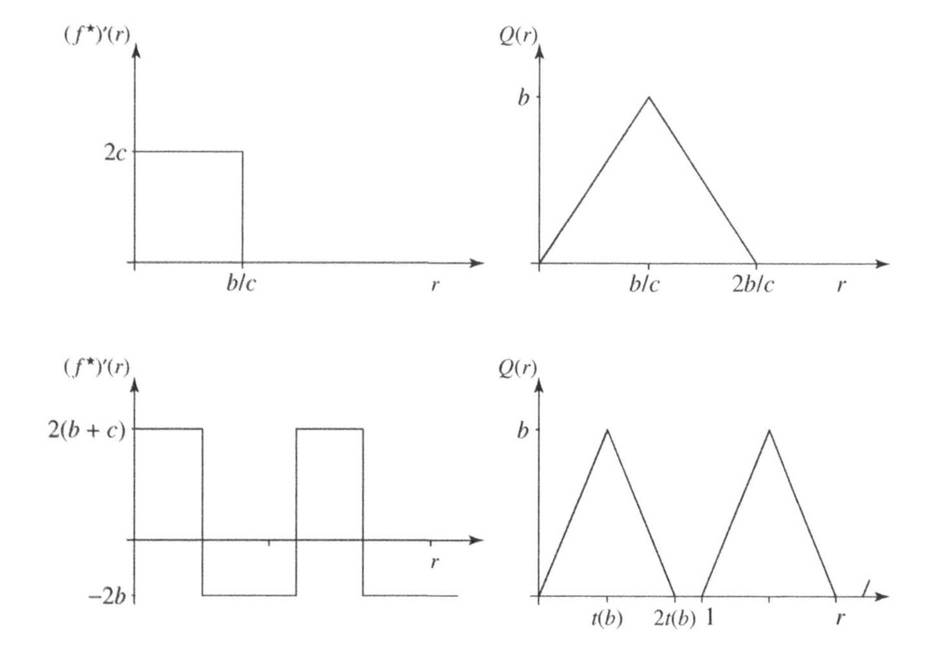

Figure 6.1: Exercise 6.1.2; the left (right) panel displays the input rate (buffer content, respectively) of the most likely path to overflow, as a function of time. The upper graphs correspond to Brownian motion, the lower graphs to Brownian Bridge.

Now we consider the properties of the most likely path, see the graphs in Figure 6.2. A direct calculation yields that, for $r \in [0, t(b)]$,

$$f^{\star}(r) = \frac{1}{2}\left((t(b))^{2H} + r^{2H} - (t(b) - r)^{2H}\right) \cdot \frac{b + ct(b)}{(t(b))^{2H}}.$$

The case $H = \frac{1}{2}$ was already dealt with in Exercise 6.1.2.

- Now consider the case with positive correlation, i.e., $H \in (\frac{1}{2}, 1)$. It is readily verified that $(f^{\star})'(0) = (f^{\star})'(t(b)) = c$, whereas

$$(f^{\star})'\left(\frac{t(b)}{2}\right) = c \cdot 2^{2-2H} > c.$$

- In case $H \in (0, \frac{1}{2})$, interestingly, $(f^{\star})'(0+) = (f^{\star})'(t(b)-) = \infty$, whereas, due to the negative correlations, $(f^{\star})'(0-) = (f^{\star})'(t(b)+) = -\infty$. Hence, the most likely path of the input rate is discontinuous. Again we have that the input rate in the middle of the interval $[0, t(b)]$ is $c \cdot 2^{2-2H} > c$. ◇

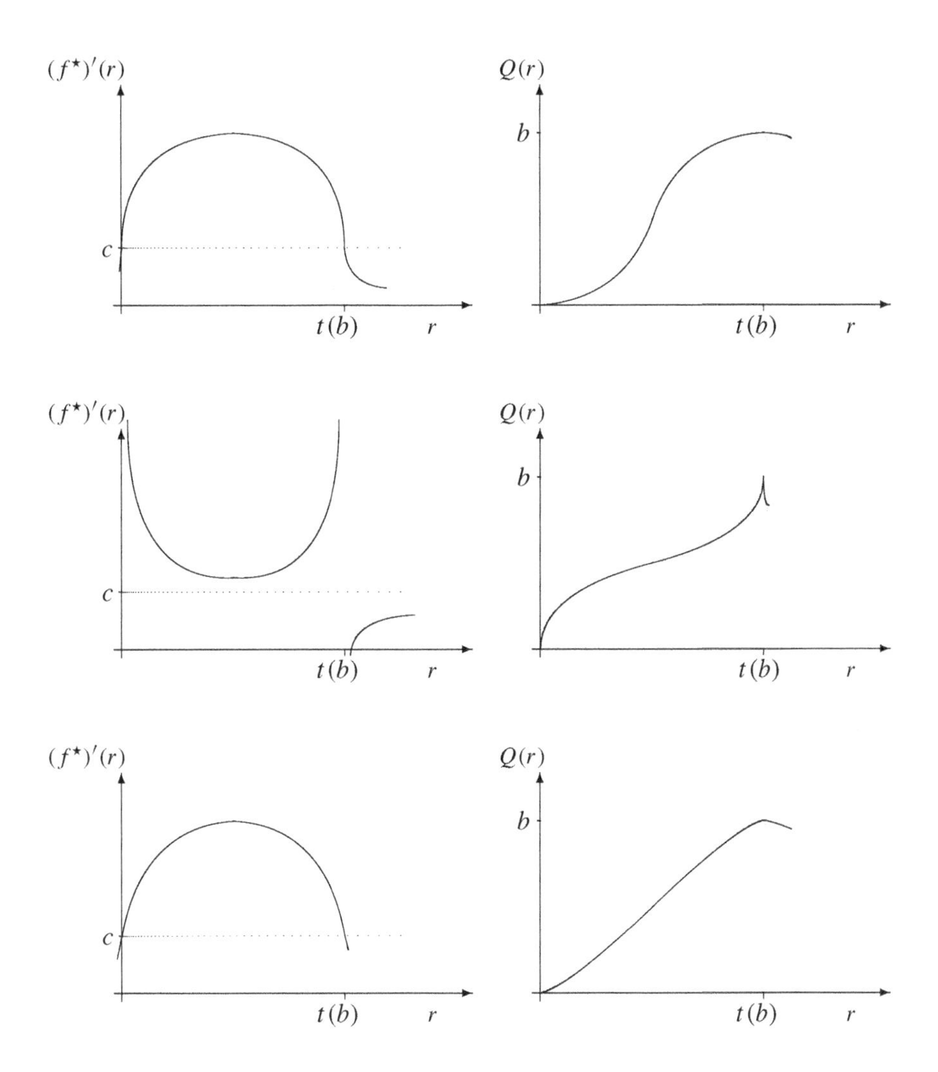

Figure 6.2: Exercises 6.1.3 and 6.1.5; the left (right) panel displays the input rate (buffer content, respectively) of the most likely path to overflow, as a function of time. The upper graphs correspond to fBm with $H > 1/2$, the middle graphs to fBm with $H < 1/2$, lower graphs to iOU.

Behavior of loss curve in 0. In the above examples, we found that the loss curve starts in 0, i.e., $I_c^{(f)}(0) = 0$. This is not true in general. To study this issue, consider variance curves of the type

$$v(t) = pt^2 - qt^3 + O(t^4), \quad \text{for } t \downarrow 0. \tag{6.5}$$

It is clear that $p \geq 0$ (as variances are nonnegative). Also, because Cauchy-Schwartz implies that $v(2t) \leq 4v(t)$ (see Exercise 2.3.2), we find that $q \geq 0$. By parameterizing $t \equiv \alpha\sqrt{b}$, we obtain

$$\begin{aligned} I_c^{(f)}(b) &= \inf_{t\geq 0} \frac{(b+ct)^2}{2pt^2 - 2qt^3 + O(t^4)} \\ &= \inf_{\alpha\geq 0} \frac{(\sqrt{b}+c\alpha)^2}{2p\alpha^2 - 2q\alpha^3\sqrt{b} + O(b)} \\ &= \frac{c^2}{2p} + \inf_{\alpha\geq 0}\left(\frac{c}{p\alpha} + \frac{c^2 q\alpha}{2p^2}\right)\sqrt{b} + O(b). \end{aligned}$$

The latter infimum is attained by $\alpha = \sqrt{2p/cq}$, so that $t(b) = \sqrt{2p/q}\cdot\sqrt{b/c}$. This yields, for $b \downarrow 0$,

$$I_c^{(f)}(b) = \frac{c^2}{2p} + \sqrt{2q}\cdot\left(\frac{c}{p}\right)^{3/2}\cdot\sqrt{b} + O(b). \tag{6.6}$$

Hence, in particular, $I_c^{(f)}(0) = c^2/2p$. We can therefore approximate the probability of a nonempty buffer by

$$\mathbb{P}(Q_n > 0) \approx \exp\left(-n\cdot\frac{c^2}{2p}\right).$$

Notice that a Gaussian process $A(\cdot)$ with a variance function that satisfies Equation (6.5) is *smooth* in the sense of Definition 2.4.4:

$$\lim_{\epsilon\downarrow 0}\frac{\mathbb{C}(t,\epsilon)}{v(\epsilon)} = \lim_{\epsilon\downarrow 0}\frac{\epsilon^2 v(t)}{2v(\epsilon)} = \frac{v(t)}{p} > 0.$$

Generally speaking, for smooth processes there is a notion of traffic rate, in the sense that $\mathbb{P}(Q_n > 0)$ can be interpreted as the probability that the aggregate input rate of the n sources exceeds nc at an arbitrary point in time. For nonsmooth processes (such as fBm) this probability has decay rate 0, but for smooth processes there is a strictly positive decay rate $c^2/2p$.

This result can be related to Corollary 4 in [5]. There a Gaussian process $A(t) = \int_0^t Z(s)\,ds$ is considered, with $Z(\cdot)$ a *stationary* Gaussian process; $R(t) := \mathbb{E}(Z(0)Z(t))$. It can be checked that $p = R(0)$; here it is used that $v''(t) = 2R(t)$, due to

$$v(t) = 2\int_0^t\int_0^s R(u)\,du\,ds,$$

cf. Exercise 3.2.2. We conclude that for this class of integrated Gaussian processes, we have that $I_c^{(f)}(0) = c^2/2p > 0$, with $p = R(0)$.

Exercise 6.1.4 (i) Consider the Gaussian counterpart of the M/G/∞ input model, as in Exercise 2.5.2. Prove that the variance function is of the form (6.5).

(ii) Consider an on-off source, with on-times having mean $\mu_{\rm on}$ and off-times having mean $\mu_{\rm off}$. Prove that the variance function is of the form (6.5).

Solution. (i) From Exercise 2.5.2 we know that

$$v''(t) = 2\lambda \int_t^\infty (1 - F_D(s))\,\mathrm{d}s.$$

This yields immediately that $p = \frac{1}{2}v''(0) = \lambda\delta$ and $q = -\frac{1}{6}v'''(0) = -\frac{1}{3}\lambda$.

(ii) Denote the distribution of the on-times (off-times) by $F_{\rm on}(\cdot)$ ($F_{\rm off}(\cdot)$, respectively). Furthermore,

$$\mu := \mu_{\rm on} + \mu_{\rm off}; \quad p_{\rm on} := \mu_{\rm on}/\mu; \quad p_{\rm off} := 1 - p_{\rm on}.$$

The probability that an arbitrary source is on at time 0 is denoted by $p_{\rm on}$; it is standard that the residual on-time has density $(1 - F_{\rm on}(\cdot))/\mu_{\rm on}$. Now focus on the kth moment of $A(t)$, for t small. Up to terms of order t^{k+2}, in self-evident notation,

$$\begin{aligned}\mathbb{E}_{\rm on}(A(t))^k &= \int_0^t \frac{(1 - F_{\rm on}(s))}{\mu_{\rm on}} \cdot s^k\,\mathrm{d}s + t^k \cdot \left(1 - \frac{t}{\mu_{\rm on}}\right) \\ &= t^k - \frac{t^{k+1}}{\mu_{\rm on}} + \frac{1}{k+1} \cdot \frac{t^{k+1}}{\mu_{\rm on}} = t^k - \frac{k}{k+1} \cdot \frac{t^{k+1}}{\mu_{\rm on}}.\end{aligned}$$

Similarly,

$$\mathbb{E}_{\rm off}(A(t))^k = \frac{1}{k+1} \cdot \frac{t^{k+1}}{\mu_{\rm off}},$$

leading to

$$\mathbb{E}(A(t))^k = p_{\rm on} \cdot t^k - \frac{1}{\mu} \cdot \frac{k-1}{k+1} \cdot t^{k+1}.$$

Now a direct calculation yields that, for $t \downarrow 0$, $\mathbb{V}\mathrm{ar}A(t) = p_{\rm on}(1 - p_{\rm on})t^2 - (3\mu)^{-1}t^3 + O(t^4)$. Hence $p := p_{\rm on}(1 - p_{\rm on})$ and $q = 1/(3\mu)$. We observe that, for t small, $\mathbb{V}\mathrm{ar}A(t)$ depends on the distribution of the on-times and off-times only through $p_{\rm on}$ and μ, or, equivalently, only through $\mu_{\rm on}$ and $\mu_{\rm off}$. As a consequence, Equation (6.6) entails that, for small b, the loss curve $I_c^{(\mathrm{f})}(b)$ depends on the on-times and off-times only through $\mu_{\rm on}$ and $\mu_{\rm off}$. This insensitivity result is in line with the findings of [190]. ◇

Exercise 6.1.5 Show that, in case of iOU, $I_c^{(\mathrm{f})}(b)$ is (i) proportional to a square root for $b \downarrow 0$ (and hence highly concave), and (ii) linear for $b \to \infty$.

Solution. First a few general remarks. The most likely timescale of overflow $t(b)$ solves

$$2c(t - 1 + e^{-t}) = (b + ct)(1 - e^{-t}); \tag{6.7}$$

this equation cannot be solved explicitly. It follows that

$$\begin{aligned}(f^\star)'(r) &= \left(1 - e^{-t(b)/2}\cosh\left(r - \frac{1}{2}t(b)\right)\right) \cdot \frac{b + ct(b)}{t(b) - 1 + \exp(-t(b))} \\ &= \left(1 - e^{-t(b)/2}\cosh\left(r - \frac{1}{2}t(b)\right)\right) \cdot \frac{2c}{1 - \exp(-t(b))},\end{aligned} \tag{6.8}$$

where the last equation is due to Equation (6.7). As a consequence, $(f^\star)'(0) = (f^\star)'(t(b)) = c$, whereas

$$(f^\star)'\left(\frac{t(b)}{2}\right) = 2c \cdot \frac{1 - e^{-t(b)/2}}{1 - e^{-t(b)}}$$

(where it is readily verified that the latter expression is larger than c). As observed in [5], the hyperbolic cosine was also obtained in [283] for the (non-Gaussian) model with exponential on-off sources; this could be expected, as iOU is the Gaussian counterpart of this model.

(i) First we concentrate on small b. Parameterize $t \equiv \alpha\sqrt{b}$. It is a matter of applying Equation (6.6), with $p = \frac{1}{2}$ and $q = \frac{1}{6}$, to obtain that the infimum over α is attained at $\sqrt{6/c}$, yielding $t(b) \approx \sqrt{6b/c}$. So the loss curve is highly concave in b (it behaves as the square root $c^2 + \frac{2}{3}\sqrt{6b} + O(b)$).

Notice that the same 'square-root behavior' is found for exponential on-off sources, and also for the (non-Gaussian) M/G/∞ model with exponential jobs, see [202, 283]. Again this makes sense, as iOU is the Gaussian counterpart of both models.

(ii) For large b, the optimum over t in Equation (6.2) is attained for large t, such that $v(t)$ is approximated by $t - 1$. Doing the calculations, we obtain

$$I_c^{(\mathrm{f})}(b) - 2c(b + c) \to 0 \quad \text{as } b \to \infty,$$

with $t(b) \approx b/c + 2$; cf. also Equation (6.7). Apparently, for large b, $I_c^{(\mathrm{f})}(b)$ becomes nearly linear. This is perhaps also what could be expected: for large t, $v(t) \approx t - 1$, which resembles the variance function of Brownian motion; the shape of the loss curve of iOU for large b indicates that even the exponential decay (as a function of the buffer size) is inherited from the Brownian motion case.

For b large, it can be checked that traffic is generated roughly at a constant rate $2c$, see Equation (6.8). In other words, during the path to overflow, the input rate goes from c to $2c$ quickly, stays there for a long while, and moves back to c immediately before $t(b)$. This is a crucial difference with the case of fBm input:

for fBm it holds that for any $\alpha \in (0, 1)$ the ratio between $(f^\star)'(\alpha t(b))$ and c is constant in b. ◇

Shape of the loss curve; correlation structure. Exercises 6.1.2, 6.1.3, and 6.1.5 show an interesting phenomenon. They indicate that a convex (concave) variance curve yields a concave (convex) loss curve. This can be explained as follows. In Chapter 2, we defined

$$\mathbb{C}(t, \epsilon) := \mathbb{C}\mathrm{ov}(A(0, \epsilon), A(t, t + \epsilon)) = \tfrac{1}{2}(v(t + \epsilon) - 2v(t) + v(t - \epsilon)),$$

and saw that, for $\epsilon \downarrow 0$, this looks like $\epsilon^2 v''(t)/2$. In other words, convexity (concavity) of $v(\cdot)$ expresses positive (negative) correlation.

If there is no correlation in the traffic (Brownian motion), we saw that $I_c^{(\mathrm{f})}(b)$ is linear in b (and the overflow probability is purely exponential in b). Positive correlations (corresponding to a convex variance function) seem to make $I_c^{(\mathrm{f})}(b)$ concave, and hence the overflow probability decays subexponentially in b. This could of course be expected: the positive correlations make it harder for the queue to cope with the arrival process, so there should be slower decay than under Brownian motion. It is also clear that in the opposite case (negative correlations, expressed by a concave variance function) the correlation structure is in some sense 'benign', so it is not surprising that then $I_c^{(\mathrm{f})}(b)$ is convex, and as a result the overflow probability decays superexponentially.

This type of duality between the variance function $v(\cdot)$ and the loss curve $I_c^{(\mathrm{f})}(\cdot)$ can be made precise. This is the subject of the next section.

6.2 Duality between loss curve and variance function

The goal of this section is to study the relationship between the correlation structure of the sources and the shape of the curve $I_c^{(\mathrm{f})}(\cdot)$. Based on the observations from the previous section, we expect that there is some connection between positive (negative) correlations and concavity (convexity) of the loss curve. This connection is explored in detail in this section.

First we describe, on an intuitive level, what convexity and concavity of the loss curve mean. Evidently, $I_c^{(\mathrm{f})}(\cdot)$ is increasing. It is important to notice that, clearly, the steeper $I_c^{(\mathrm{f})}(\cdot)$ at some buffer size b, the higher the marginal benefits of an additional unit of buffering (where 'benefits' are in terms of reducing the overflow probability). If $I_c^{(\mathrm{f})}(\cdot)$ is *convex*, then adding buffering capacity is getting more and more beneficial; if $I_c^{(\mathrm{f})}(\cdot)$ is concave, then the benefit of buffering becomes smaller and smaller. This motivates the examination of the characteristics of the shape of the loss curve $I_c^{(\mathrm{f})}(\cdot)$.

A key notion in our analysis is the most likely timescale of overflow $t(b)$, i.e., the most likely duration of a busy period preceding overflow over buffer level nb, as introduced in the previous section. The main contribution of this section is that we show that the curve $I_c^{(f)}(\cdot)$ is convex (concave) in b, *if and only if* the Gaussian input exhibits negative (positive) correlations on the timescale $t(b)$. All proofs are elementary, and add insight into the marginal benefits of buffering, i.e., the nature of $I_c^{(f)}(\cdot)$ (in terms of its derivative and second derivative with respect to the buffer size b).

As we are dealing with convexity and concavity of $v(\cdot)$, we have to assume that this is formally justified. We also impose a (mild) technical assumption on $v(\cdot)$ that guarantees uniqueness of $t^\star(b)$ for all b. Define the standard deviation function by $\varsigma(t) := \sqrt{v(t)}$.

Assumption 6.2.1 *The following two assumptions are imposed on the variance function:* (i) $v(\cdot) \in C_2([0,\infty))$, *(ii)* $\varsigma(\cdot)$ *is strictly increasing and strictly concave.*

Later in this section, we illustrate our results by means of four analytic examples and two numerical examples. The traffic models of the two numerical examples do not obey (some of) the requirements included in Assumption 6.2.1; the corresponding numerical results show the impact of this.

Lemma 6.2.2 *Assumption 6.2.1 entails that, for any b, minimization Equation* (6.2) *has a* unique *minimizer $t(b)$. In fact, $t(b)$ is the unique solution to*

$$F(b,t) := 2c\,v(t) - (b+ct)v'(t) = 0, \quad or \quad b = c\left(2\,\frac{v(t)}{v'(t)} - t\right). \tag{6.9}$$

Proof. First rewrite the minimization Equation (6.2) as

$$\inf_{t\geq 0}\frac{m^2(t)}{2}, \quad \text{with } m(t) := \frac{b+ct}{\varsigma(t)}.$$

Define $\phi(t) := \varsigma(t)/\varsigma'(t) - t$. Since

$$m'(t) = \frac{c\varsigma(t) - (b+ct)\varsigma'(t)}{\varsigma^2(t)},$$

and because of element (ii) of Assumption 6.2.1, it suffices to prove that (i) for each $b > 0$ and $c > 0$

$$\phi(t) = \frac{b}{c} \tag{6.10}$$

has a root $t(b)$, and (ii) $\phi(\cdot)$ is strictly increasing.

Due to $v(t)/t^{\alpha} \to 0$ for some $\alpha < 2$, it follows that $\lim_{t\to\infty} m(t) = \infty$ for each $b, c > 0$. Moreover, since $\varsigma(0) = 0$, it follows that $\lim_{t\to 0} m(t) = \infty$ for each $b, c > 0$. As a consequence, Equation (6.10) has at least one solution. Moreover

$$\phi'(t) = \frac{(\varsigma'(t))^2 - \varsigma(t)\varsigma''(t)}{(\varsigma'(t))^2} - 1 = -\frac{\varsigma(t)\varsigma''(t)}{(\varsigma'(t))^2} > 0,$$

since $\varsigma''(t) < 0$ due to the strict concavity of $\varsigma(\cdot)$, cf. element (ii) of Assumption 6.2.1. Thus $\phi(\cdot)$ is strictly increasing. This completes the proof. □

Our main result on the relation between the shape of the decay rate function $I_c^{(\mathrm{f})}(\cdot)$, and the correlation structure of the Gaussian sources, is stated in Theorem 6.2.5. We first prove two lemmas.

The first lemma says that the most likely epoch of overflow $t(b)$ is an increasing function of the buffer size b.

Lemma 6.2.3 *$t(\cdot) \in C_1([0,\infty))$, and is strictly increasing.*

Proof. Recall the fact that $t(b)$ is the *unique* solution to Equation (6.9). In conjunction with $v(\cdot) \in C_2([0,\infty))$ and $v'(\cdot) > 0$ (Assumption 6.2.1), we conclude that $t(\cdot)$ is continuous. From Equation (6.9), we see that

$$t'(b) = -\frac{\partial F/\partial b}{\partial F/\partial t} = \frac{v'(t(b))}{cv'(t(b)) - (b + ct(b))v''(t(b))}$$
$$= \frac{1}{c} \cdot \left(1 - 2\,\frac{v(t(b))v''(t(b))}{v'(t(b))^2}\right)^{-1}, \tag{6.11}$$

such that the continuity of $t(\cdot)$, together with $v(\cdot) \in C_2([0,\infty))$, implies that $t'(\cdot)$ is continuous, too.

Assumption 6.2.1 states that, for all $t \geq 0$,

$$\frac{\mathrm{d}^2}{\mathrm{d}t^2}\sqrt{v(t)} < 0 \iff 2\,\frac{v(t)v''(t)}{v'(t)^2} < 1,$$

thus proving the lemma. □

As we have seen in Chapters 4 and 5, $I_c^{(\mathrm{f})}(b)$ can be written as the variational problem

$$I_c^{(\mathrm{f})}(b) = \inf_{t\geq 0}\sup_{\theta}\left(\theta(b+ct) - \log \mathbb{E}e^{\theta A(t)}\right). \tag{6.12}$$

The optimizing θ reads

$$\theta_t(b) := \frac{b+ct}{v(t)}. \tag{6.13}$$

The second lemma states a relation between the derivative of the loss curve and the tilting parameter of the Fenchel–Legendre transform in Equation (6.12). Here we use the shorthand notation $\theta(b) \equiv \theta_{t(b)}(b)$.

Lemma 6.2.4 *For all $b > 0$, it holds that $(I_c^{(\mathrm{f})})'(b) = \theta(b)$.*

Proof. Recalling that $t(b)$ is the optimizing t, differentiating Equation (6.2) with respect to b yields

$$(I_c^{(\mathrm{f})})'(b) = \left(\frac{b + c\,t(b)}{v(t(b))}\right) - t'(b)\left(\frac{b + c\,t(b)}{2v^2(t(b))}\right)\big((b + ct(b))v'(t(b)) - 2c\,v(t(b))\big).$$

Now note that this equals $\theta(b)$, due to Equations (6.9) and (6.13). □

The main result of this section can be proven now. It describes the duality relation between the shape of $I_c^{(\mathrm{f})}(\cdot)$ and the correlation structure (which is uniquely determined by $v(\cdot)$). More specifically, it is shown that the curve $I_c^{(\mathrm{f})}(\cdot)$ is convex at some buffer size b if and only if there are negative correlations on the timescale $t(b)$ on which the overflow most likely takes place.

Theorem 6.2.5 *For all $b > 0$,*

$$(I_c^{(\mathrm{f})})''(b) \geq 0 \iff v''(t(b)) \leq 0.$$

Proof. Due to Lemma 6.2.4, $(I_c^{(\mathrm{f})})''(b) = \theta'(b)$. Trivial calculus yields

$$\theta'(b) = \frac{v(t(b))(1 + c\,t'(b)) - 2c\,t'(b)v(t(b))}{v^2(t(b))} = \frac{1 - c\,t'(b)}{v(t(b))},$$

where the last equality is due to Equation (6.9). As $v(t)$ is nonnegative for any $t > 0$, conclude that $(I_c^{(\mathrm{f})})''(b) \geq 0$ is equivalent to $c\,t'(b) \leq 1$. So we are left to prove that $c\,t'(b) \leq 1$ is equivalent to $v''(t(b)) \leq 0$.

To show this equivalence, note that relation Equation (6.11) yields

$$t'(b) = \frac{1}{c}\left(1 - \left(t(b) + \frac{b}{c}\right)\frac{v''(t(b))}{v'(t(b))}\right)^{-1}.$$

Now recall that $t(b) \geq 0$, $t'(b) \geq 0$ (due to Lemma 6.2.3) and $v'(t(b)) \geq 0$. Conclude that $c\,t'(b) \leq 1$ is equivalent to $v''(t(b)) \leq 0$. □

Obviously, for all b and t, it holds that $I_c^{(\mathrm{f})}(b) \leq (b + ct)^2/(2v(t))$. Noticing that both $v(\cdot)$ and $I_c^{(\mathrm{f})}(\cdot)$ are nonnegative, this results in the following interesting corollary.

Corollary 6.2.6 *For all $t > 0$, it holds that*

$$v(t) \leq \inf_{b>0} \frac{(b + ct)^2}{2I_c^{(\mathrm{f})}(b)}. \tag{6.14}$$

In Chapter 15, we show that in many situations Equation (6.14) is actually an equality, thus constituting a *duality* result between $I_c^{(\mathrm{f})}(\cdot)$ and $v(\cdot)$.

Analytic examples. We now present several different Gaussian input models that illustrate Theorem 6.2.5. The first highlights a model in which the type of correlation is determined by the choice of a model parameter. Example 2 relates to negatively correlated input traffic, whereas Examples 3 and 4 focus on positively correlated traffic. All examples add some specific extra insights.

1. *iOU.* First verify that $v(t) = t - 1 + e^{-t}$ satisfies Assumption 6.2.1. It is easy to see that $v(\cdot)$ is convex, so we will have 'decreasing marginal buffering benefits', i.e., $I_c^{(\mathrm{f})}(\cdot)$ is concave due to Theorem 6.2.5. This example shows the relation between the 'level of positive correlation' and the shape of $I_c^{(\mathrm{f})}(\cdot)$. The strong convexity for small t indicates strong positive correlation on short timescale, whereas this positive correlation becomes weaker and weaker as the timescale increases (reflected by the asymptotically linear shape of $v(\cdot)$ for t large).

   First we concentrate on small b. We have seen in Exercise 6.1.5 that $I_c^{(\mathrm{f})}(b) = c^2 + \frac{2}{3}\sqrt{6bc^3} + O(b)$, where $t(b) \approx \sqrt{6b/c}$. So $I_c^{(\mathrm{f})}(\cdot)$ is highly concave for b small (i.e., behaving as a square root), expressing the strong positive correlations on a short timescale.

   For large b, we found that $I_c^{(\mathrm{f})}(b) - 2c(b+c) \to 0$, with $t(b) \approx b/c + 2$. Apparently, for large b, $I_c^{(\mathrm{f})}(\cdot)$ becomes nearly linear, as expected by the weak correlation on long timescales.

2. *fBm.* For $H < \frac{1}{2}$ this function is (uniformly) concave, indicating negative correlations, whereas $H > \frac{1}{2}$ entails that $v(\cdot)$ is convex corresponding to positive correlations – for $H = \frac{1}{2}$, the increments are independent. Assumption 6.2.1 is fulfilled; notice that $\sqrt{v(t)} = t^H$, which is concave. The results of Exercise 6.1.3 show that $I_c^{(\mathrm{f})}(\cdot)$ is indeed convex (concave) when the Hurst parameter is smaller (larger) than $\frac{1}{2}$, as expected on the basis of Theorem 6.2.5.

3. *Brownian bridge.* With $v(t) = t(1-t)$, Assumption 6.2.1 is not satisfied, but it is not hard to verify that we can restrict ourselves to $t \in [0, \frac{1}{2})$, rather than $t > 0$, on which $\varsigma(\cdot) = \sqrt{v(\cdot)}$ is concave. Direct arguments show that Theorem 6.2.5 applies.

   As we have derived in Exercise 6.1.2, $t(b) = b(c - 1 + 2b)^{-1}$, resulting in the loss curve $I_c^{(\mathrm{f})}(b) = 2b(b+c)$. As expected from Theorem 6.2.5, the concavity of $v(\cdot)$ indeed translates into $I_c^{(\mathrm{f})}(\cdot)$ being convex. In other words, due to the negative correlations, the marginal benefits of buffering are relatively high.

4. *M/G/∞ input model with Pareto jobs.* Consider the Gaussian source model in which the variance function is given by Equation (2.5), which obeys the

requirements stated in Assumption 6.2.1. As in the case of iOU, we obtain a highly concave shape at $b \downarrow 0$:

$$I_c^{(\mathrm{f})}(b) = \frac{\alpha - 1}{2}(c-\mu)^2 + \frac{1}{3}\sqrt{6b(\alpha-1)^3(c-\mu)^3} + O(b).$$

The interesting part, however, is $b \to \infty$. If $\alpha > 2$, then $v(t)$ is essentially linear in t (as in the iOU example above), yielding

$$t(b) \approx \frac{b}{c-\mu} - \frac{2}{3-\alpha}, \quad \text{and}$$

$$I_c^{(\mathrm{f})}(b) - 2(\alpha-1)(\alpha-2)(c-\mu)\left(b - \frac{c-\mu}{3-\alpha}\right) \to 0.$$

If $\alpha \in (1, 2)$, then $v(t)$ is roughly of the order $t^{3-\alpha}$. It can be verified easily that $t(b)$ looks like $b\mathbb{E}D(c-\mu)^{-1}$, implying that $I_c^{(\mathrm{f})}(b) = O(b^{\alpha-1})$. We conclude that for these α, the curve remains highly concave, also for large b.

Numerical examples. We now present two numerical examples that show interesting nontrivial behavior. The former focuses on sources with positive correlations on a short timescale, and negative correlations on a longer timescale, whereas the latter model displays the opposite behavior. In both examples, Assumption 6.2.1 is only 'partly met' – the impact of this is reflected by the numerical results.

- *Deterministic on-off sources.* Consider a source that alternates between being silent and transmitting (at a constant rate of, say, 1). Let the on- and off-periods be deterministic, with lengths σ and τ, respectively. Hence the source is purely periodic, with periods of length $\sigma + \tau$; the position of the start of the transmission phase, within the period of the source, has a uniform distribution on $[0, \sigma + \tau]$. We assume that $\sigma \leq \tau$; a similar reasoning applies to the case $\tau < \sigma$.

  This type of sources has been used to describe the 'worst-case traffic' [95, 163, 168, 178] that can go through a traffic policer (or access regulator); the traffic contract parameters (usually the so-called sustainable traffic rate, the peak traffic rate, and the maximum burst size) can be mapped on the source's on-time σ, off-time τ, and peak rate (which is, in the setting of the present example, normalized to 1).

  The variance $v(t)$ of such a source is given by

$$-\frac{t^3}{3(\sigma+\tau)} + \frac{\sigma t^2}{\sigma+\tau} - \frac{\sigma^2 t^2}{(\sigma+\tau)^2} \quad \text{for } t \in [0, \sigma);$$

$$-\frac{\sigma^3}{3(\sigma+\tau)} + \frac{\sigma^2 t}{\sigma+\tau} - \frac{\sigma^2 t^2}{(\sigma+\tau)^2} \quad \text{for } t \in [\sigma, \tau);$$

$$-\frac{\sigma^3}{3(\sigma+\tau)}+\frac{\sigma^2 t}{\sigma+\tau}+\frac{(t-\tau)^3}{3(\sigma+\tau)}-\frac{\sigma^2 t^2}{(\sigma+\tau)^2} \quad \text{for } t \in [\tau, \sigma+\tau).$$

We now focus on the Gaussian counterpart of such a source, i.e., $A(t)$ has a normal distribution with mean $\sigma\,(\sigma+\tau)^{-1}$ and variance $v(t)$. Notice that the source exhibits positive correlations on a timescale below σ (as expressed by $v''(t) \geq 0$ for $t \in [0, \sigma)$), whereas on a longer timescale traffic is negatively correlated (as $v''(t) \leq 0$ for $t \in [\sigma, \tau)$). Similarly to the case of periodic arrivals, when evaluating the right-hand side of Equation (6.2) we can restrict ourselves to $t \in [0, (\sigma+\tau)/2)$. Notice, however, that $v(\cdot) \in C_1([0, \infty))$ rather than $C_2([0, \infty))$; hence Assumption 6.2.1 is *not* met.

In the example, we let n of these on-off sources feed into a buffered resource. We use the parameters $\sigma = 1$ and $\tau = 2$, and hence the average traffic rate generated by a single source is $\frac{1}{3}$. We choose the normalized link rate $\frac{1}{2}$. To transform the process into a queue fed by *centered* sources, we therefore take $c = \frac{1}{2} - \frac{1}{3} = \frac{1}{6}$.

The function $v(\cdot)$ shifts from convexity to concavity at $t = 1$. We see in Figure 6.3 that $I_c^{(\mathrm{f})}(\cdot)$ is indeed concave as long as $t(b)$ is below 1, whereas it become convex if $t(b)$ exceeds 1, as implied by Theorem 6.2.5; the transition takes place at $b = 0.168$. The fact that $v(\cdot) \in C_1([0, \infty))$ – instead of $C_2([0, \infty))$ – is reflected by the fact that $t(\cdot)$ is continuous, but not differentiable.

- *A combined packet/burst level model.* In this example, we consider an on-off source with an explicitly modeled packet and burst level, as was introduced in Section 2.5. In the on-times packets (of fixed size, say 1) are sent periodically. We normalize time such that the corresponding packet interarrival time is 1. Both the on- and off-times are distributed geometrically, with means p^{-1} and

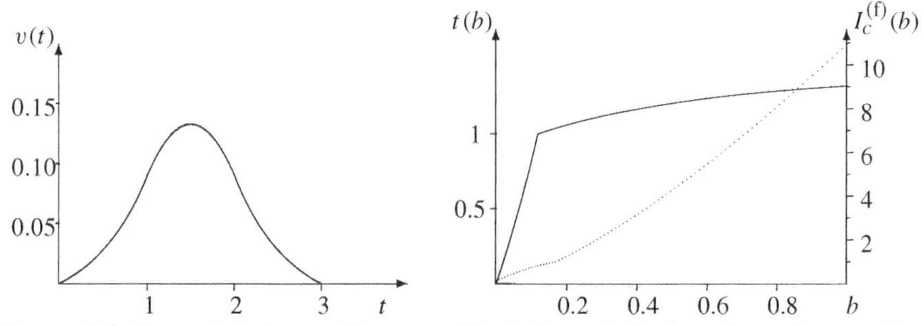

Figure 6.3: Deterministic on-off sources. The left panel displays the variance $v(\cdot)$ as a function of t. The right panel depicts the decay rate $I_c^{(\mathrm{f})}(\cdot)$ – the dotted line – and the time to overflow $t(\cdot)$ – the solid line – both as functions of the buffer size b.

q^{-1}, respectively. The mgf $\mathbb{E}\exp(\theta A(t))$ was given explicitly in Section 2.5, from which the moments of $A(t)$ can be derived through differentiation. This enables the computation of $v(t)$. As before, consider the Gaussian counterpart of such a source.

In the example, we choose $p = 0.01$ and $q = 0.02$. This gives $\mu = \frac{2}{3}$. Let the normalized link speed be $\frac{4}{5}$, such that we have to use $c = \frac{2}{15}$. The variance function shows the negative (positive) correlations on a short (longer) timescale. Notice that Assumption 6.2.1 is *not* met, see the left panel of Figure 6.4 – in the right panel we see how this affects the relation between $v(\cdot)$ and $I_c^{(f)}(\cdot)$.

Figure 6.4 shows an interesting phenomenon. For small buffers, overflow is mainly caused by 'colliding packets': overflow happens within the packet interarrival time (i.e., 1). For larger buffers, it becomes more likely that the number of sources that is in the on-time is higher than on average – then the time to overflow is a multiple of packet interarrival times. A more detailed description of these packet and burst effects is given in [189], cf. also Example 6 in [206].

The curve $I_c^{(f)}(\cdot)$ is convex for small b, due to the negative correlations on a short timescale, and has some 'angle' (in the example at a critical buffer level of $b = 0.130$) as soon as the positive correlations kick in. The fact that $v(\cdot) \in C_0([0,\infty))$ – instead of $C_2([0,\infty))$ – is reflected by the fact that $t(\cdot)$ is discontinuous.

The following exercise is inspired by Example 3.1 in [109]. Although it relates to slotted time rather than continuous time, it gives a result that has an interesting interpretation in the light of Theorem 6.2.5.

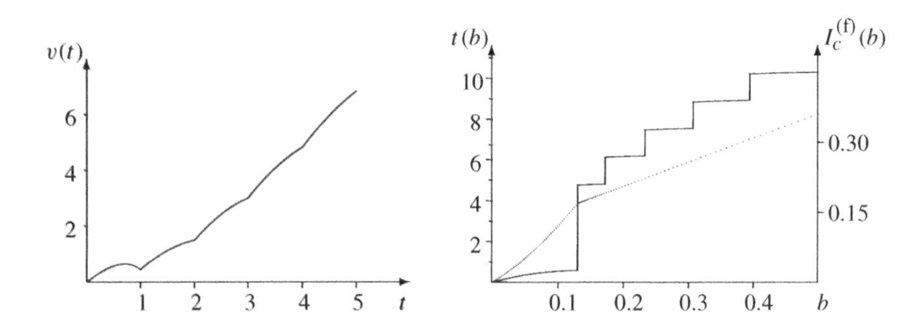

Figure 6.4: Combined packet/burst level model. The left panel displays the variance $v(\cdot)$ as a function of t. The right panel depicts the decay rate $I_c^{(f)}(\cdot)$ – the dotted line – and the time to overflow $t(\cdot)$ – the solid line – both as functions of the buffer size b.

Exercise 6.2.7 Consider the following process in slotted time: with $n \in \mathbb{N}$,

$$X_n := a \cdot X_{n-1} + \sqrt{1-a^2} \cdot \epsilon_n,$$

where $a \in (-1, 1)$ and $\epsilon_1, \epsilon_2, \ldots$ i.i.d. $\mathcal{N}(0, 1)$ random variables. Let $A\{1, n\} := \sum_{i=1}^{n} X_i$.

(i) Show that $A\{1, n\} =_{\rm d} \mathcal{N}(0, v(n))$, with

$$v(n) := n \cdot \frac{1+a}{1-a} - 2a\frac{1-a^n}{(1-a)^2}.$$

(ii) Show that there are finite α, β, such that

$$I_c^{(\rm f)}(b) - \alpha b \to \beta, \text{ as } b \to \infty.$$

(iii) Interpret the sign of β.

Solution. First observe that $\mathbb{C}\mathrm{ov}(X_0, X_n) = \mathbb{C}\mathrm{ov}(X_0, a^n X_0) = a^n$. If $a \in (0, 1)$, then the correlations are positive; if $a \in (-1, 0)$, then the correlations between consecutive increments are negative. Hence,

$$v(n) = \mathbb{V}\mathrm{ar}\left(\sum_{i=1}^{n} X_i\right) = n + 2\sum_{i=2}^{n}\sum_{j=1}^{i-1} \mathbb{C}\mathrm{ov}(X_i, X_j)$$

$$= n + 2\sum_{i=2}^{n}\sum_{j=1}^{i-1} a^{i-j} = n \cdot \frac{1+a}{1-a} - 2a\frac{1-a^n}{(1-a)^2}.$$

(ii) This is a consequence of [40, Theorem 3] (but it can be proven directly as well), which states the following. Let $\theta^\star$ denote the unique positive solution to

$$\lim_{n\to\infty} \frac{1}{n} \log \mathbb{E} \exp(\theta A\{1, n\}) = c\theta.$$

Furthermore, suppose that $-\lim_{n\to\infty} \log \mathbb{E} \exp(\theta^\star A\{1, n\} - \theta^\star cn) =: \nu$ exists. Then it holds that the loss curve is asymptotically linear: $I_c^{(\rm f)}(b) - \theta^\star b \to \nu$, for $b \to \infty$.

In our setting, it turns out that

$$\alpha \equiv \alpha(a) = \theta^\star := 2c\left(\frac{1-a}{1+a}\right); \quad \beta \equiv \beta(a) = \nu := -4c^2\frac{2a}{(1+a)^2}.$$

As could be expected, the slope of $I_c^{(\rm f)}(b)$ is relatively large in case of negative correlations (i.e., $a \in (-1, 0)$), and relatively small in case of positive correlations (i.e., $a \in (0, 1)$).

(iii) We see that $\beta > 0$ iff $a < 0$. This is in line with the concave (convex) shape of $I_c^{(\rm f)}(b)$ in case of positive (negative) correlations. ◇

6.3 The buffer-bandwidth curve is convex

A network provider has essentially two types of resources that he can deploy to meet the customers' performance requirements. When he chooses to increase the amount of buffer available in the network element, this clearly has a positive impact on the loss probability (albeit at the expense of incurring additional delay); the alternative is to increase the queue's service capacity (which reduces both the loss probability and the delay).

In other words: to achieve a certain predefined loss probability, say ϵ, the provider has to choose with which buffer size and link capacity this target is achieved. It is clear that the two types of resources trade off, and the goal of this section is to further analyze this. Once we have an explicit characterization of the trade-off between buffer and capacity for a given loss probability ϵ, and knowing the prices of both resources, the 'optimal' (i.e., most cost-effective) values can be selected.

In this section, we rely on the many-sources framework introduced earlier in this chapter: we have n sources sharing a network element with service rate nc and buffer threshold nb, with the performance objective $p_n(b, c) \leq \epsilon$. Relying on the (very crude) approximation $p_n(b, c) \approx \exp(-nI_c^{(\mathrm{f})}(b))$, our objective becomes $I_c^{(\mathrm{f})}(b) \geq \delta$, where the identification $e^{-n\delta} = \epsilon$ is used (such that $\delta > 0$). In other words, all values b, c such that

$$\inf_{t \geq 0} \frac{(b + ct)^2}{2v(t)} \geq \delta$$

satisfy the performance requirement.

Interestingly, the many-sources framework allows us to find the minimally required link capacity c for a given buffer b and loss constraint δ, as follows. By definition,

$$c_b(\delta) \equiv c_b := \inf \left\{ c \mid \inf_{t \geq 0} \frac{(b + ct)^2}{2v(t)} \geq \delta \right\}.$$

It is clear, however, that if the infimum of a function $f(t)$ over t is larger than (or equal to) δ, then *for all* t it should hold that $f(t) \geq \delta$. In other words:

$$c_b = \inf \left\{ c \mid \forall t \geq 0 : \frac{(b + ct)^2}{2v(t)} \geq \delta \right\}.$$

Isolating the c, this further reduces to

$$c_b = \inf \left\{ c \mid \forall t \geq 0 : c \geq \frac{\sqrt{2\delta v(t)} - b}{t} \right\}$$

$$= \inf \left\{ c \mid c \geq \sup_{t \geq 0} \frac{\sqrt{2\delta v(t)} - b}{t} \right\} = \sup_{t \geq 0} \frac{\sqrt{2\delta v(t)} - b}{t}. \qquad (6.15)$$

Similarly, the minimally required b (for given c, δ) can be computed:

$$b_c = \sup_{t \geq 0} \left(\sqrt{2\delta v(t)} - ct \right).$$

Exercise 6.3.1 Compute c_b and b_c for fBm.

Solution. Applying the results above, we have that $c_b = \inf_{t \geq 0} f(t)$, with

$$f(t) := \sqrt{2\delta}\, t^{H-1} - \frac{b}{t}.$$

It is clear that $f(t) \to -\infty$ as $t \downarrow 0$; also $f(t) \to 0$ as $t \to \infty$. At the same time it can be verified that $f'(\cdot)$ has one zero, and $f''(\cdot)$ changes sign just once. In other words: we find the unique maximum by solving $f'(t) = 0$. This yields

$$t = \left(\frac{b}{\sqrt{2\delta}(1-H)} \right)^{1/H}.$$

Inserting this in the objective function yields

$$c_b = H(2\delta)^{1/2H} \left(\frac{b}{1-H} \right)^{1-1/H}.$$

We see that c and b trade off 'hyperbolically'. Similarly,

$$b_c = (1-H)(2\delta)^{-1/(2H-2)} \left(\frac{c}{H} \right)^{H/(H-1)}.$$

The above calculations reveal that, along the trade-off curve, $b^{1-H}c^{H}$ remains constant (where this constant depends on H and δ). Economists would say that buffer and bandwidth are substitute commodities that trade off according to a so-called *Cobb–Douglas curve*; background on Cobb–Douglas production functions can be found in nearly any economics textbook, see for instance [203, p. 73]. ◇

In the previous exercise we saw that the resources traded off in a convex way, in the sense that, for given δ, c_b is a convex function of b. In fact, this holds in general, as can be proven surprisingly directly.

Proposition 6.3.2 *The required link capacity $c_b(\delta) \equiv c_b$ for given buffer b and decay rate δ, as given by Equation* (6.15), *is a convex function.*

Proof. Evidently, the objective function in Equation (6.15), i.e., $\sqrt{2\delta v(t)}/t - b/t$, is linear in b. The maximum of linear functions is convex. □

Exercise 6.3.3 (i) Consider fBm traffic, and require that the decay rate of the loss probability is at least δ. Impose the following cost structure: the cost per unit buffer is κ_b, and the cost per unit capacity is κ_c. Determine the optimal buffer size $b^\star$ and capacity $c^\star$.

(ii) Characterize the solution for a general function $c_b(\delta)$.

Solution. (i) We saw that, to obtain a decay rate δ, the resources b and c are such that $b^{1-H}c^H$ is constant; this constant, say φ, depends on H and δ. Consequently, the problem we have to solve is:

$$\min_{b\geq 0, c\geq 0} \kappa_b b + \kappa_c c \quad \text{subject to} \quad b^{1-H}c^H = \varphi.$$

Due to the convex form of the constraint, this can be solved immediately through Lagrangian optimization. It is easily verified that, taking for ease $\varphi \equiv 1$,

$$c^\star = \left(\frac{\kappa_b}{\kappa_c} \cdot \frac{H}{1-H}\right)^{1-H}, \quad b^\star = \left(\frac{\kappa_c}{\kappa_b} \cdot \frac{1-H}{H}\right)^{H}.$$

(ii) Elementary convex analysis yields that we have to find the b for which the derivative of $c_b(\delta)$ is κ_b/κ_c, i.e., $b^\star$ is solved from

$$\frac{\kappa_b}{\kappa_c} = -\left(\frac{\partial}{\partial b} c_b(\delta)\right);$$

$c^\star$ then equals $c_{b^\star}(\delta)$. ◇

Bibliographical notes

The first to examine queuing under the many-sources scaling was Weiss [283], who considered a queue fed by many exponential on-off sources, see also the contributions by Botvich and Duffield [40], Courcoubetis and Weber [55], Mandjes *et al.* [188, 190, 195] and Wischik *et al.* [109, 292].

Addie, Mannersalo, and Norros [5] wrote a nice introduction on logarithmic asymptotics for single Gaussian queues under the many-sources scaling. [5] also includes a series of interesting examples, some of which are used throughout this chapter.

The duality results on the loss curve were derived by Mandjes [185]. Kumaran and Mandjes [162] established the convexity of the buffer-bandwidth trade-off curve in the many-sources context (i.e., not just for Gaussian sources); see also Kumaran, Mandjes, and Stolyar [164] for a similar result in a different setting.

Chapter 7

Exact many-sources asymptotics

In the previous chapter, we addressed the logarithmic asymptotics (in the many-sources scaling) of $p_n(b, c) := \mathbb{P}(Q_n \geq nb)$ and a number of interesting ramifications. The main result that we found there was Theorem 6.1.1, which we could prove as a direct application of Schilder's theorem. Theorem 6.1.1 characterizes the exponential decay rate (in n) of $p_n(b, c)$, and can alternatively be written as

$$p_n(b, c) = \phi(n) \exp\left(-nI_c^{(f)}(b)\right). \tag{7.1}$$

We found that the exponential decay rate $I_c^{(f)}(b)$ equals $\frac{1}{2}\inf_{t\geq 0}(m(t))^2$, with

$$m(t) := \frac{b + ct}{\varsigma(t)}, \tag{7.2}$$

where, as before, $\varsigma(t) := \sqrt{v(t)}$. Apart from the exponential part, Representation (7.1) also involves a 'subexponential' part, i.e., a function $\phi(\cdot)$ that satisfies $\log \phi(n) = o(n)$. As explained earlier, this hardly specifies $\phi(\cdot)$: for instance, $\phi(n) = n^{\alpha}$ and even $\phi(n) = \exp(n^{1-\epsilon})$ (for some small, positive ϵ) are still possible. Hence, logarithmic asymptotics are useful to get first insights, but it would be desirable to know the exact asymptotics as well.

Therefore, the goal of this chapter is to find the exact asymptotics of $p_n(b, c)$. It turns out that it is relatively straightforward to characterize these exact asymptotics in the case of slotted time. The result is stated in Section 7.1, and proved in Section 7.2 by elementary arguments. The case of continuous time requires considerably heavier machinery. Again we first state, in Section 7.3, the result, and then provide the proofs, in Section 7.4.

Large deviations for Gaussian queues M. Mandjes

7.1 Slotted time: results

So far, the focus of this book has been on Gaussian sources and Gaussian queues in continuous time. In some cases, however, working in slotted (i.e., discrete) time offers attractive advantages. When considering exact asymptotics in the many-sources regime, for instance, the proofs turn out to be rather elementary and need no heavy machinery. The most advanced tool needed is the Bahadur-Rao result (Proposition 4.1.6 and Exercise 4.1.7) for normally distributed random variables.

In slotted time, the probability of our interest reads

$$p_n(b,c) = \mathbb{P}\left(\exists t \in \mathbb{N} : \sum_{i=1}^{n} A_i\{-t,-1\} \geq nb + nct\right),$$

where we recall that $A_i\{s,t\}$ is defined as the amount of traffic generated by the ith source in slots s up to t, with $s < t$. Evidently, using time-reversibility arguments, we can replace $A_i\{-t,-1\}$ in the previous display by $A_i\{1,t\}$.

As in continuous time, a crucial role is played by the function $m(\cdot)$ defined in Equation (7.2); recall that $\frac{1}{2}(m(t))^2$ can be interpreted as the exponent in the normal density (with mean 0 and variance $v(t)$) evaluated in $b + ct$. We assume that $m(\cdot)$ has a unique minimizer in $\mathbb{N}$; notice that this does not follow from Assumption 6.2.1 anymore. We therefore impose the uniqueness explicitly.

Assumption 7.1.1 *Assume that*

$$t^\star := \arg\min_{t\in\mathbb{N}} m(t)$$

is unique.

In our analysis, it turns out that the contribution of the 'most likely timescale' $t^\star$ is dominant. In other words, the probability of our interest is asymptotically (i.e., for large values of n) equal to

$$p_{n,t}(b,c) := \mathbb{P}\left(\sum_{i=1}^{n} A_i\{1,t\} \geq nb + nct\right), \tag{7.3}$$

for $t = t^\star$. To simplify notation, we also introduce

$$f_n(t) := \left(\frac{1}{\sqrt{2\pi}}\frac{1}{\sqrt{n}\,m(t)}\exp\left(-\frac{1}{2}n(m(t))^2\right)\right).$$

Theorem 7.1.2 Exact asymptotics, slotted time. *If Assumption 7.1.1 holds, then, for any $b, c > 0$,*

$$\lim_{n\to\infty} p_n(b,c)\,/\,f_n(t^\star) = 1.$$

The above theorem says that, in slotted time, the structure of 'Bahadur-Rao' carries over to queues: for positive numbers γ, δ, we have that the overflow probability $p_n(b, c)$ behaves asymptotically as $\gamma n^{-1/2} e^{-n\delta}$. In the second part of the chapter, which is devoted to the corresponding continuous-time queue, we find that this structure changes: then we still find an exponential term (as follows from Theorem 6.1.1), but the polynomial function is not necessarily of the form $n^{-1/2}$ anymore.

7.2 Slotted time: proofs

The proof of Theorem 7.1.2 consists of an easy lower bound, and a somewhat more involved upper bound.

Lower bound. As mentioned, the proof of the lower bound of Theorem 7.1.2 is relatively straightforward; it heavily relies on the 'principle of the largest term'. The validity of the first part of the following lemma follows immediately from Exercise 4.1.7 (i.e., Bahadur-Rao asymptotics for normal random variables). The second part follows from the uniqueness of $t^\star$.

Lemma 7.2.1 *(i) For any $t \in \mathbb{N}$,*

$$\lim_{n\to\infty} \frac{p_{n,t}(b, c)}{f_n(t)} = 1.$$

(ii) Moreover, for $t \neq t^\star$, we have that under Assumption 7.1.1, $f_n(t)/f_n(t^\star) \to 0$ as $n \to \infty$.

The proof of the lower bound follows from Lemma 7.2.1, as follows.

Proof of lower bound. For all $t \in \mathbb{N}$, we have that $p_n(b, c) \geq p_{n,t}(b, c)$. Choose $t = t^\star$, and use part (i) of Lemma 7.2.1. □

Upper bound. The upper bound is more involved, as we have to find a way to bound the probability of an infinite union of events. The advantage of slotted time is that this concerns a *countable* union (rather than an uncountable union, as would have been the case in continuous time), and countable unions can be bounded from above by sums. The idea is then to split this sum into two parts: one that contains $t^\star$, and the 'tail'. The contribution of the tail-terms is then controlled by applying Lemmas 7.2.2 and 7.2.3.

Lemma 7.2.2 *There exists $\epsilon > 0$ such that, for all $t > T$ (where $T > t^\star$), it holds that $\frac{1}{2}(m(t))^2 > t^{2\epsilon}$.*

Proof. This is an immediate consequence of Assumption 4.2.2, as this entails that there is a T and an $\alpha < 2$ such that for all $t > T$ it holds that $v(t) < t^\alpha$.

Now use the trivial lower bound $(b+ct)^2 > c^2t^2$, and we find that the statement holds. □

As a consequence of Lemma 7.2.2, we have that the Chernoff bound (i.e., Proposition 4.1.2) implies that, for $t > T$,

$$p_{n,t}(b,c) = \mathbb{P}\left(\sum_{i=1}^{n} A_i\{1,t\} \geq nb + nct\right) \leq \exp\left(-\frac{1}{2}n(m(t))^2\right) \leq e^{-nt^{2\epsilon}}. \quad (7.4)$$

Now that we can bound, for $t > T$, each of the probabilities $p_{n,t}(b,c)$, our next goal is to find an upper bound to the *sum* of these probabilities. This is done in Lemma 7.2.3. Define $q := 1/(2\epsilon)$. Let k be the largest natural number such that $q-1-k \in (0,1]$. Moreover, we introduce two notions that play an important role in the lemma:

$$\gamma_q := q-1-k, \text{ and } \beta_q := \frac{(q-1)\cdots(q-k)}{\gamma_q^k e^{\gamma_q}}. \quad (7.5)$$

By definition $\gamma_q \in (0,1]$ and $\beta_q > 0$.

Lemma 7.2.3 *If* $\epsilon \in [1/2, \infty)$, *then*

$$\sum_{t=T+1}^{\infty} e^{-nt^{2\epsilon}} \leq \frac{q}{n} e^{-nT^{2\epsilon}}.$$

If $\epsilon \in (0, 1/2)$, *then*

$$\sum_{t=T+1}^{\infty} e^{-nt^{2\epsilon}} \leq \left(\frac{q\beta_q}{n-\gamma_q}\right) e^{-(n-\gamma_q)T^{2\epsilon}}.$$

Proof. First consider the case $\epsilon \geq 1/2$. Since $q \leq 1$ and $T \in \mathbb{N}$, we can bound the sum as follows:

$$\begin{aligned}\int_T^{\infty} \exp\left(-nt^{1/q}\right) \mathrm{d}t &= q\int_{T^{1/q}}^{\infty} \exp(-ny) y^{q-1}\,\mathrm{d}y \\ &\leq q\left(T^{1/q}\right)^{q-1} \int_{T^{1/q}}^{\infty} \exp(-ny)\,\mathrm{d}y \\ &= \frac{q}{n}\left(T^{1/q}\right)^{q-1} \exp\left(-nT^{1/q}\right) \leq \frac{q}{n} e^{-nT^{2\epsilon}};\end{aligned}$$

here, the fact that it is used (in the first inequality) that y^{q-1} is decreasing in y, and (in the second inequality) that, for any $T \in \mathbb{N}$, it holds that $T^{1-1/q} \leq 1$.

For the case $\epsilon \in (0, 1/2)$ considerably more work needs to be done, but the arguments are elementary. First note that $q > 1$, which is crucial throughout the proof. Recall that $k \geq 0$ denotes the largest integer such that $q - 1 - k \in (0, 1]$. As before, we have by a simple substitution,

$$\int_T^\infty \exp\left(-nt^{1/q}\right) \mathrm{d}t = q \int_{T^{1/q}}^\infty \exp(-ny) y^{q-1} \mathrm{d}y. \tag{7.6}$$

The idea is to select $\beta, \gamma \in (0, \infty)$ such that

$$y^{q-1} \leq \beta e^{\gamma y} \tag{7.7}$$

for all $y \in \mathbb{R}_+$. We now discuss how these parameters can be chosen.

If $q \in (1, 2]$, then $p_q : y \mapsto y^{q-1}$ is concave. Since p_q is differentiable at 1 with derivative $q - 1$, by [255, Theorem 25.1] we have for all $y \in \mathbb{R}_+$,

$$y^{q-1} \leq 1 + (q-1)(y-1). \tag{7.8}$$

Similarly, since $y \mapsto \beta e^{\gamma y}$ is convex and differentiable at 1 with derivative $\beta\gamma e^\gamma$, we have for all $y \in \mathbb{R}_+$,

$$\beta e^{\gamma y} \geq \beta e^\gamma + \beta\gamma e^\gamma (y-1). \tag{7.9}$$

By comparing Equation (7.8) with Equation (7.9), we see that $y^{q-1} \leq \beta e^{\gamma y}$ upon choosing $\gamma = q - 1$ and $\beta = e^{-\gamma}$.

To find β, γ such that Equation (7.7) holds for $q \in (k+1, k+2]$ where $k > 0$, the key observation is that this inequality is always satisfied for $y = 0$. Therefore, it suffices to choose β, γ such that the derivative of the left-hand side of Equation (7.7) does not exceed the right-hand side. By applying this idea k times, one readily observes that it suffices to require that β, γ satisfy the inequality

$$\beta\gamma^k e^{\gamma y} \geq (q-1)\cdots(q-k) y^{q-k-1}. \tag{7.10}$$

Note that the right-hand side of Equation (7.10) is concave as a function of y since $q - k - 1 \in (0, 1]$, and that the left-hand side is convex as a function of y. Therefore, we are in a similar situation as we were for $k = 0$. In this case, we choose β and γ such that

$$\begin{aligned} \beta\gamma^k e^\gamma &= (q-1)\cdots(q-k) \\ \beta\gamma^{k+1} e^\gamma &= (q-1)\cdots(q-k)(q-k-1). \end{aligned}$$

Note that β and γ as defined in Equation (7.5) solve this system of equations uniquely. As before, [255, Theorem 25.1] is applied twice to see that, for $y \in \mathbb{R}_+$,

$$\begin{aligned} &(q-1)\cdots(q-k) y^{q-k-1} \\ &\leq (q-1)\cdots(q-k) + (q-1)\cdots(q-k)(q-k-1)(y-1) \\ &= \beta\gamma^k e^\gamma + \beta\gamma^{k+1} e^\gamma (y-1) \leq \beta\gamma^k e^{\gamma y}. \end{aligned}$$

Now that we have found simple bounds on y^{q-1}, the assertion in the lemma follows upon combining these bounds with Equation (7.6):

$$\int_T^\infty \exp\left(-nt^{1/q}\right)\mathrm{d}t \leq q\beta \int_{T^{1/q}}^\infty \exp\left(-(n-\gamma)y\right)\mathrm{d}y$$
$$= \left(\frac{q\beta}{n-\gamma}\right)\exp\left(-(n-\gamma)T^{1/q}\right),$$

as desired. □

Proof of upper bound. Clearly, it holds that

$$p_n(b,c) \leq \sum_{t=1}^\infty p_{n,t}(b,c) = \sum_{t=1}^T p_{n,t}(b,c) + \sum_{t=T+1}^\infty p_{n,t}(b,c),$$

due to the union bound.

Consider the first sum. Because of Lemma 7.2.1, we see immediately that $\sum_{t=1}^T p_{n,t}(b,c)/f_n(t^\star) \to 1$ as $n \to \infty$, for any finite $T \geq t^\star$, as long as Assumption 7.1.1 is fulfilled.

Now consider the second sum. Choose T so large that for all $t > T$ it holds that $\frac{1}{2}(m(t))^2 > t^{2\epsilon}$ (which is possible due to Lemma 7.2.2) and also that $T^{2\epsilon} \geq \frac{1}{2}(m(t^\star))^2 + \delta$, for some $\delta > 0$. Inequality (7.4) and Lemma 7.2.3 entail that there is a positive constant κ such that

$$\sum_{t=T+1}^\infty p_{n,t}(b,c) \leq \sum_{t=T+1}^\infty e^{-nt^{2\epsilon}}$$
$$\leq \kappa e^{-nT^{2\epsilon}} \leq \exp\left(-\frac{1}{2}n(m(t^\star))^2 - n\delta\right).$$

Now divide by $f_n(t^\star)$, and let $n \to \infty$, to obtain 0. This completes the proof of the upper bound. □

7.3 Continuous time: results

After having established the exact asymptotics for the single queue in slotted time, we now turn to continuous time. Unfortunately, the techniques required are not that elementary anymore, and more sophisticated machinery is required. Interestingly, we observe that the asymptotics are not necessarily of the form $\gamma n^{-1/2}e^{-n\delta}$ (which was the case in slotted time), but rather of the form $\gamma' n^{-\zeta}e^{-n\delta}$, for a ζ that could differ from $1/2$. We have to impose an additional assumption.

Assumption 7.3.1 *The variance function $v(\cdot)$ behaves as a polynomial for small t: for some constant $A_0 > 0$ and $\gamma > 0$,*

$$\lim_{t\downarrow 0} \frac{v(t)}{t^\gamma} = A_0.$$

In the following result, which is the main theorem of this section, we impose, besides the Assumptions 6.2.1 and 7.3.1, also Assumption 4.2.2 (which we had supposed to be valid throughout the book). Recall that Assumption 6.2.1 implies that, due to Lemma 6.2.2 there is a unique t that minimizes $m(t)$; define the unique minimizer again by

$$t^\star := \arg\min_{t\geq 0} m(t).$$

It is immediately clear that $t^\star$ solves

$$c\varsigma(t^\star) = (b + ct^\star)\varsigma'(t^\star), \tag{7.11}$$

which is equivalent to Equation (6.9).

Let $\Psi(\cdot)$ be defined as the complementary distribution function of a standard normal random variable:

$$\Psi(x) := 1 - \Phi(x) = \int_x^\infty \frac{1}{\sqrt{2\pi}} e^{-y^2/2}\,\mathrm{d}y.$$

From Exercise 4.1.7 we know that

$$\lim_{x\to\infty} \Psi(x)\cdot\left(x\sqrt{2\pi}e^{x^2/2}\right) = 1. \tag{7.12}$$

The following result describes the exact asymptotics in continuous time. The result involves the Pickands constant as defined in Section 5.6.

Theorem 7.3.2 Exact asymptotics, continuous time. *Assume that Assumptions 4.2.2, 6.2.1, and 7.3.1 are satisfied.*

(i) If $0 < \gamma < 2$, then, for $n \to \infty$ and any $b, c > 0$,

$$p_n(b,c) \Big/ \frac{\mathcal{H}_\gamma}{\sqrt{\pi}} \left(\frac{2\varsigma(t^\star)}{-\varsigma''(t^\star)}\right)^{1/2} \left(\frac{A_0}{2\varsigma^2(t^\star)}\right)^{1/\gamma}$$

$$\cdot\left(m(t^\star)\right)^{\frac{2}{\gamma}-1}\cdot n^{\frac{1}{\gamma}-\frac{1}{2}}\Psi\left(m(t^\star)\sqrt{n}\right) \to 1.$$

(ii) If $\gamma = 2$, then, for $n \to \infty$ and any $b, c > 0$,

$$p_n(b,c) \Big/ \sqrt{1 + \frac{(\varsigma'(t^\star))^2 - (\varsigma'(0))^2}{\varsigma(t^\star)\varsigma''(t^\star)}} \cdot \Psi\left(m(t^\star)\sqrt{n}\right) \to 1\,.$$

Due to Equation (7.12), the above results imply that, $p_n(b, c)$ is asymptotically proportional to

$$n^{1/\gamma-1} \exp(-nm^2(t^\star)/2)$$

(as a function of n). We emphasize that this is essentially different from the discrete-time asymptotics (where the polynomial part is $n^{-1/2}$ rather than $n^{1/\gamma-1}$). Recalling that $\gamma \in (0, 2]$, we have that $n^{1/\gamma-1}$ is larger than $n^{-1/2}$; this qualitative argument confirms the evident property that the overflow probabilities in continuous-time are larger than in discrete time. In fact, our continuous-time result gives insight into the probability of excursions *between the grid points* exceeding nb.

The following exercise specializes Theorem 7.3.2 to the important special case of fBm.

Exercise 7.3.3 Determine the asymptotics of $p_n(b, c)$ in the case of fBm input.

Solution. It is easily seen that Assumptions 4.2.2, 6.2.1, and 7.3.1 are met, with $\gamma \equiv 2H$ and $A_0 \equiv 1$. As we found before,

$$t^\star = \frac{H}{1-H} \cdot \frac{b}{c};$$

$$m := m(t^\star) = \left(\frac{b}{1-H}\right)^{1-H} \left(\frac{c}{H}\right)^H .$$

It is now a matter of inserting these expressions in the formulas of Theorem 7.3.2. This leads to

$$p_n(b, c) \Bigg/ \frac{\mathcal{H}_{2H}}{\pi} \cdot \frac{1}{\sqrt{H(1-H)}} \cdot \left(\frac{m}{\sqrt{2}}\right)^{\frac{1}{H}-1} \cdot n^{1/2H-1/2} \Psi(m\sqrt{n}) \to 1 ,$$

as $n \to \infty$.

Interestingly, the polynomial function $n^{1/2H-1/2}$ decays more slowly than $1/\sqrt{n}$, which makes sense, as the probability of exceeding nb is larger in continuous time than in slotted time. ◇

7.4 Continuous time: proofs

In this section, we present the proof of Theorem 7.3.2. We first introduce some general notation.

- In the sequel we denote the variance function and the standard deviation function of a Gaussian process $X(\cdot)$ by $v_X(\cdot)$ and $\varsigma_X(\cdot)$, respectively; evidently $v_A(\cdot) \equiv v(\cdot)$ and $\varsigma_A(\cdot) \equiv \varsigma(\cdot)$.

- We also introduce the so-called 'standardized process' by adding a bar: $\overline{X}(t) = X(t)/\varsigma_X(t)$ for some Gaussian process $X(\cdot)$. The standardized process has the feature that $\mathbb{V}\mathrm{ar}\overline{X}(t) = 1$ for all t.

Moreover, for a given $R > 0$ and $\alpha \in (0, 2]$, let

$$\mathcal{H}_\alpha^R := \lim_{S\to\infty} \mathbb{E}\exp\left(\sup_{t\in[-S,S]} \left(B_{\alpha/2}(t) - (1+R)|t|^\alpha\right)\right);$$

$$\mathcal{F}_\alpha^R := \lim_{S\to\infty} \mathbb{E}\exp\left(\sup_{t\in[0,S]} \left(B_{\alpha/2}(t) - (1+R)t^\alpha\right)\right),$$

cf. the definition of the Pickands constant in Section 5.6.

The idea of the proof of Theorems 7.3.2 is based on an appropriate use of Theorem 1 in [245]. Since this result plays the crucial role in the following analysis, we present it in the form that is suitable for us; see also Theorem 2.2 in [158].

To make the notation somewhat more concise, we write $f(n) \sim g(n)$ to denote 'asymptotic equivalence', i.e., $f(n)/g(n) \to 1$ as $n \to \infty$.

Theorem 7.4.1 *Let* $(\xi(t))_{t\in[0,T]}$ *be a centered Gaussian process with continuous sample paths a.s. and variance function* $v_\xi(\cdot)$ *such that the maximum of* $\varsigma_\xi(\cdot)$ *on* $[0, T]$ *is attained at a unique point* $t^\star$ *with* $\varsigma_\xi(t^\star) = 1$. *Make the following assumptions:*

(a) $\varsigma_\xi(\cdot)$ is polynomial in a neighborhood of $t^\star$: *there exist* $\kappa, \beta > 0$ *such that*

$$1 - \sigma_\xi(t + t^\star) = \kappa|t|^\beta(1 + o(1)) \quad \text{as } t \to 0;$$

(b) Local stationarity: *there exist* $D, \alpha > 0$ *such that*

$$1 - \mathbb{C}\mathrm{ov}\left(\overline{\xi}(t), \overline{\xi}(s)\right) = D|t-s|^\alpha + o\left(|t-s|^\alpha\right) \quad \text{as } s, t \to t^\star;$$

(c) Regularity: *there exist* $C, \alpha_1 > 0$ *such that, for* $s, t \in [0, T]$,

$$\mathbb{E}\left(\xi(t) - \xi(s)\right)^2 \leq C|t-s|^{\alpha_1}.$$

Then the following statements hold.

(i) If $\beta > \alpha$, *with* $\mathcal{G}_{\alpha,\beta} := \mathcal{H}_\alpha \Gamma(1/\beta) D^{1/\alpha}\beta^{-1}A^{-1/\beta}$, *as* $u \to \infty$,

$$\mathbb{P}\left(\sup_{t\in[0,T]} \xi(t) > u\right) \sim u^{\frac{2}{\alpha}-\frac{2}{\beta}}\Psi(u) \cdot \begin{cases} 2\cdot\mathcal{G}_{\alpha,\beta} & \text{if } t^\star \in (0,T); \\ \mathcal{G}_{\alpha,\beta} & \text{if } t^\star = 0 \text{ or } t^\star = T. \end{cases}$$

(ii) If $\beta = \alpha$, *with* $R := A/D$, *as* $u \to \infty$,

$$\mathbb{P}\left(\sup_{t\in[0,T]} \xi(t) > u\right) \sim \Psi(u) \cdot \begin{cases} \mathcal{H}_\alpha^R & \text{if } t^\star \in (0,T); \\ \mathcal{F}_\alpha^R & \text{if } t^\star = 0 \text{ or } t^\star = T. \end{cases}$$

(iii) If $\beta < \alpha$, as $u \to \infty$,

$$\mathbb{P}\left(\sup_{t\in[0,T]} \xi(t) > u\right) \sim \Psi(u).$$

Before presenting the proof of Theorem 7.3.2 we first give some technical lemmas.

Lemma 7.4.2 *Let $A(\cdot)$ satisfy Assumption 6.2.1. Then, for each $T \in [0, t^\star)$,*

$$(i)\quad 1 - \frac{m(t^\star)}{m(t+t^\star)} = -\frac{\varsigma''(t^\star)}{2\varsigma(t^\star)} \cdot t^2 \cdot (1 + o(1)) \quad \text{as } t \to 0;$$

$$(ii)\quad 1 - \frac{m(T)}{m(T-t)} = -\frac{m'(T)}{m(T)} \cdot t \cdot (1 + o(1)) \quad \text{as } t \to 0.$$

Proof. First we prove (i). Because of Assumption 6.2.1, we can represent $\varsigma(\cdot)$ around $t^\star$ by a second-order Taylor expansion:

$$\begin{aligned}
1 - \frac{m(t^\star)}{m(t+t^\star)} &= \frac{\varsigma(t^\star)(b + c(t+t^\star)) - \varsigma(t+t^\star)(b+ct^\star)}{\varsigma(t^\star)(b+c(t+t^\star))} \\
&= \frac{\varsigma(t^\star)(b+c(t+t^\star)) - \left(\varsigma(t^\star) + \varsigma'(t^\star)t + \varsigma''(t^\star+\theta(t))\frac{t^2}{2}\right)(b+ct^\star)}{\varsigma(t^\star)(b+c(t+t^\star))} \\
&= -\frac{\varsigma''(t^\star+\theta(t))}{2\varsigma(t^\star)} \cdot \frac{b+ct^\star}{b+c(t+t^\star)} \cdot t^2, \qquad (7.13)
\end{aligned}$$

where $\theta(t) \in [0, t]$, and Equation (7.13) is due to Equation (7.11). Hence, taking $t \to 0$, we have completed the proof of (i).

Similarly, to prove (ii) notice that

$$\begin{aligned}
1 - \frac{m(T)}{m(T-t)} &= \frac{\varsigma(T)\cdot(b+c(T-t)) - \varsigma(T-t)\cdot(b+cT)}{\varsigma(T)(b+c(T-t))} \\
&= \frac{\varsigma(T)\cdot(b+c(T-t)) - (\varsigma(T) - \varsigma'(T+\theta(t))t)\cdot(b+cT)}{\varsigma(T)(b+c(T-t))} \\
&= -\frac{\varsigma(T)c - (b+cT)\cdot\varsigma'(T+\theta(t))}{\varsigma(T)(b+c(T-t))} \cdot t
\end{aligned}$$

where $\theta(t) \in [0, t]$. Now notice that

$$\frac{\varsigma(T)c - (b+cT)\cdot\varsigma'(T)}{\varsigma(T)(b+cT)} = \frac{m'(T)}{m(T)}$$

and $m'(T) < 0$ for $T < t^\star$ (apply Lemma 6.2.2). This completes the proof. □

Lemma 7.4.3 *Let $A(\cdot)$ satisfy Assumptions 6.2.1 and 7.3.1. Then the following statements hold.*

(i) If $\gamma < 2$, then, for each $T > 0$,

$$1 - \mathbb{C}\text{ov}\left(\overline{A}(s), \overline{A}(t)\right) = \frac{A_0}{2\varsigma^2(T)}|t-s|^\gamma + o\left(|t-s|^\gamma\right) \quad \text{as } s,t \to T;$$

(ii) if $\gamma = 2$, then, for each $T > 0$,

$$1 - \mathbb{C}\text{ov}\left(\overline{A}(s), \overline{A}(t)\right) = \frac{(\varsigma'(0))^2 - (\varsigma'(T))^2}{2\varsigma^2(T)}|t-s|^2 + o\left(|t-s|^2\right)$$
$$\text{as } s,t \to T.$$

Proof. Case (i) is due to Lemma 3.2 in [69]. It follows directly from Assumption 7.3.1 and

$$\begin{aligned}\lim_{s,t\to T} \frac{1 - \mathbb{C}\text{ov}\left(\overline{A}(s), \overline{A}(t)\right)}{|t-s|^\gamma} &= \lim_{s,t\to T}\left(\frac{\varsigma^2(t-s)}{2\varsigma(s)\varsigma(t)} - \frac{(\varsigma(t)-\varsigma(s))^2}{2\varsigma(t)\varsigma(s)}\right)|t-s|^{-\gamma} \\ &= \lim_{s,t\to T}\left(\frac{\varsigma^2(t-s)}{2\varsigma(s)\varsigma(t)}\right)|t-s|^{-\gamma} = \frac{A_0}{2\varsigma^2(T)},\end{aligned}$$

assuming, without loss of generality, $t > s$.

Notice that this reasoning does not go through for $\gamma = 2$, as in that case

$$\lim_{s,t\to T} \frac{(\varsigma(t)-\varsigma(s))^2}{2\varsigma(t)\varsigma(s)}|t-s|^{-2} = \frac{(\varsigma'(T))^2}{2\varsigma^2(T)} > 0.$$

In order to prove the case (ii), note that

$$\begin{aligned}\lim_{s,t\to T} \frac{1 - \mathbb{C}\text{ov}\left(\overline{A}(s), \overline{A}(t)\right)}{|t-s|^2} &= \frac{1}{2\varsigma^2(T)} \lim_{s,t\to T} \frac{\varsigma^2(|t-s|) - (\varsigma(t)-\varsigma(s))^2}{|t-s|^2} \\ &= \frac{(\varsigma'(0))^2 - (\varsigma'(T))^2}{2\varsigma^2(T)}.\end{aligned}$$

This completes the proof. □

Lemma 7.4.4 *Let $A(\cdot)$ satisfy Assumption 7.3.1. Then for each $T > 0$ there exist constants $G_T > 0$ and $\alpha_1 > 0$ such that*

$$\mathbb{E}\left(\overline{A}(t)\frac{m(t^\star)}{m(t)} - \overline{A}(s)\frac{m(t^\star)}{m(s)}\right)^2 \le G_T \cdot |t-s|^{\alpha_1} \tag{7.14}$$

for all $s, t \in [0, T]$.

Proof. Let $T > 0$ be given. It is straightforward to show that the left-hand side of Equation (7.14) equals

$$m^2(t^\star)\,\mathbb{E}\left(\frac{A(t)}{b+ct} - \frac{A(s)}{b+cs}\right)^2$$
$$= \frac{m^2(t^\star)}{(b+ct)^2(b+cs)^2}\mathbb{E}\left((b+cs)A(t) - (b+ct)A(s)\right)^2.$$

This expression is majorized as follows:

$$\leq \frac{m^2(t^\star)}{b^4}\mathbb{E}\left((A(t)-A(s))(b+cs) - A(s)(ct-cs)\right)^2$$
$$\leq \frac{m^2(t^\star)}{b^4}\left(2\mathbb{E}\left((A(t)-A(s))(b+cs)\right)^2 + 2\mathbb{E}\left(A(s)(ct-cs)\right)^2\right) \tag{7.15}$$
$$= \frac{m^2(t^\star)}{b^4}\left(2v(|t-s|)\cdot(b+cs)^2 + 2v(s)\cdot c^2|t-s|^2\right),$$

where Equation (7.15) follows from the fact that $(x-y)^2 \leq 2x^2 + 2y^2$ for all x, y. Now application of Assumption 7.3.1 yields that there exists a constant $G_T > 0$ such that the above is not larger than $G_T \cdot |t-s|^\gamma$ for all $s, t \in [0, T]$. □

We are now in a position to present the basic idea behind our proof. By a number of simple Transformations, we first rewrite the probabilities $p_n(b, c)$ in terms of the framework of Theorem 7.4.1. For the moment, we focus on the 'transient' probabilities

$$p_n^T(b, c) := \mathbb{P}\left(\exists t \in [0, T] : \sum_{i=1}^n A_i(t) \geq nb + nct\right),$$

with $T > t^\star$; later we also take into account the contribution of $t > T$ (that is, we show that this contribution is negligible).

Observe that

$$p_n^T(b, c) = \mathbb{P}\left(\sup_{t\in[0,T]}\left(\sum_{i=1}^n A_i(t) - nct\right) > nb\right)$$
$$= \mathbb{P}\left(\sup_{t\in[0,T]}\frac{\sum_{i=1}^n A_i(t)}{\sqrt{n}}\cdot\frac{1}{b+ct} > \sqrt{n}\right)$$
$$= \mathbb{P}\left(\sup_{t\in[0,T]} A(t)\cdot\frac{1}{b+ct} > \sqrt{n}\right) \tag{7.16}$$

$$= \mathbb{P}\left(\sup_{t\in[0,T]} \frac{A(t)}{\varsigma(t)} \cdot \frac{m(t^\star)}{m(t)} > m(t^\star)\sqrt{n}\right),$$

where Equation (7.16) follows from the fact that $A_i(\cdot)$ are Gaussian i.i.d. copies of $A(\cdot)$. Let

$$\mu(t) := \frac{A(t)}{\varsigma(t)} \cdot \frac{m(t^\star)}{m(t)}.$$

Notice that we are in the framework of Theorem 7.4.1: the process $\mu(\cdot)$ has standard deviation function $\varsigma_\mu(\cdot)$ with $\varsigma_\mu(t) = m(t^\star)/m(t)$; evidently, $\varsigma_\mu^2(\cdot)$ has $t^\star$ as unique maximizer (see Lemma 6.2.2) and $\max_{t\in[0,T]} \varsigma_\mu(t) = 1$. So, to prove Theorem 7.3.2 (for $p_n^T(b,c)$ rather than $p_n(b,c)$), we have to check if Assumptions (a), (b), and (c) of Theorem 7.4.1 apply.

Lemma 7.4.5 *With $p_n(b,c)$ replaced by $p_n^T(b,c)$, and $T > t^\star$, Theorem 7.3.2 holds.*

Proof. Noting that

$$\mathbb{C}\mathrm{ov}(\overline{\mu}(s), \overline{\mu}(t)) = \mathbb{C}\mathrm{ov}(\overline{A}(s), \overline{A}(t)),$$

and following Lemmas 6.2.2, 7.4.2, and 7.4.3, we infer that Assumptions (a) and (b) of Theorem 7.4.1 are satisfied with $\alpha \equiv \gamma$, $\beta \equiv 2$, and $\kappa \equiv -\varsigma''(t^\star)/2\varsigma(t^\star)$. Moreover, from Lemma 7.4.4 it follows that Assumption (c) is satisfied.

- If $0 < \gamma < 2$, then Lemma 7.4.3 states that $D \equiv A_0/2\varsigma^2(t^\star)$. Since $T > t^\star$, Theorem 7.4.1 gives the result immediately, using that $\Gamma(1/2) = \sqrt{\pi}$.
- If $\gamma = 2$, then from Lemma 7.4.3

  $$D \equiv \frac{(\varsigma'(0))^2 - (\varsigma'(t^\star))^2}{2\varsigma^2(t^\star)}.$$

  Thus, using an argumentation similar to the case $0 < \gamma < 2$, we obtain

  $$\mathbb{P}\left(\sup_{t\in[0,T]} \mu(t) > m(t^\star)\sqrt{n}\right) \sim \Psi\left(m(t^\star)\sqrt{n}\right) \cdot \mathcal{H}_\gamma^R,$$

  where

  $$R \equiv -\left(\frac{\varsigma''(t^\star)}{2\varsigma(t^\star)}\right) \cdot \left(\frac{(\varsigma'(0))^2 - (\varsigma'(t^\star))^2}{2\varsigma^2(t^\star)}\right)^{-1} = \frac{\varsigma(t^\star)\varsigma''(t^\star)}{(\varsigma'(t^\star))^2 - (\varsigma'(0))^2}.$$

Now notice that $\mathcal{H}_{\gamma}^{R} = \sqrt{1+R^{-1}}$, according to [158]. This concludes our proof. □

To prove Theorem 7.3.2, we need the following standard inequality.

Theorem 7.4.6 Borell's inequality. *Let $A(\cdot)$ be a centered Gaussian process, and let $||A|| := \sup_{t\in T} A(t)$ and $||v|| := \sup_{t\in T} v(t)$. Then,*

$$\mathbb{P}(|\ ||A|| - \mathbb{E}||A||\ | > u) \le 2\exp\left(-\frac{u^2}{2||v||}\right).$$

Proof of Theorem 7.3.2. Since for each $T > 0, n \ge 0$ we have $p_n^T(b, c) \le p_n(b, c)$, it is enough to show that

$$\limsup_{n\to\infty} \frac{p_n(b, c)}{p_n^T(b, c)} \le 1$$

for $T > t^\star$.

Note that, using the same argumentation as in the proof of Lemma 7.4.5,

$$p_n(b, c) = \mathbb{P}\left(\sup_{t\in[0,\infty)} \frac{A(t)}{\varsigma(t)} \cdot \frac{m(t^\star)}{m(t)} > m(t^\star)\sqrt{n}\right).$$

Using the union bound, we have

$$p_n(b, c) \le p_n^T(b, c) + \mathbb{P}\left(\sup_{t\in[T,\infty)} \frac{A(t)}{\varsigma(t)} \cdot \frac{m(t^\star)}{m(t)} > m(t^\star)\sqrt{n}\right).$$

Due to Lemma 6.2.2 there exists a $T > t^\star$ such that $m(t) > 2m(t^\star)$ for $t > T$, yielding

$$\begin{aligned}
\mathbb{P}\left(\sup_{t\in[T,\infty)} \frac{\eta(t)}{\varsigma(t)} \cdot \frac{m(t^\star)}{m(t)} > m(t^\star)\sqrt{n}\right) &\le \mathbb{P}\left(\sup_{t\in[T,\infty)} \frac{\eta(t)}{\varsigma(t)} > 2m(t^\star)\sqrt{n}\right) \\
&\le 2\Psi\left(2m(t^\star)\sqrt{n} - K\right) \qquad (7.17) \\
&= o\left(p_n^T(b, c)\right)
\end{aligned}$$

as $n \to \infty$, where Equation (7.17) follows from Borell's inequality, and K is a constant. Hence, the contribution of $t > T$ can be neglected. This completes the proof. □

Bibliographical notes

The analysis of the slotted-time case was inspired by the work by Likhanov and Mazumdar [177]. This reference also concentrates on exact asymptotics under the many-sources scaling, but allows considerably more general sources. Because we restrict ourselves to Gaussian sources, it turns out that the proofs simplify substantially.

The continuous-time case was analyzed in detail by Dębicki and Mandjes [63]. It is noted that that latter paper also found the exact asymptotics of the transient overflow probability $p_n^T(b, c)$ (rather than just the steady-state probability $p_n(b, c)$); also attention is paid to the special case of integrated Gaussian sources.

Chapter 8

Simulation

In many situations, one is interested in a probability that cannot be explicitly calculated. The previous chapters presented asymptotics that can be used as approximations if the scaling parameter involved (buffer B, or number of sources n) grows large. A disadvantage of using these asymptotics is, however, that it is usually not clear up front whether a certain asymptotic regime has already kicked in.

In these situations, one could resort to estimating the probabilities of interest 'empirically', i.e., through simulation. An inherent problem of estimating tail probabilities is that the corresponding event is *rare*, and therefore it takes long to obtain a sufficient number of useful observations; in general, one could say that the number of runs needed to obtain an estimate with predefined accuracy and confidence is inversely proportional to the probability to be estimated.

A technique devised to circumvent this problem is called *importance sampling*: draw the samples under a distribution that is different from the actual one, and then weigh the simulation output by the so-called likelihood ratios to recover unbiasedness. This chapter discusses importance sampling techniques for estimating $p_n(b, c)$.

A prerequisite for using importance sampling techniques, however, is that one should be able to simulate Gaussian queues. So far, there are hardly any satisfactory approaches for simulation of continuous-time Gaussian queues. In fact, from a conceptual point of view, there are two serious problems.

- In the first place, it is not known how continuous-time processes that are so highly irregular may be stimulated. Of course, one solution to this is to just consider the value of the process on a discrete grid, say $\epsilon\mathbb{N}$; one would hope that when ϵ is small enough, the discrepancy with the probability of our interest becomes small. Comparing the asymptotics in slotted time (which are always of the form $\gamma n^{-1/2}e^{-n\delta}$) with those for continuous time (which are of

Large deviations for Gaussian queues M. Mandjes

the form $\gamma' n^{-\zeta} e^{-n\delta}$, for a ζ not necessarily equal to 1/2), however, already indicates that this reasoning is dangerous. In fact, this argument shows that the limits of n and ϵ do not commute. When the events involved are not extremely rare, approximation of a continuous-time Gaussian queue by its slotted counterpart seems reasonable, though.

- However, when we accept that the continuous time parameter can be replaced by a discrete time parameter, a second problem still needs to be solved. To determine whether

$$\sup_{t \in \mathbb{N}} \frac{1}{n} \sum_{i=1}^{n} A_i\{1, t\} - ct \geq b$$

the process has to be simulated over an *infinite* time horizon; as in Section 7.1 we use the notation $A\{1, t\}$ to denote the arrivals in $\{1, \ldots, t\}$. Accepting a minor bias, however, this horizon can be truncated. Interestingly, the truncation horizon can be computed relying on Lemma 7.2.3.

This chapter is organized as follows. First we determine, in Section 8.1, the simulation horizon for a given maximum bias. Then Section 8.2 describes in detail how importance sampling works. A method to quantify the effectiveness of importance sampling estimator is based on the so-called 'asymptotic efficiency'; this concept is introduced in Section 8.3. Finally, Section 8.4 presents a number of asymptotically efficient importance sampling algorithms for estimating $p_n(b.c)$.

8.1 Determining the simulation horizon

Suppose that we truncate the infinite simulation horizon to some finite T. Then we approximate $p_n(b, c)$ by

$$p_n^T(b, c) = \mathbb{P}\left(\exists t \in \{1, \ldots, T\} : \sum_{i=1}^{n} A_i\{1, t\} \geq nb + nct\right). \tag{8.1}$$

In the previous chapter, we have already seen that, for n large, most of the probability mass is centered around $t^\star = \arg\min m(t)$, with $m(\cdot)$ defined by Equation (7.2). This means that, when choosing T sufficiently large (i.e., well above $t^\star$), we can make sure that the approximation error is small. To investigate this error, let

$$\tau_n := \inf\left\{t \in \mathbb{N} : \sum_{i=1}^{n} A_i\{1, t\} \geq nb + nct\right\}$$

denote the epoch of the first buffer overflow, so that $p_n(b, c) = \mathbb{P}(\tau_n < \infty)$. When approximating $p_n(b, c)$ by $\mathbb{P}(\tau_n \leq T)$, the contribution of $\mathbb{P}(T < \tau_n < \infty)$ is discarded. Consequently, the resulting estimator is *biased*: it has a mean that is *smaller* than $p_n(b,c)$. To control the error made, we can choose T sufficiently large, such that

$$\frac{\mathbb{P}(T < \tau_n < \infty)}{p_n(b,c)} < \delta, \tag{8.2}$$

for some predefined $\delta > 0$. When δ is chosen small, the truncation is clearly of minor impact. The question left is, how can we find T such that Equation (8.2) holds, for given b and c, bias δ and Gaussian sources (characterized through their variance function $v(\cdot)$)?

Unfortunately, the requirement in Equation (8.2) does not directly translate into an explicit expression for the simulation horizon T. We tackle this problem below by establishing tractable bounds on the probabilities $\mathbb{P}(T < \tau_n < \infty)$ and $p_n(b,c)$ (more precisely: an upper bound on $\mathbb{P}(T < \tau_n < \infty)$ and a lower bound on $p_n(b,c)$). These bounds then eventually give an explicit expression for T.

- The lower bound on $p_n(b,c)$ is as follows. Define

$$g_n(t) := \left(\sqrt{\frac{n(m(t))^2}{2}} + \sqrt{\frac{n(m(t))^2}{2} + 2}\right)^{-1}.$$

For any $t \in \mathbb{N}$, recalling the definition of $p_{n,t}(b,c)$ from Equation (7.3),

$$\begin{aligned} p_n(b,c) \geq p_{n,t}(b,c) &= \int_{m(t)\sqrt{n}}^{\infty} \frac{1}{\sqrt{2\pi}} e^{-t^2/2}\,\mathrm{d}t \\ &\geq \frac{1}{\sqrt{\pi}} g_n(t) \exp\left(-\frac{n(m(t))^2}{2}\right), \end{aligned} \tag{8.3}$$

where the last inequality is a standard bound for the standard normal cumulative density function (see [218] for related inequalities and references). In order to find a sharp lower bound, we insert the t that maximizes the exponential part, which is the t that minimizes the exponent. In other words, we choose $t = t^\star$, i.e., the minimizer of $m(t)$.

- Analogously to the upper bound used to derive the exact asymptotics, we have (apply the Chernoff bound)

$$\begin{aligned} \mathbb{P}(T < \tau_n < \infty) &= \sum_{t=T+1}^{\infty} \mathbb{P}(\tau_n = t) \leq \sum_{t=T+1}^{\infty} p_{n,t}(b,c) \\ &\leq \sum_{t=T+1}^{\infty} \exp\left(-\frac{n(m(t))^2}{2}\right). \end{aligned} \tag{8.4}$$

In the present generality, it is difficult to bound this quantity further. We could proceed by focusing on a specific correlation structure, for instance, fBm with $v(t) = t^{2H}$, for $H \in (0, 1)$. Instead, we focus on the somewhat more general situation that the variance function can be bounded (from above) by a polynomial: $v(t) \leq Ct^{2H}$, for some $H \in (0, 1)$ and $C \in (0, \infty)$; cf. Assumption 4.2.2.

For instance, if $v(\cdot)$ is regularly varying [34] with index α, then $v(t)$ can be bounded from above (for t sufficiently large) by Potter's bound $Ct^{\alpha+\zeta}$, for any $C > 1$ and $\zeta > 0$; see [34, Theorem 1.5.6]. Obviously, it is desirable to choose the horizon as small as possible under the restriction that Equation (8.2) holds; for this, C and H should be chosen as small as possible.

Under $v(t) \leq Ct^{2H}$, we can bound Equation (8.4) as follows:

$$\sum_{t=T+1}^{\infty} \exp\left(-\frac{n(m(t))^2}{2}\right) \leq \sum_{t=T+1}^{\infty} \exp\left(-n\frac{c^2}{2C}t^{2-2H}\right)$$

$$\leq \int_{T}^{\infty} \exp\left(-n\frac{c^2}{2C}t^{2\epsilon}\right) \mathrm{d}t, \tag{8.5}$$

with $\epsilon := 1 - H$.

It appears that the natural way of finding an upper bound to Equation (8.5) depends critically on the value of ϵ. Consequently, the cases $\epsilon \geq 1/2$ and $\epsilon \in (0, 1/2)$ should be considered separately. Set $C_0 := c^2/(2C)$ and $q := 1/(2\epsilon)$ for notational convenience; recall the definitions of γ_q and β_q from Section 7.2. The upper bound is now a direct consequence of Lemma 7.2.3.

Now using Inequality (8.3) to bound $p_n(b, c)$ from below (choose $t = t^\star$), and Lemma 7.2.3 to bound $\mathbb{P}(\tau_n > T)$ from above, we obtain the following result.

Proposition 8.1.1 *Suppose that $v(t) \leq Ct^{2H}$, for some $H \in (0, 1)$ and $C \in (0, \infty)$. For $\epsilon \in [1/2, \infty)$, let $T(n)$ be the largest integer smaller than*

$$\left(-\frac{1}{nC_0}\log\left(\frac{1}{q\sqrt{\pi}} \cdot nC_0\delta \cdot g_n(t^\star)\exp\left(-\frac{n(m(t^\star))^2}{2}\right)\right)\right)^q,$$

and for $\epsilon \in (0, 1/2)$, let $T(n)$ be the largest integer smaller than

$$\left(-\frac{1}{nC_0}\log\left(\frac{1}{q\beta_q\sqrt{\pi}} \cdot (n-\gamma_q)C_0^q\delta \cdot g_n(t^\star)\exp\left(-\frac{n(m(t^\star))^2}{2}\right)\right)\right)^q.$$

Then, with T chosen as $T(n)$, the error as defined in Equation (8.2) does not exceed δ.

Moreover,

$$\overline{T} := \lim_{n\to\infty} T(n) = \left(\frac{1}{2C_0}(m(t^\star))^2\right)^q.$$

We recall that $t^\star$ could be interpreted as the most likely epoch of overflow. Given that overflow occurs, most of the probability mass will be around $t^\star$. Hence, it is not surprising that $\overline{T} > t^\star$:

$$\frac{1}{2C_0}(m(t^\star))^2 = \frac{(b+ct^\star)^2}{2v(t^\star)} \Big/ \frac{c^2}{2C} > (t^\star)^{2\epsilon} = (t^\star)^{1/q}.$$

8.2 Importance sampling algorithms

Now that we have found a procedure to find the simulation horizon T for a given δ, the next question is how the resulting probability $\mathbb{P}(\tau_n \leq T)$ should be estimated. We have already argued that $\mathbb{P}(\tau_n \leq T)$ is a rare-event probability, and consequently it is time consuming to estimate it in a direct way. One of the remedies is called *importance sampling*.

The idea behind importance sampling can be described as follows. For ease, we first focus on estimating the event that $p_n := \mathbb{P}\big(\sum_{i=1}^n X_i > na\big)$ with the X_i i.i.d. standard normal random variables, and $a > 0$. This is of course a somewhat artificial example, as we already know that $p_n = 1 - \Phi(a\sqrt{n})$, but it is particularly suitable to illustrate the difference between importance sampling and the direct method.

- *Direct simulation.* Perform N i.i.d. experiments; if in the jth experiment indeed $\sum_{i=1}^n X_i > na$, then put $I_n^{(j)} := 1$, and otherwise put the value as 0. The direct method estimates p_n by the sample mean $\sum_{j=1}^N I_n^{(j)}/N$.

  The next question is, how big should N be to make sure that the estimate's accuracy (defined as the width of the confidence interval, divided by the estimate) is below some predefined fraction (for instance 10%)? Because the width of the confidence interval is proportional to the standard deviation of the estimator, i.e.,

$$\sqrt{\mathbb{V}\mathrm{ar}\left(\frac{1}{N}\sum_{j=1}^N I_n^{(j)}\right)} = \sqrt{\frac{1}{N}\mathbb{V}\mathrm{ar} I_n} = \sqrt{\frac{p_n(1-p_n)}{N}} \approx \sqrt{\frac{p_n}{N}},$$

  it immediately follows that the number of experiments needed is inversely proportional to p_n. But $1/p_n$ is growing very rapidly in n: according to 'Bahadur-Rao', it behaves as

$$N_n \sim a\sqrt{2\pi n}\, e^{na^2/2}.$$

- *Importance sampling.* An alternative would be to use importance sampling: sample the X_i from the $\mathcal{N}(a, 1)$ distribution (to which we associate the measure $\mathbb{Q}$), rather than the $\mathcal{N}(0, 1)$ distribution (to which we associate $\mathbb{P}$).

Then the event of interest is clearly not rare anymore: a substantial fraction of the experiments will be such that $I_n^{(j)} = 1$. From the identity

$$\mathbb{P}(E) = \int 1\{E\}\,\mathrm{d}\mathbb{P} = \int 1\{E\}\left(\frac{\mathrm{d}\mathbb{P}}{\mathrm{d}\mathbb{Q}}\right)\mathrm{d}\mathbb{Q},$$

with $1\{E\}$ the indicator function of an event E, we see that if we do so, we have to weigh the $I_n^{(j)}$ with a *likelihood ratio* (cf. Radon–Nikodým derivative) $L_n^{(j)} := \mathrm{d}\mathbb{P}/\mathrm{d}\mathbb{Q}$, which measures, in the jth experiment, the relative likelihood of the realization under measure $\mathbb{P}$ to what it is under $\mathbb{Q}$. As a consequence, with $L_n^{(j)}$ being the likelihood ratio in experiment j, the estimator

$$\frac{1}{N}\sum_{j=1}^{N} L_n^{(j)} I_n^{(j)}$$

is unbiased (i.e., its mean is p_n).

It is not hard to verify that for a single experiment based on $X_1, \ldots, X_n$, this likelihood ratio equals

$$L_n = \left(\frac{1}{\sqrt{2\pi n}}e^{-(ny)^2/(2n)}\right) \bigg/ \left(\frac{1}{\sqrt{2\pi n}}e^{-(ny-na)^2/(2n)}\right) = e^{-nya+na^2/2}$$

when $\sum_{i=1}^{n} X_i = ny$. As before, the number of runs needed to guarantee some predefined accuracy will be roughly proportional to the ratio of the standard deviation of a single experiment, i.e., the square root of (in self-evident notation)

$$\mathbb{V}\mathrm{ar}_{\mathbb{Q}}(L_n I_n) = \mathbb{E}_{\mathbb{Q}}(L_n^2 I_n) - (\mathbb{E}_{\mathbb{Q}}(L_n I_n))^2, \quad \text{with} \quad I_n := 1\left\{\sum_{i=1}^{n} X_i \geq na\right\},$$

and the probability p_n. We now determine the asymptotic behavior of $\mathbb{V}\mathrm{ar}_{\mathbb{Q}}$ $(L_n I_n)$ (as n grows large). From this we can quantify the gain (in terms of the number of runs needed), compared to direct simulation.

Exercise 8.2.1 Show that, with Z distributed $\mathcal{N}(na, n)$, and $\delta > 0$,

$$\lim_{n\to\infty} \mathbb{E}\left(1\{Z \geq na\}e^{-\delta Z}\right)\cdot\left(\sqrt{2\pi n}\,\delta e^{na\delta}\right) = 1.$$

Solution. It is easy, though tedious, to verify that

$$\mathbb{E}\left(1\{Z \geq na\}e^{-\delta Z}\right) = e^{-na\delta+n\delta^2/2}\Psi(\sqrt{n}\delta),$$

where $\Psi(\cdot)$ is the complementary distribution function of a standard normal random variable, which satisfies

$$\lim_{x\to\infty} \Psi(x)x\sqrt{2\pi}e^{x^2/2} = 1;$$

use the inequalities in Equation (4.3). This substantiates the claim. 

Notice that under $\mathbb{Q}$ we have that $\sum_{i=1}^{n} X_i$ is distributed $\mathcal{N}(na, n)$. Therefore, we find by applying Exercise 8.2.1 that both $\mathbb{E}_{\mathbb{Q}}(L_n^2 I_n)$ and $(\mathbb{E}_{\mathbb{Q}}(L_n I_n))^2$ decay roughly as $\exp(-na^2)$, but the former decreases slightly slower than the latter:

$$\mathbb{E}_{\mathbb{Q}}(L_n^2 I_n) \cdot \sqrt{2\pi n}(2a)e^{na^2} \to 1\,; \quad (\mathbb{E}_{\mathbb{Q}}(L_n I_n))^2 \cdot (2\pi n)\, a^2\, e^{na^2} \to 1$$

(the latter statement was in fact already clear: realize that $\mathbb{E}_{\mathbb{Q}}(L_n I_n) = p_n$ and apply 'Bahadur-Rao'). As a consequence, for n large, $\mathbb{V}\mathrm{ar}_{\mathbb{Q}}(L_n I_n)$ behaves as $\mathbb{E}_{\mathbb{Q}}(L_n^2 I_n)$. Consequently, the ratio of the square root of $\mathbb{V}\mathrm{ar}_{\mathbb{Q}}(LI)$ and p_n is roughly of the form

$$\sqrt{\frac{a}{2}}\sqrt[4]{2\pi n}.$$

In other words, the number of runs needed to achieve a predefined accuracy grows extremely slowly in n, viz. as $\sqrt{n}$:

$$N_n \sim \frac{1}{2}a\sqrt{2\pi n}.$$

This growth rate is substantially slower than that corresponding to direct simulation, where the number of runs is essentially exponential in n. We see that importance sampling is preferred over direct simulation.

The above illustrates how importance sampling works, and that a substantial speed-up can be achieved. The example focused on the probability $\mathbb{P}(\mathcal{N}(0, n) \geq na)$ of a sample mean of i.i.d. normal random variables attaining some extreme value. Moreover, as we have seen in Section 7.1 that $p_n(b, c)$ behaves more or less as

$$\mathbb{P}(\mathcal{N}(0, nv(t^\star)) \geq nb + nct^\star),$$

we may wonder whether this method also works to estimate $p_n(b, c)$. We discuss this in Section 8.4. We first introduce in Section 8.3 a useful criterion that measures the efficiency of importance sampling procedures.

8.3 Asymptotic efficiency

A careful inspection of the analysis of the example of Section 8.2 indicates that importance sampling worked so well because the square root of $\mathbb{V}\mathrm{ar}(L_n I_n)$ was more or less of the same order as p_n (namely $\exp(-na^2/2)$, up to some polynomial functions). Stated differently, $\mathbb{E}_{\mathbb{Q}}(L_n^2 I_n)$ is of about the same order as $(\mathbb{E}_{\mathbb{Q}}(L_n I_n))^2 = p_n^2$ (namely $\exp(-na^2)$). Noticing that $\mathbb{E}_{\mathbb{Q}}(L_n^2 I_n) \geq (\mathbb{E}_{\mathbb{Q}}(L_n I_n))^2 = p_n^2$, a criterion for optimality could be the following.

Definition 8.3.1 Asymptotic efficiency. *A measure* $\mathbb{Q}$ *is* asymptotically efficient *if*

$$\lim_{n\to\infty} \frac{1}{n} \log \mathbb{E}_{\mathbb{Q}}(L_n^2 I_n) = \lim_{n\to\infty} \frac{1}{n} \log(\mathbb{E}_{\mathbb{Q}}(L_n I_n))^2 \tag{8.6}$$

(where the right-hand side is, by definition, equal to $2 \cdot \lim_{n\to\infty} n^{-1} \log p_n$*).*

It is obvious that the left-hand side of Equation (8.6) is always larger than, or equal to, the right-hand side. Hence, if there is equality, then we have found a measure $\mathbb{Q}$ that is in some sense optimal. (Notice that this implies that, in fact, it only needs to be shown that the left-hand side is not larger than the right-hand side, to conclude asymptotic efficiency). We remark that in the literature there are various other optimality criteria.

Clearly, the importance sampling procedure that we proposed for the example in Section 8.2, based on the new measure $\mathbb{Q}$, satisfies the asymptotic efficiency criterion: both sides of Equation (8.6) equal $-a^2$. A quick way to prove asymptotic efficiency for this example is as follows. The fact that the right-hand side in Equation (8.6) equals $-a^2$ follows directly from 'Cramér'. Now concentrate on the left-hand side:

$$\begin{aligned}
&\lim_{n\to\infty} \frac{1}{n} \log \mathbb{E}_{\mathbb{Q}}(L_n^2 I_n) \\
&\quad = \lim_{n\to\infty} \frac{1}{n} \log \mathbb{E}\left(\exp\left(-2a \sum_{i=1}^{n} X_i + na^2\right) 1\left\{\sum_{i=1}^{n} X_i \geq na\right\}\right) \\
&\quad \leq \lim_{n\to\infty} \frac{1}{n} \log\left(\exp(-2na^2 + na^2)\right) = -a^2,
\end{aligned}$$

as desired. We say that the new measure is of *exponentially bounded likelihood ratio*: with respect to the event of interest, it holds that $L_n \leq e^{-n\delta}$ for some $\delta > 0$ (or, equivalently, $L_n I_n \leq e^{-n\delta}$).

8.4 Efficient estimation of the overflow probability

In the previous sections we explained how to truncate the time range $\mathbb{N}$ to $\{1, \ldots, T\}$; we introduced the concept of importance sampling (with, as an example, the estimation of the rare-event probability $\mathbb{P}(\mathcal{N}(0, n) \geq na)$) and gave a criterion to assess the quality of a new measure $\mathbb{Q}$. Bearing in mind the finding of Section 7.1 that $p_n^T(b,c)$ is asymptotically equal to (for n large)

$$f_n(t^\star) = \mathbb{P}(\mathcal{N}(0, nv(t^\star)) \geq nb + nct^\star),$$

we now attempt to devise an asymptotically efficient algorithm to estimate $p_n^T(b,c)$. Recall also that

$$\lim_{n\to\infty} \frac{1}{n} \log p_n^T(b,c) = -\frac{1}{2}(m(t^\star))^2. \tag{8.7}$$

Direct simulation. The first problem is that it is not clear up front how to simulate Gaussian processes. Notice however that Gaussian processes in slotted time, on a finite time horizon $\{1, \ldots, T\}$, correspond in fact to a multivariate normal distribution (of dimension T), and how to sample from such a distribution is, in principle, known. As we assume Gaussian processes with stationary increments, this multivariate normal distribution has a special form (the correlation between two slots only depends on the distance between these two slots, rather than their positions). Due to this property, traces of fixed length, say, T can be sampled relatively fast [59]; in the special case of fBm, the effort required to sample a trace of length T is roughly proportional to $T \log T$. For more reflections on these methods, see also [81]. Interesting (and fast, but not exact) algorithms are given in [227, 238].

When being able to sample the Gaussian process, a direct method would be to simulate $\sum_{i=1}^n A_i\{1,t\}$ for $t = 1, \ldots, T$ several times, count the number of runs in which that process exceeds the straight line $nb + nct$, and divide this by the number of runs. We recall that, as argued in Section 8.2, this method will be slow, because the event under consideration is rare.

A first proposal. As suggested by the example in Section 8.2, we have to change the distribution of $\sum_{i=1}^n A_i\{1,t^\star\}$. Under $\mathbb{P}$ this distribution was $\mathcal{N}(0, nv(t^\star))$, and the example indicates that under $\mathbb{Q}$ it should become $\mathcal{N}(nb + nct^\star, nv(t^\star))$. This tells us how to change the measure of $\sum_{i=1}^n A_i\{1,t^\star\}$, but it remains unclear what, under this new measure, the distribution of $\sum_{i=1}^n A_i\{1,t\}$ should be for all other $t \in \{1, \ldots, T\}$.

A reasonable choice seems to be to associate $\mathbb{Q}$ with a Gaussian process $G_n(\cdot)$ of which the value at time s has distribution $\mathcal{N}(n\mu(s), nv(s))$, where $\mu(\cdot)$ is defined by

$$\mu(s) := \mathbb{E}(A\{1,s\} \mid A\{1,t^\star\} = b + ct^\star) = \frac{\Gamma(s,t^\star)}{v(t^\star)}(b + ct^\star);$$

here it recalled that $\Gamma(s,t) = \mathbb{C}\text{ov}(A\{1,s\}, A\{1,t\})$ (verify that $G_n(\cdot)$ does not have stationary increments!). In other words, we have that the vector

$$\left(\sum_{i=1}^n A_i\{1,1\}, \ldots, \sum_{i=1}^n A_i\{1,T\}\right)$$

is distributed T-variate normal, with mean vector $(\mu(1), \ldots, \mu(T))$ and covariance matrix $\Sigma := (\Gamma(s,t))_{s,t=1}^{T}$.

This choice is motivated by the fact that, under this specific choice of $\mu(s)$, the mean of the process $n^{-1}G_n(\cdot)$ coincides with the most likely path to overflow; notice that $\mu(t^\star) = b + ct^\star$, as expected. The variance function is left unchanged, in line with the findings of Section 8.2 (hence for fBm again an algorithm of complexity $T \log T$ can be used).

Each run simulates the Gaussian process that was associated with $\mathbb{Q}$ until either (i) $G_n(s)$ exceeds $nb + ncs$ for some $s \in \{0, \ldots, T\}$ (and then we put $I_n := 1$), or (ii) time $T + 1$ is reached (and then we put $I_n := 0$). In case (i) the likelihood at the stopping epoch s is

$$L_n = \exp\left(-ny\frac{\mu(s)}{v(s)} + n\frac{(\mu(s))^2}{2v(s)}\right)$$

when $ny = \sum_{i=1}^{n} A_i\{1, s\}$.

Suppose first that $G_n(s)$ exceeds $nb + ncs$ *for the first time* at time $t^\star$. Then we indeed have that

$$L_n I_n \leq \exp\left(-n(b + ct^\star)\frac{\mu(t^\star)}{v(t^\star)} + n\frac{(\mu(t^\star))^2}{2v(t^\star)}\right) = \exp\left(-\frac{n}{2}(m(t^\star))^2\right);$$

in conjunction with Equation (8.7), this would give asymptotic efficiency. There is, however, also the possibility of $G_n(s)$ exceeding $nb + ncs$ (for the first time) at an $s \neq t^\star$. It turns out that then $L_n I_n$ cannot be bounded so nicely. In fact, as a result of this effect, the procedure proposed above is *not* asymptotically efficient.

A more refined method. Introduce the following probabilities:

$$\overline{p}_{n,t}(b, c) := \mathbb{P}\left(\left\{\sum_{i=1}^{n} A_i\{1, t\} \geq nb + nct\right\} \cap \left\{\sup_{s\in\{1,\ldots,t-1\}} \sum_{i=1}^{n} A_i\{1, s\} < nb + nct\right\}\right);$$

hence, $\overline{p}_{n,t}(b, c)$ is interpreted as the probability of exceeding nb at time t *for the first time*. As these events are disjoint, we have

$$p_n^T(b, c) = \sum_{t=1}^{T} \overline{p}_{n,t}(b, c).$$

The idea of the refined method is to simulate ‘timeslot-by-timeslot’, i.e., estimate each of these T events separately by performing T ‘subexperiments’. This procedure was proposed in [37].

In subexperiment t, a Gaussian process $G_{n,t}(\cdot)$ of which the value at time s has distribution $\mathcal{N}(n\mu_t(s), nv(s))$, where $\mu_t(\cdot)$ is defined by

$$\mu_t(s) := \mathbb{E}(A\{1,s\} \mid A\{1,t\} = b + ct) = \frac{\Gamma(s,t)}{v(t)}(b+ct);$$

call the associated measure $\mathbb{Q}_t$. We put $I_{n,t} := 1$ if $G_{n,t}(s)$ exceeds $nb + ncs$ for the first time at time $s = t$. If $I_{n,t} = 1$, then

$$L_{n,t} = \exp\left(-ny\frac{\mu_t(t)}{v(t)} + n\frac{(\mu_t(t))^2}{2v(t)}\right)$$

when $ny = \sum_{i=1}^{n} A_i\{1,t\}$; note that $\mu_t(t) = b + ct$. In this way we have enforced that

$$L_{n,t}I_{n,t} \leq \exp\left(-\frac{n}{2}(m(t))^2\right).$$

Perform this experiment (that consists of T subexperiments) N times, and use the estimator

$$\frac{1}{N}\sum_{j=1}^{N}\left(\sum_{t=1}^{T} L_{n,t}^{(j)} I_{n,t}^{(j)}\right).$$

The second moment of the contribution of a single experiment (i.e., T subexperiments) reads

$$\sum_{t=1}^{T}\mathbb{E}_{\mathbb{Q}_t}(L_{n,t}^2 I_{n,t}) + 2\sum_{t=1}^{T}\sum_{s=1}^{t-1}\mathbb{E}_{\mathbb{Q}_t}(L_{n,t}I_{n,t})\,\mathbb{E}_{\mathbb{Q}_s}(L_{n,s}I_{n,s})$$
$$\leq T^2\exp\left(-n(m(t^\star))^2\right),$$

as desired (use that $t^\star$ minimizes $m(\cdot)$, and that $t^\star < T$). We conclude that this 'partitioning approach' yields an asymptotically efficient estimator.

Other approaches. Another way to better control the likelihood ratio is to simulate 'source-by-source' (rather than 'timeslot-by-timeslot'). The idea is that the optimal change of measure that is applied to source i (for $i = 1, \ldots, n$) should depend on the paths of the sources that were already simulated (i.e., $A_j\{1,t\}$ for $j = 1, \ldots, i-1$ and $t = 1, \ldots, T$). It can be shown that this procedure is also asymptotically efficient [81, 89]; see also [90].

A second alternative is presented in [260]: in the language of this section, it is proposed that a random $t \in \{1, \ldots, T\}$ be picked up per experiment and simulation under $\mathbb{Q}_t$ be performed. Also this approach is asymptotically efficient. It depends critically on the specific model parameters, which of three asymptotically efficient approaches performs best (in terms of required simulation time to achieve a predefined accuracy).

Bibliographical notes

A standard introductory textbook on simulation is Law and Kelton [170], and an early article on importance sampling is by Hammersley and Handscomb [126]; see also Hopmans and Kleijnen [133] and Bratley, Fox, and Schrage [41]. The idea of applying a large deviations motivated change of measure is due to Siegmund [268]. Cottrell, Fort, and Malgouyres [54] is a pioneering study that uses large-deviations for importance sampling purpose. See also Bucklew [43] for a recent textbook on rare-event simulation.

Consequently, estimators based on large deviation results are natural candidates for efficient simulation. In fact, they are asymptotically efficient in many settings; see Asmussen and Rubinstein [14] and Heidelberger [131] and references therein. However, Glasserman and Wang [117] give examples showing that this need not always be the case; more specifically: if the likelihood ratio cannot be bounded on the event of interest, there is a substantial risk of the large-deviations based change of measure being *not* asymptotically efficient, cf. our first proposal in Section 8.4.

Dieker [77] gives a nice overview on methods to generate Gaussian traces (and in particular fBm); see also [80, 227]. The asymptotically efficient method described in Section 8.4 is due to Dieker and Mandjes [81]. Some related results on fast simulation of queues with Gaussian input have been reported by Michna [215] and by Huang, Devetsikiotis, Lambadaris, and Kaye [134]. Michna focuses on fBm input under the large-buffer scaling, but does not consider asymptotic efficiency of his simulation scheme (in fact, one may check that his estimator is asymptotically inefficient). Huang *et al.* also work in the large-buffer asymptotic regime, and present a constant-mean change of measure; this consequently does not correspond to the (curved) most likely path, and is therefore asymptotically inefficient in the many-sources regime. We also mention the interesting work by Baldi and Pacchiarotti [20], who use recent insights into certain Gaussian martingales, and a novel 'bridge'-approach [115].

We conclude by remarking that other efficient rare-event simulation techniques have been proposed; see the surveys [118, 171]; an interesting alternative is the so-called splitting method or RESTART [116, 280]. We also mention the powerful, recently proposed cross-entropy method, in which the optimal change of measure is found *during the simulation* [35, 258, 259].

Chapter 9

Tandem and priority queues

So far, we have analyzed the single queue operating under the first-come-first-serve discipline. This has provided useful insights, but the model is evidently an oversimplification of reality. We mention two serious limitations.

First, in real networking environments traffic streams usually traverse *concatenations* of hops (rather than just a single node). Secondly, it is envisaged that the service at these hops distinguishes between several traffic classes (by using *priority* mechanisms, or the more advanced *generalized processor sharing* discipline), cf. the Differentiated Services (DiffServ) approach proposed by the Internet Engineering Task Force [31, 155, 284]. This motivates the recent interest in performance evaluation for these more complex queuing models.

This chapter concentrates on the evaluation of tail asymptotics in queuing systems that are more advanced than a single FIFO node. More specifically, we examine in detail *tandem* queues (particularly the second queue) and *priority* queues (particularly the low-priority queue)–it turns out that the analysis of the tandem queue essentially carries over to the priority system. The results are meant as a first step toward the analysis of networks with general topology, with nodes operating under advanced scheduling disciplines such as GPS; It is the subject of the next chapter.

As in the previous chapters, the focus will be on logarithmic many-sources asymptotics. In the tandem model we assume that n i.i.d. Gaussian sources feed into a first queue, and the output of the first queue serves as input to a second queue. Then the (deterministic) service rates of both queues, as well as the buffer thresholds, are scaled by n, too, and we let n go to infinity. The first queue can be analyzed by using the techniques discussed in the previous chapters. The decay rate of the buffer content of the second queue, however, is considerably harder to analyze. Therefore, the primary goal of this chapter is to find the counterpart of Theorem 6.1.1 for the second queue in a tandem network.

Large deviations for Gaussian queues M. Mandjes

Similarly, one could consider a (strict) priority system in which n sources of type 'h' feed traffic into a high-priority (hp) queue, and n sources of type 'ℓ' into a low-priority (lp) queue. (The assumption of an equal number of hp sources and lp sources is purely a technicality; the above setting, in fact, also covers the case of unequal numbers of hp sources and lp sources, as we will argue later.) There is nc service capacity available. As long as the hp queue is nonempty, it is assigned all capacity; if the hp queue is empty, then the lp queue is allocated the unused service capacity. The hp traffic does not 'see' the lp traffic, and hence the hp queue is essentially just a single queue, that can be analyzed as described in the previous chapters. The lp queue, however, is more complicated, as it is served with a time-varying capacity. For priority systems, we aim at finding the counterpart of Theorem 6.1.1, but now for the lp queue of the priority system.

The material presented in this chapter fits in the framework of a series of articles [5, 204, 205, 206]. These papers examine queues with Gaussian sources, such as the single-node FIFO queue, but also priority queues and queues operating under GPS scheduling. For the latter types of queues, they derive heuristics for the decay rate of the overflow probabilities.

The analysis of the present chapter shows that, for priority queues, the heuristics of [5, 204, 205, 206] are typically close, but that there is a gap with the exact outcome. We derive for both the tandem and priority queue a lower bound on the decay rate of the overflow probability (that is, for tandems the overflow probability of the second queue, and for priority systems the overflow probability of the lp queue). In addition, we present an explicit condition under which this lower bound actually matches the exact value of the decay rate (we say: the lower bound is *tight*). In a number of important cases, our results indicate that the lower bound is actually tight. Notice that lower bounds on the decay rate (which correspond to upper bounds on the probability) are usually of practical interest, as typically the network has to be designed such that overflow is sufficiently rare. The analysis presented in this chapter also further motivates the heuristics of [5, 204, 205, 206].

A few words on the approach. First consider the tandem queue. As said above, the decay rate of the buffer content of the first queue follows from the previous chapter. Interestingly, however, the decay rate of the *total* amount of traffic in the system (i.e., the sum of the buffer contents of both queues) follows in a similar fashion. The fact that we know how to analyze both the first queue and the total queue is the key in our analysis of the second queue.

In the priority system a similar reasoning applies: there we know how to characterize the decay rate of the hp queue and the total queue, which enables the analysis of the lp queue. In this sense, our analysis exploits the above-mentioned similarity between priority and tandem queues. We mention that for priority systems in discrete time, different bounds were found by Wischik [292]; we comment on the relation with our results later.

The chapter is organized as follows. Section 9.1 introduces the tandem model. Section 9.2 analyzes the decay rate of the overflow probability of the second queue in a tandem system, and finds a lower bound. In Section 9.3 we present a condition under which the lower bound equals the decay rate, i.e., under which there is 'tightness'. Some reflections on the most likely path are given in Section 9.4. The results are illustrated in Section 9.5 by a number of (analytical and numerical) examples. Section 9.6 studies the priority system, addressing the decay rate of the overflow probability in the lp queue.

9.1 Tandem: model and preliminaries

Tandem model. Consider a two-queue tandem model, with (deterministic) service rate nc_1 for the first queue and nc_2 for the second queue. We assume that $c_1 > c_2$, in order to exclude the trivial case where the buffer of the second queue cannot build up.

We consider n i.i.d. Gaussian sources that feed into the first queue. Traffic of these sources that has been served at the first queue immediately flows into the second queue – we assume no additional sources to feed the second queue. We are interested in the steady-state probability of the buffer content of the second queue $Q_{2,n}$ exceeding a certain threshold nb, $b > 0$, when the number of sources gets large, or, more specifically, its logarithmic asymptotics:

$$I_c^{(\mathrm{t})}(b) := -\lim_{n\to\infty} \frac{1}{n} \log \mathbb{P}(Q_{2,n} > nb), \tag{9.1}$$

where c denotes the vector $(c_1, c_2)^{\mathrm{T}}$. Note that we assume the buffer sizes of both queues to be infinite. We remark that it is not *a priori* clear that the limit in Equation (9.1) exists; its existence is, in fact, one of the results reported in this chapter (Theorem 9.2.1).

Many-sources scaling. In this section, we show that the probability of our interest can be written in terms of the 'empirical mean process' $n^{-1}\sum_{i=1}^{n} A_i(\cdot)$. The following lemma exploits the fact that we know both a representation of the first queue $Q_{1,n}$ (in steady-state) and a representation of the *total* queue $Q_{1,n} + Q_{2,n}$ (in steady-state). Let $t^0 := b/(c_1 - c_2)$.

Lemma 9.1.1 *$\mathbb{P}(Q_{2,n} > nb)$ equals*

$$\mathbb{P}\left(\exists t > t^0 : \forall s \in (0, t) : \frac{1}{n}\sum_{i=1}^{n} A_i(-t, -s) > b + c_2 t - c_1 s\right).$$

Proof. Notice that a 'reduction principle' applies: the total queue length is unchanged when the tandem network is replaced by its slowest queue, see e.g., [15, 104];

cf. [282]. More formally, $Q_{1,n} + Q_{2,n} = \sup_{t>0}(\sum_{i=1}^{n} A_i(-t, 0) - nc_2 t)$. Consequently, we can rewrite the buffer content of the downstream queue as

$$Q_{2,n} = (Q_{1,n} + Q_{2,n}) - Q_{1,n}$$
$$= \sup_{t>0}\left(\sum_{i=1}^{n} A_i(-t, 0) - nc_2 t\right) - \sup_{s>0}\left(\sum_{i=1}^{n} A_i(-s, 0) - nc_1 s\right). \tag{9.2}$$

It was shown, see [249, Lemma 5.1], that the negative of the optimizing t in Equation (9.2) corresponds to the start of the last busy period of the total queue in which time 0 is contained; similarly, the optimizing s is the start of the last busy period of the first queue in which time 0 is contained. Notice that a positive first queue induces a positive total queue, which immediately implies that we can restrict ourselves to $s \in (0, t)$. Hence $\mathbb{P}(Q_{2,n} > nb)$ equals

$$\mathbb{P}\left(\exists t > 0 : \forall s \in (0, t) : \frac{1}{n}\sum_{i=1}^{n} A_i(-t, -s) > b + c_2 t - c_1 s\right).$$

Because for $s \uparrow t$ the requirement

$$\frac{1}{n}\sum_{i=1}^{n} A_i(-t, -s) > b + c_2 t - c_1 s$$

reads $0 > b + (c_2 - c_1)t$, we can restrict ourselves to $t > t^0$. We can iterpret t^0 as the minimum time it takes to cause overflow in the second queue (notice that the maximum net input rate of the second queue in a tandem system is $c_1 - c_2$). □

The crucial implication of the above lemma is that for analyzing $\mathbb{P}(Q_{2,n} \geq nb)$, we only have to focus on the behavior of the empirical mean process. More concretely,

$$\mathbb{P}(Q_{2,n} > nb) = p_n[\mathcal{S}^{(t)}] = \mathbb{P}\left(\frac{1}{n}\sum_{i=1}^{n} A_i(\cdot) \in \mathcal{S}^{(t)}\right), \tag{9.3}$$

where the set of 'overflow paths' $\mathcal{S}^{(t)}$ is given by

$$\mathcal{S}^{(t)} := \{f \in \Omega : \exists t > t^0, \forall s \in (0, t) : f(-s) - f(-t) > b + c_2 t - c_1 s\};$$

recall the definition of Ω from Section 4.2.

Remark 9.1.2 A straightforward time-shift shows that the probability that the empirical mean process is in $\mathcal{S}^{(t)}$ coincides with the probability that it is in $\mathcal{T}$, with

$$\mathcal{T} := \{f \in \Omega : \exists t > t^0, \forall s \in (0, t) : f(s) > b + c_2 t - c_1(t - s)\}. \tag{9.4}$$

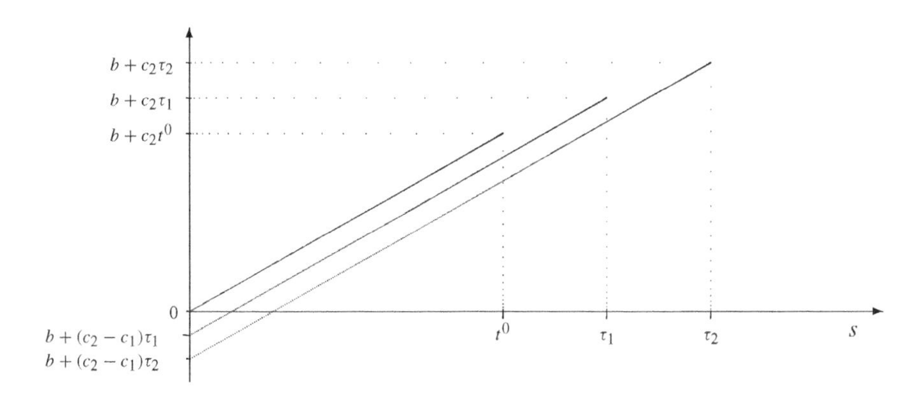

Figure 9.1: Graphical representation of the overflow set. For different values of t, the curve $b + c_2 t - c_1(t - s)$ has been drawn. Overflow occurs if there is a $t > t^0$ such that the empirical mean process lies, for $s \in (0, t)$, above the corresponding curve.

However, the set $\mathcal{T}$ is somewhat easier to interpret, see Figure 9.1. For different values of t (i.e., $\tau_2 > \tau_1 > t^0 = b/(c_1 - c_2)$), the line $b + c_2 t - c_1(t - s)$ has been drawn. The empirical mean process $n^{-1} \sum_{i=1}^{n} A_i(\cdot)$ is in $\mathcal{T}$ if there is a $t > t^0$ such that for all $s \in (0, t)$ it stays above the line $b + c_2 t - c_1(t - s)$. Notice that $\mathcal{T}$ resembles the set corresponding to the probability of long busy periods in a single queue, as studied in [223]. $\diamond$

Remark 9.1.3 As indicated above, our results are for centered sources, but, as before, they can be translated easily into results for noncentered sources, cf. Remark 5.2.1. Then the traffic generated by Gaussian source i in the interval $[s, t)$ is $A(s, t) + \mu(t - s)$, where $A(s, t)$ corresponds to a centered source; here $0 < \mu < \min\{c_1, c_2\}$ and $s < t$. Let $q(\mu, c_1, c_2)$ be the probability that the second queue exceeds nb, given that input rate μ and service rates c_1 and c_2 are in force. From Equation (9.2), it follows immediately that

$$q(\mu, c_1, c_2) = q(0, c_1 - \mu, c_2 - \mu),$$

and hence we can restrict ourselves to centered sources. $\diamond$

9.2 Tandem: lower bound on the decay rate

In this section, we start analyzing the logarithmic asymptotics of $\mathbb{P}(Q_{2,n} > nb)$. More specifically, we use 'Schilder' (Theorem 4.2.3) to formulate the decay rate as a variational problem, and then we find a lower bound on this decay rate.

Decay rate as a variational problem. We now consider the decay rate as in Equation (9.1) of $\mathbb{P}(Q_{2,n} > nb)$. From Equation (9.3) we know that $\mathbb{P}(Q_{2,n} > nb)$

can be rewritten as the probability that the empirical mean process is in $\mathcal{S}^{(t)}$ (which is an open subset of Ω). The existence of the decay rate is now a consequence of Schilder's theorem, by showing (the plausible fact) that $\mathcal{S}^{(t)}$ is an $\mathbb{I}$-continuity set, i.e., that the infima of $\mathbb{I}(\cdot)$ over $\mathcal{S}^{(t)}$ and its closure, say $\overline{\mathcal{S}^{(t)}}$, match. This proof of $\mathcal{S}^{(t)}$ being an $\mathbb{I}$-continuity set is beyond the scope of this book, and can be found in Appendix A of [201].

Theorem 9.2.1

$$I_c^{(t)}(b) = \inf_{f \in \overline{\mathcal{S}^{(t)}}} \mathbb{I}(f) = \inf_{f \in \mathcal{S}^{(t)}} \mathbb{I}(f).$$

Lower bound on the decay rate. Our next goal is to derive a tractable lower bound on $I_c^{(t)}(b)$. This is presented in Theorem 9.2.2.

Observe that

$$\mathcal{S}^{(t)} = \bigcup_{t > t^0} \bigcap_{s \in (0,t)} \mathcal{S}_{s,t}^{(t)} \quad \text{with} \quad \mathcal{S}_{s,t}^{(t)} := \{f \in \Omega : f(-s) - f(-t) > b + c_2 t - c_1 s\}.$$

Hence we are interested in the decay rate of the union of intersections. The decay rate of a union of events is simply the minimum of the decay rates of the individual events, as we have seen several times earlier. The decay rate of an intersection, however, is not standard. In the next theorem, we find a straightforward lower bound on this decay rate. Define

$$\mathcal{U}_{s,t} := \{f \in \Omega : -f(-t) \geq b + c_2 t;\ f(-s) - f(-t) \geq b + c_2 t - c_1 s\}.$$

Theorem 9.2.2 *The following lower bound applies:*

$$I_c^{(t)}(b) \geq \inf_{t > t^0} \sup_{s \in (0,t)} \inf_{f \in \mathcal{U}_{s,t}} \mathbb{I}(f). \tag{9.5}$$

Proof. Clearly,

$$I_c^{(t)}(b) = \inf_{t > t^0} \inf_{f \in \bigcap_{s \in (0,t)} \mathcal{S}_{s,t}^{(t)}} \mathbb{I}(f).$$

Now fix t and consider the inner infimum. If $f(-s) - f(-t) > b + c_2 t - c_1 s$ for all $s \in (0, t)$, even then (f is continuous) $f(-s) - f(-t) \geq b + c_2 t - c_1 s$ for all $s \in [0, t]$. Hence,

$$\bigcap_{s \in (0,t)} \mathcal{S}_{s,t}^{(t)} \subseteq \bigcap_{s \in [0,t]} \mathcal{U}_{s,t} \subseteq \mathcal{U}_{r,t}$$

for all $r \in (0, t)$, and consequently

$$\inf_{f \in \bigcap_{s \in (0,t)} \mathcal{S}_{s,t}^{(1)}} \mathbb{I}(f) \geq \inf_{f \in \mathcal{U}_{r,t}} \mathbb{I}(f).$$

Now take the supremum over r in the right-hand side. □

Theorem 9.2.2 contains an infimum over $f \in \mathcal{U}_{s,t}$. In the next lemma, we show how this infimum can be computed.

Before stating this lemma, we first introduce some additional notations. Recalling Theorem 4.1.8, the bivariate large-deviations rate function of

$$\left(\sum_{i=1}^{n} \frac{A_i(-t, 0)}{n}, \sum_{i=1}^{n} \frac{A_i(-t, -s)}{n} \right)$$

is, for $y, z \in \mathbb{R}$ and $t > 0$, $s \in (0, t)$, given by $\Lambda(y, z) := \frac{1}{2}\,(y, z)\ \Sigma(t-s, t)^{-1}\ (y, z)^{\mathrm{T}}$, with

$$\Sigma(s, t) := \begin{pmatrix} v(t) & \Gamma(s, t) \\ \Gamma(s, t) & v(s) \end{pmatrix}.$$

We also define the following quantity, which plays a key role in our analysis:

$$\begin{aligned} k(s, t) &:= \mathbb{E}(A(-s, 0) \mid A(-t, 0) = b + c_2 t) \\ &= \mathbb{E}(A(s) \mid A(t) = b + c_2 t) = \frac{\Gamma(s, t)}{v(t)}(b + c_2 t). \end{aligned} \tag{9.6}$$

Recall Assumption 6.2.1: the standard deviation function was supposed to be $\mathcal{C}^2([0, \infty))$ and strictly increasing and strictly concave.

Lemma 9.2.3 *Under Assumption 6.2.1, for $t > t^0$ and $s \in (0, t)$,*

$$\inf_{f \in \mathcal{U}_{s,t}} \mathbb{I}(f) = \Upsilon(s, t) := \begin{cases} \Lambda(b + c_2 t, b + c_2 t - c_1 s), & \text{if } k(s, t) > c_1 s; \\ (b + c_2 t)^2 / 2v(t), & \text{if } k(s, t) \leq c_1 s. \end{cases}$$

Proof. Observe that

$$\begin{aligned} p_n[\mathcal{U}_{s,t}] &\equiv \mathbb{P}\left(\sum_{i=1}^{n} \frac{A_i(\cdot)}{n} \in \mathcal{U}_{s,t} \right) \\ &= \mathbb{P}\left(\sum_{i=1}^{n} \frac{A_i(-t, 0)}{n} \geq b + c_2 t; \sum_{i=1}^{n} \frac{A_i(-t, -s)}{n} \geq b + c_2 t - c_1 s \right). \end{aligned} \tag{9.7}$$

We conclude that we can use Theorem 4.1.8 ('bivariate Cramér') to find the decay rate of $p_n[\mathcal{U}_{s,t}]$. We obtain

$$\inf_{f \in \mathcal{U}_{s,t}} \mathbb{I}(f) = \inf \Lambda(y, z),$$

where the last infimum is over $y \geq b + c_2 t$ and $z \geq b + c_2 t - c_1 s$. Using that $\Lambda(\cdot, \cdot)$ is convex, this problem can be solved in the standard manner, as in Exercise 4.1.9. It is easily verified that the contour of Λ that touches the line $y = b + c_2 t$ does so at z-value

$$z^\star := \frac{\Gamma(t-s,t)}{v(t)}(b + c_2 t);$$

also, the contour that touches $z = b + c_2 t - c_1 s$ does so at y-value

$$y^\star := \frac{\Gamma(t-s,t)}{v(t-s)}(b + c_2 t - c_1 s).$$

We first show that it cannot be that $y^\star > b + c_2 t$, as follows. If $y^\star > b + c_2 t$, then the optimum would be attained at $(y^\star, b + c_2 t - c_1 s)$. Straightforward computations, however, show that $y^\star > b + c_2 t$ would imply that (use $\Gamma(t, t-s) \leq \sqrt{v(t)v(t-s)}$)

$$\left(\sqrt{v(t)} - \sqrt{v(t-s)}\right)(b + c_2 t) > \sqrt{v(t)}\, c_1 s. \tag{9.8}$$

This inequality is not fulfilled for $s = 0$ ($0 \not> 0$) nor for $s = t$ ($b + c_2 t \not> c_1 t$ for $t > t_0$). As the left-hand side of Equation (9.8) is convex (in s) due to Assumption 6.2.1, whereas the right-hand side is linear (in s), there is no $s \in (0, t)$ for which the inequality holds. Conclude that $y^\star > b + c_2 t$ can be ruled out.

Two cases are left:

(A) Suppose $z^\star > b + c_2 t - c_1 s$, or, equivalently, $k(s,t) \leq c_1 s$. Then the optimum is attained in $(b + c_2 t, z^\star)$, with rate function $(b + c_2 t)^2 / 2v(t)$, independent of s.

(B) In the remaining case (where $y^\star \leq b + c_2 t$ and $z^\star \leq b + c_2 t - c_1 s$) the optimum is attained at the $(b + c_2 t, b + c_2 t - c_1 s)$, i.e., the 'corner point'. This happens if $k(s,t) > c_1 s$, and gives the desired decay rate.

This proves the statement. As an aside we mention that if $k(s,t) = c_1 s$, then both regimes coincide: $\Lambda(b + c_2 t, b + c_2 t - c_1 s) = (b + c_2 t)^2 / 2v(t)$. □

Exercise 9.2.4 Prove that, under Assumption 6.2.1, the following lower bound applies:

$$I_c^{(t)}(b) \geq \inf_{t > t^0} \sup_{s \in (0,t)} \Upsilon(s,t).$$

Solution. Straightforward. 

Interpretation of the lower bound. The above results have a helpful interpretation, leading to two regimes for values of c_1. For c_1 smaller than some critical link rate $c_1^\star$, we show in Corollary 9.2.6 that the lower bound of Exercise 9.2.4 can be simplified considerably.

We start by drawing a parallel with the single-node result. There, t has to be found such that

$$L_c(t) := \frac{(b + ct - \mathbb{E}A(t))^2}{2\mathbb{V}\mathrm{ar}A(t)}$$

is minimized. Let t_c denote an optimizing argument t. $L_c(t)$ can be interpreted as the cost of generating $b + ct$ in an interval of length t, and t_c as the time duration yielding the 'lowest cost'.

Now we turn to our tandem setting, and in particular to the result of Lemma 9.2.3. Computing the minimum of $\Lambda(y, z)$ over its admissible region, we saw that, under Assumption 6.2.1, in both cases the optimizing y was equal to $y = b + c_2 t$. On the contrary, for the optimizing z there were two possible regimes.

Now recall the Representation (9.6) of $k(s, t)$ as a conditional mean, and Equation (9.7). The result in Lemma 9.2.3 essentially states that in the regime $k(s, t) \leq c_1 s$ the most likely realization of $\sum_{i=1}^n A_i(-t, 0) \geq nb + nc_2 t$ yields $\sum_{i=1}^n A_i(-t, -s) \geq nb + nc_2 t - nc_1 s$ (with high probability, n large). In the other regime, $k(s, t) > c_1 s$, the most likely realization of $\sum_{i=1}^n A_i(-t, 0) \geq nb + nc_2 t$ does not automatically yield $\sum_{i=1}^n A(-t, -s) \geq nb + nc_2 t - nc_1 s$ (with high probability, n large); in this case, fulfilling the second constraint in Equation (9.7) requires additional 'cost'.

The next decomposition result follows immediately from Lemma 9.2.3 and the above.

Corollary 9.2.5 *For $s \in (0, t)$, we have $\Upsilon(s, t) = L_{c_2}(t) + L(s \mid t)$, with*

$$L(s \mid t) := \frac{\max^2\{\mathbb{E}(A(s) \mid A(t) = b + c_2 t) - c_1 s, 0\}}{2\mathbb{V}\mathrm{ar}(A(s) \mid A(t) = b + c_2 t)}$$
$$= \frac{\max^2\{k(s, t) - c_1 s, 0\}}{2\mathbb{V}\mathrm{ar}(A(s) \mid A(t) = b + c_2 t)}. \tag{9.9}$$

Similar to the interpretation of the single-node result, we can interpret $\Upsilon(s, t)$ as the cost of generating the required amount of traffic. Denoting by $s^\star$ and $t^\star$ optimizing arguments in Exercise 9.2.4, the intuition is as follows:

- 'Cost component' $L_{c_2}(t)$ is needed to generate $b + c_2 t$ in the interval $(-t, 0]$. By taking the *infimum* over t (to get $t^\star$) we find the *most likely* epoch to meet the constraint.
- 'Cost component' $L(s \mid t)$ is required to make sure that no more than $c_1 s$ is generated in the interval $(-s, 0]$, *conditional* on the event $A(-t, 0) =$

$b + c_2 t$. We can interpret $s^\star$ as the epoch at which *most* effort has to be done to fulfill this requirement. This is, of course, reflected by the fact that in Equation 9.2.4 we have to take the *supremum* over all s in $(0, t)$. Evidently, if $k(s, t) \leq c_1 s$ for all $s \in (0, t)$, this cost component is 0.

For large values of c_1, $k(s, t)$ will be smaller than $c_1 s$ for all $s \in (0, t)$ (note that $k(s, t)$ does not depend on c_1). As argued above, in this case, the second term in Corollary 9.2.5 vanishes. If this holds for the t that minimizes the first term, i.e., t_{c_2}, then

$$\inf_{t > t^0} \sup_{s \in (0,t)} \Upsilon(s, t) = L_{c_2}(t_{c_2}). \tag{9.10}$$

This clearly holds for all c_1 larger than

$$\begin{aligned} c_1^\star &:= \inf\{c_1 \mid \forall s \in (0, t_{c_2}) : k(s, t_{c_2}) \leq c_1 s\} \\ &= \inf\left\{c_1 \mid \forall s \in (0, t_{c_2}) : c_1 \geq \frac{k(s, t_{c_2})}{s}\right\} \\ &= \sup_{s \in (0, t_{c_2})} \frac{k(s, t_{c_2})}{s}. \end{aligned}$$

Corollary 9.2.6 *For all* $c_1 \geq c_1^\star$, *Equation* (9.10) *applies.*

Based on the above arguments, we conclude the following:

- It implies that, for c_1 larger than the critical link speed $c_1^\star$, the lower bound on $I_c^{(t)}(b)$ of Exercise 9.2.4 coincides with the result of a single-node queue with service rate c_2. The intuition behind this is that, essentially, in this regime all traffic entering the first queue is served immediately, and goes directly into the second queue; traffic is not 'reshaped' by the first queue.
- On the contrary, however, if $c_1 < c_1^\star$ then the first queue *does* play a role in delaying and reshaping the traffic before entering the second queue.

This dichotomy will be further explored in the next section.

9.3 Tandem: tightness of the decay rate

The result of Exercise 9.2.4 yields a lower bound on the decay rate $I_c^{(t)}(b)$. Of course, such a bound is only useful if it is relatively close to the actual decay rate, or, even better, coincides with it. In the latter case, we say that the lower bound is *tight*.

In Section 9.2, we have derived a lower bound on $I_c^{(t)}(b)$ by replacing the decay rate of an intersection of events by the decay rate of the least likely of these. It is important to observe that if the optimum path in this least likely set happens to be in all the sets of the intersection, then the lower bound is tight.

More specifically, let $s^\star$ and $t^\star$ be optimizers in the lower bound of Corollary 9.2.4. Clearly, we can prove tightness of the lower bound by showing that the most probable path in $\mathcal{U}_{s^\star,t^\star}$ is in $\mathcal{S}^{(t)}$ (or $\overline{\mathcal{S}^{(t)}}$, use Theorem 9.2.1). In our analysis, we distinguish between the two cases (or *regimes*) (A) $c_1 \geq c_1^\star$, and (B) $c_1 < c_1^\star$.

Regime (A): c_1 larger than the critical service rate. In this situation, we know from Exercise 9.2.6 that the lower bound in Corollary 9.2.4 reduces to the decay rate in a single queue. The following result follows easily.

Theorem 9.3.1 *Under Assumption 6.2.1, if $c_1 \geq c_1^\star$, then*

$$I_c^{(t)}(b) = \inf_{t > t^0} \sup_{s \in (0,t)} \Upsilon(s,t) = L_{c_2}(t_{c_2}),$$

and a most probable path in $\mathcal{S}^{(t)}$ is

$$f^\star(r) = -\mathbb{E}(A(r,0) \mid A(-t_{c_2},0) = b + c_2 t_{c_2}). \tag{9.11}$$

Proof. As shown in Section 9.2, in this regime $t^\star = t_{c_2}$, whereas the choice of $s^\star$ is irrelevant (as $c_1 \geq c_1^\star$ implies $L(s \mid t^\star) = 0$ for all $s \in (0, t^\star)$). Notice that it is now sufficient to show that $f^\star \in \mathcal{S}^{(t)}$, or $f^\star \in \overline{\mathcal{S}^{(t)}}$ (use Theorem 9.2.1). We claim that $f^\star(\cdot) \in \overline{\mathcal{S}^{(t)}}$, or more precisely, that there exists $t \geq t^0$ such that for all $s \in (0,t)$ it holds that $f^\star(-s) - f^\star(-t) \geq b + c_2 t - c_1 s$. This follows because, by definition of $c_1^\star$, for all $s \in (0, t^\star)$,

$$\begin{aligned} f^\star(-s) - f^\star(-t^\star) &= \mathbb{E}(A(-t^\star,-s) \mid A(-t^\star,0) = b + c_2 t^\star) \\ &= b + c_2 t^\star - k(s,t^\star) \geq b + c_2 t^\star - c_1 s. \end{aligned}$$

This completes the proof. □

We want to stress that the above theorem holds for all Gaussian processes, regardless of the specific shape of the variance function. Consequently, the result is also valid for lrd processes, such as fBm (with $H > \frac{1}{2}$).

Regime (B): c_1 smaller than the critical service rate. We follow the same approach as in Regime (A): first we derive (in Lemma 9.3.3) a most probable path in $\mathcal{U}_{s^\star,t^\star}$, and then we verify (in Theorem 9.3.4) whether this path is in $\mathcal{S}^{(t)}$. It turns out that we have to impose certain additional conditions to make the lower bound of Corollary 9.2.4 tight. We proceed by two technical lemmas; the proof of Lemma 9.3.2 is given in Appendix B of [201].

Lemma 9.3.2 *Under Assumption 6.2.1, if $c_1 < c_1^\star$, then $k(s^\star,t^\star) \geq c_1 s^\star$.*

Lemma 9.3.3 *If $k(s,t) \geq c_1 s$, then a most probable path in $\mathcal{U}_{s,t}$ is*

$$f(r) = -\mathbb{E}(A(r,0) \mid A(-t,0) = b + c_2 t, A(-s,0) = c_1 s), \tag{9.12}$$

with the corresponding decay rate $I_c^{(t)}(b) = \Lambda(b + c_2 t, b + c_2 t - c_1 s)$.

Proof. Using standard properties of conditional multivariate normal random variables, we see that $f(r)$ equals

$$-\theta_1^\star(s,t)\Gamma(-r,t) - \theta_2^\star(s,t)\Gamma(-r,s),$$

with

$$\begin{pmatrix} \theta_1^\star(s,t) \\ \theta_2^\star(s,t) \end{pmatrix} := \Sigma(s,t)^{-1} \begin{pmatrix} b + c_2 t \\ c_1 s \end{pmatrix}. \tag{9.13}$$

We finish the proof by applying Lemma 9.2.3, and observing that

$$\frac{1}{2}||f||_R^2 = \Upsilon(s,t) = \Lambda(b + c_2 t, b + c_2 t - c_1 s),$$

which is a matter of straightforward calculus. □

Before presenting our tightness result for the case $c_1 < c_1^\star$, we first introduce some new notations.

- For $r_1, r_2 < 0$,

  $$\overline{\mathbb{E}}A(r_1,r_2) := \mathbb{E}(A(r_1,r_2) \mid A(-t^\star,0) = b + c_2 t^\star),$$

  with $\overline{\mathbb{V}\mathrm{ar}}(\cdot)$ and $\overline{\mathbb{C}\mathrm{ov}}(\cdot,\cdot)$ defined similarly. Also, $\overline{v}(r_1) := \overline{\mathbb{V}\mathrm{ar}}A(r_1,0)$ and $\overline{\Gamma}(r_1,r_2) := \overline{\mathbb{C}\mathrm{ov}}(A(r_1,0), A(r_2,0))$.

- For $r \in (-t^\star, 0)$ we define the functions

  $$\overline{m}(r) := \frac{\overline{\mathbb{E}}A(r,0) + c_1 r}{\sqrt{\overline{v}(r)}}, \quad m(r) := \frac{\overline{m}(r)}{\overline{m}(-s^\star)}, \quad \rho(r) := \frac{\overline{\Gamma}(r,-s^\star)}{\sqrt{\overline{v}(r)\,\overline{v}(-s^\star)}}.$$

Theorem 9.3.4 *Suppose*

$$m(-s) \leq \rho(-s) \quad \textit{for all } s \in (0,t^\star). \tag{9.14}$$

Under Assumption 6.2.1, if $c_1 < c_1^\star$, then

$$I_c^{(t)}(b) = \inf_{t > t^0} \sup_{s \in (0,t)} \Upsilon(s,t) = \Lambda(b + c_2 t^\star, b + c_2 t^\star - c_1 s^\star),$$

and a most probable path is

$$f^\star(r) = -\mathbb{E}(A(r,0) \mid A(-t^\star,0) = b + c_2 t^\star, A(-s^\star,0) = c_1 s^\star).$$

Proof. As in Theorem 9.3.1, we have to show that $f^\star(\cdot)$ is in $\overline{\mathcal{S}^{(t)}}$. This is done as follows.

$$\begin{aligned} f^\star(-s) - f^\star(-t^\star) &= \mathbb{E}(A(-t^\star, -s) \mid A(-t^\star, 0) = b + c_2 t^\star, A(-s^\star, 0) = c_1 s^\star) \\ &= b + c_2 t^\star - \overline{\mathbb{E}}(A(-s, 0) \mid A(-s^\star, 0) = c_1 s^\star) \\ &= b + c_2 t^\star - \overline{\mathbb{E}}A(-s, 0) - \frac{\overline{\Gamma}(-s, -s^\star)}{\overline{v}(-s^\star)} \left(c_1 s^\star - \overline{\mathbb{E}}A(-s^\star, 0) \right). \end{aligned}$$

Now it is easily seen that Equation (9.14) implies that

$$f^\star(-s) - f^\star(-t^\star) \geq b + c_2 t^\star - c_1 s$$

for all $s \in (0, t^\star)$. Due to Lemma 9.3.2, we have that $k(s^\star, t^\star) \geq c_1 s^\star$. With Lemma 9.3.3, the expression for $I_c^{(t)}(b)$ follows. □

Although the condition (9.14), required in Theorem 9.3.4, is stated in terms of the model parameters, as well as known statistics of the arrival process, it could be a tedious task to verify it in a specific situation. The next lemma presents a somewhat more transparent *necessary* condition for (9.14).

The intuition behind the lemma is the following. Observe that both $\rho(\cdot)$ and $m(\cdot)$ attain a maximum 1 at $r = -s^\star$. For $\rho(\cdot)$ this follows from the observation that $\rho(r)$ is a correlation coefficient; for $m(\cdot)$ from Corollary 9.2.5 and Lemma 9.3.2. Then a necessary condition (9.14) is that in $s^\star$ the curve $m(\cdot)$ is 'more concave' than $\rho(\cdot)$. The proof of the lemma is given in Appendix C of [201].

Lemma 9.3.5 *A necessary condition for* (9.14) *is*

$$m''(-s^\star) \leq \rho''(-s^\star), \tag{9.15}$$

or equivalently,

$$\theta_1^\star(s^\star, t^\star)\,(v''(t^\star - s^\star) - v''(s^\star)) \;+\; \theta_2^\star(s^\star, t^\star)\,(v''(0) - v''(s^\star)) \;\geq\; 0. \tag{9.16}$$

Condition (9.16) has an insightful interpretation, which will be given in the next section.

9.4 Tandem: properties of the input rate path

So far, we have analyzed paths $f(\cdot)$ of the *cumulative* amount of traffic injected into the system. In this section we turn our attention to the first derivative $g(\cdot) := f'(\cdot)$ of $f(\cdot)$, which can be interpreted as the path of the *input rate* of the queuing system.

As before, we have to consider two regimes: (A) $c_1 \geq c_1^\star$, and (B) $c_1 < c_1^\star$; let Assumption 6.2.1 be in force. Consider the paths $f^\star$ as identified in Theorems 9.3.1

and 9.3.4, and, more specifically, their derivative $g^\star(\cdot) := (f^\star)'(\cdot)$. In case (A), with $t^\star = t_{c_2}$, and $r \in (-t^\star, 0)$,

$$g^\star(r) = \frac{b + c_2 t^\star}{2v(t^\star)}\,(v'(r + t^\star) + v'(-r)),$$

whereas in case (B) it turns out that, with $r \in (-t^\star, -s^\star]$,

$$g^\star(r) = \frac{v'(r + t^\star) + v'(-r)}{2}\;\theta_1^\star(s^\star, t^\star) + \frac{-v'(-r - s^\star) + v'(-r)}{2}\;\theta_2^\star(s^\star, t^\star),$$

and with $r \in [-s^\star, 0)$,

$$g^\star(r) = \frac{v'(r + t^\star) + v'(-r)}{2}\;\theta_1^\star(s^\star, t^\star) + \frac{v'(r + s^\star) + v'(-r)}{2}\;\theta_2^\star(s^\star, t^\star).$$

If $v'(0) = 0$, we prove below that the path $g^\star(\cdot)$ has some nice properties. Notice that the requirement $v'(0) = 0$ holds for many Gaussian processes. It is not valid for standard Brownian motion, since then $v(t) = t$, but the special structure (independent increments!) of Brownian motion allows an explicit analysis. Fractional Brownian motion (fBm), with $v(t) = t^{2H}$, has $v'(0) = 0$ only for $H \in (\frac{1}{2}, 1]$. These examples are treated in Section 9.5.

Proposition 9.4.1 *If $c_1 \geq c_1^\star$ and $v'(0) = 0$, then $g^\star(0) = g^\star(-t^\star) = c_2$.*

Proof. Notice that $t^\star$ satisfies

$$2c_2\,\frac{v(t^\star)}{v'(t^\star)} = b + c_2 t^\star.$$

The statement follows immediately from $v'(0) = 0$. (As an aside, we mention that $g^\star(\cdot)$ is symmetric in $-t^\star/2$.) □

Just as we exploited properties of $t^\star$ in the proof of Proposition 9.4.1, we need conditions for $s^\star$ and $t^\star$ in the regime $c_1 < c_1^\star$. These are derived in the next lemma.

Exercise 9.4.2 Show that, if $c_1 < c_1^\star$, then $s^\star$ and $t^\star$ satisfy the following equations:

$$2c_2 = \theta_1^\star(s^\star, t^\star)\, v'(t^\star) + \theta_2^\star(s^\star, t^\star)\,(v'(t^\star) - v'(t^\star - s^\star));$$
$$2c_1 = \theta_2^\star(s^\star, t^\star)\, v'(s^\star) + \theta_1(s^\star, t^\star)\,(v'(s^\star) + v'(t^\star - s^\star)).$$

Solution. By Lemma 9.3.2, $k(s^\star, t^\star) \geq c_1 s^\star$. Observe that we can rewrite $\Upsilon(s, t) = \Lambda(b + c_2 t, b + c_2 t - c_1 s) =$

$$\theta^{\mathrm{T}} x(s, t) - \frac{1}{2}\theta^{\mathrm{T}}\Sigma(s, t)\theta, \quad \text{where } x(s, t) := \begin{pmatrix} b + c_2 t \\ c_1 s \end{pmatrix}; \tag{9.17}$$

here we abbreviate $\theta \equiv (\theta_1^\star(s,t), \theta_2^\star(s,t))^{\mathrm{T}}$. We write ∂_t and ∂_s for the partial derivatives with respect to t and s, respectively. The optimal $s^\star$ and $t^\star$ necessarily satisfy the first-order conditions, obtained by differentiating Equation (9.17) to t and s, and equating them to 0. Direct calculations yield

$$\begin{pmatrix} \theta_1 c_2 \\ \theta_2 c_1 \end{pmatrix} = \begin{pmatrix} \partial_t \theta_1 & \partial_t \theta_2 \\ \partial_s \theta_1 & \partial_s \theta_2 \end{pmatrix} (\Sigma(s,t)\theta - x(s,t)) + \begin{pmatrix} \frac{1}{2}\theta_1^2 v'(t) + \partial_t \Gamma(s,t)\theta_1\theta_2 \\ \frac{1}{2}\theta_2^2 v'(s) + \partial_s \Gamma(s,t)\theta_1\theta_2 \end{pmatrix}.$$

Equation (9.13) provides $x(s,t) = \Sigma(s,t)\theta$. Now the statement follows directly. ◇

Proposition 9.4.3 *If* $c_1 < c_1^\star$ *and* $v'(0) = 0$, *then* (i) $g^\star(-t^\star) = c_2$, *and* (ii) $g^\star(-s^\star) = c_1$. *Also, the necessary condition* (9.16) *is equivalent to* $(g^\star)'(-s^\star) \geq 0$.

Proof. Claims (i) and (ii) follow directly from $v'(0) = 0$ and Exercise 9.4.2. The last statement follows directly after some calculations. □

Proposition 9.4.3 can be interpreted as follows. The second queue starts a busy period at time $-t^\star$. During this trajectory, the first queue starts to fill at time $-s^\star$ and is empty again at time 0, if the conditions of Theorem 9.3.4 apply. It is also easily seen that the necessary condition (9.16) has the appealing interpretation that $(g^\star)'(-s^\star) \geq 0$: the input rate path should be increasing at time $-s^\star$.

Some remarks. We end this section by giving two hints on extensions. The first shows how to make the lower bound tighter, while the second indicates how to extend the analysis to multiple-node tandem systems.

Remark 9.4.4 In our lower bound, we replace the intersection over $s \in (0,t)$ by the *least likely event* of the intersection. Under condition (9.16) the occurrence of the least likely event *implies all the other events in the intersection*, with high probability (in the sense that $f^\star \in \mathcal{U}_{s^\star,t^\star}$ implies that $f^\star \in \mathcal{U}_{s,t^\star}$ for all $s \in (0,t^\star)$). The examples in Section 9.5 show that the Condition (9.14) is met for many 'standard' Gaussian models, but not always.

If there is no tightness, a better lower bound can be obtained by approximating the intersection by more than just one event:

$$I_c^{(\mathrm{t})}(b) \geq \inf_{t>t^0} \sup_{\bar{s}\in(0,t)^m} \inf_{f\in\mathcal{U}_{\bar{s},t}} \mathbb{I}(f),$$

where $\bar{s} = (s_1, \cdots, s_m)$, and the 'multiple-constraints set' $\mathcal{U}_{\bar{s},t}$ is defined by

$$\mathcal{U}_{\bar{s},t} := \left\{ \begin{array}{r} f \in \Omega : -f(-t) \geq b + c_2 t;\ f(-s_i) - f(-t) \geq b + c_2 t - c_1 s_i, \\ \text{for } i = 1, \ldots, m \end{array} \right\}.$$

Obviously, the lower bound becomes tighter when increasing m. ◇

Remark 9.4.5 The approach we have followed in this section to analyze the two-node tandem network, can be easily applied to an m-node tandem network, with strictly decreasing service rates, i.e., $c_1 > \ldots > c_m$–nodes i for which $c_i \leq c_{i+1}$ can be ignored, cf. [15, 104, 147]. Note that $\sum_{i=1}^{k} Q_{i,n}$ is equivalent to the single queue in which the sources feed into a buffer that is emptied at rate c_k. This means that we have the characteristics of both $\sum_{i=1}^{m-1} Q_{i,n}$ and $\sum_{i=1}^{m} Q_{i,n}$, which enables the analysis of $Q_{m,n}$, just as in the two-node tandem case. ◇

9.5 Tandem: examples

As argued extensively in the earlier chapters, one of the reasons for considering Gaussian input processes, is that they cover a broad range of correlation structures. More specifically, choosing the variance function appropriately, we can make the input process exhibiting for instance lrd behavior. In this section, we do the computations for various variance functions. We also discuss in detail the condition in Theorem 9.3.4.

Brownian motion. As seen before, $v(t) = t$ yields $t_{c_2} = b/c_2$. According to Corollary 9.2.6, $c_1^\star$ is the largest value of c_1 such that for all $s \in (0, t_{c_2})$,

$$\frac{s}{t_{c_2}}(b + c_2 t_{c_2}) - c_1 s \leq 0,$$

i.e., $c_1^\star = 2c_2$. Hence, using Theorem 9.3.1, we have for $c_1 \geq 2c_2$ that $I_c^{(t)}(b) = 2bc_2$, with a constant input rate $g^\star(r) = 2c_2$ for $r \in (-b/c_2, 0)$ and $g^\star(r) = 0$ elsewhere.

Now we turn to the case where $c_1 < 2c_2$. The optimizing $s^\star$ and $t^\star$ are determined by solving the first-order equations for s and t, see Theorem 9.3.4. We immediately obtain that $t^\star = b/(c_1 - c_2)$ and $s^\star = 0$. We conclude that for this regime the service rate of the first queue *does* play a role. The most probable input rate path reads $g^\star(r) = c_1$, for $r \in (-t^\star, 0)$ and $g^\star(r) = 0$ elsewhere. It is easily verified that the most probable path $f^\star(\cdot)$ is in $\bar{\mathcal{S}}^{(t)}$, making the decay rate as found in Theorem 9.3.4 tight. In other words,

$$I_c^{(t)}(b) = \Lambda(b + c_2 t^\star, b + c_2 t^\star - c_1 s^\star) = \frac{bc_1^2}{2(c_1 - c_2)}.$$

Observe that, interestingly, Brownian motion apparently changes its rate instantaneously, as reflected by the most likely input rate path. This is evidently an immediate consequence of the independence of the increments.

Fractional Brownian motion. Choosing $v(t) = t^{2H}$ gives

$$t_{c_2} = \frac{b}{c_2}\frac{H}{1 - H}.$$

By Theorem 9.3.1,

$$I_c^{(t)}(b) = \frac{1}{2}\left(\frac{b}{1-H}\right)^{2-2H}\left(\frac{c_2}{H}\right)^{2H}$$

for all $c_1 \geq c_1^\star$. Unfortunately, for general H there does not exist a closed-form expression for $c_1^\star$.

Exercise 9.5.1 Note that, in general, $c_1^\star$ is a function of b. Prove that for fBm $c_1^\star$ does *not* depend on b.

Solution. Straightforward calculus yields that

$$c_1^\star = \frac{c_2}{2H}\left(\sup_{\alpha\in(0,1)} \frac{1+\alpha^{2H}-(1-\alpha)^{2H}}{\alpha}\right);$$

observe that the self-similarity entails that for fBm $c_1^\star$ does *not* depend on b. ◇

Now turn to the case $c_1 < c_1^\star$. Lemma 9.3.5 states that Condition (9.16) is a necessary condition for tightness to hold. Observe that $v''(t) = (2H-1)2Ht^{2H-2}$ and hence $v''(0) = \infty$. It is easily checked that $\theta_2^\star(s^\star, t^\star) \leq 0$, which implies that in this case (9.16) is not satisfied. Therefore the lower bound on $I_c^{(t)}(b)$ is *not* tight.

M/G/∞ input. In Section 2.5 we introduced the Gaussian counterpart of the M/G/∞ model; let λ be the arrival rate and δ the mean job size. For fBm we could *a priori* rule out tightness of the lower bound due to $v''(0) = \infty$, see Lemma 9.3.5. For M/G/∞ inputs Exercise 2.5.2 shows that $v''(0)$ is finite (even for heavy-tailed job durations D). It implies that condition (9.14) needs to be checked to verify tightness.

Now we consider some examples of session-length distributions. In all the examples we take $b = 0.5$, $\lambda = 0.125$, $\delta = 2$ and $c_2 = 1$.

- *Exponential.* In this case, $v(\cdot)$ is given by Equation (2.5). Notice that $v(\cdot)$ tends to a straight line for large t (corresponding to short-range dependence). Numerical computations then give $c_1^\star = 1.195$. Taking $c_1 = 1.1$ results in $s^\star = 4.756$, $t^\star = 5.169$ and $m(r)$, $\rho(r)$ as given in Figure 9.3. The upper panel of Figure 9.3 shows $m(r)$ and $\rho(r)$ for $r \in (-t^\star, 0)$, whereas the lower panel magnifies the graph around $-s^\star$. We see that indeed $m(\cdot) \leq \rho(\cdot)$ on the desired interval, so the decay rate is tight. A corresponding input rate path is given in Figure 9.2, which satisfies the properties as indicated in Proposition 9.4.3.

- *Hyperexponential.* In case D has a hyperexponential distribution, with probability $p_i \in (0,1)$ it behaves as an exponential random variable with mean ν_i^{-1}, with $i = 1, 2$ and $p_1 + p_2 = 1$. It is easily verified that

$$v(t) = 2\lambda\frac{p_1}{\nu_1^3}\left(\nu_1 t - 1 + e^{-\nu_1 t}\right) + 2\lambda\frac{p_2}{\nu_2^3}\left(\nu_2 t - 1 + e^{-\nu_2 t}\right),$$

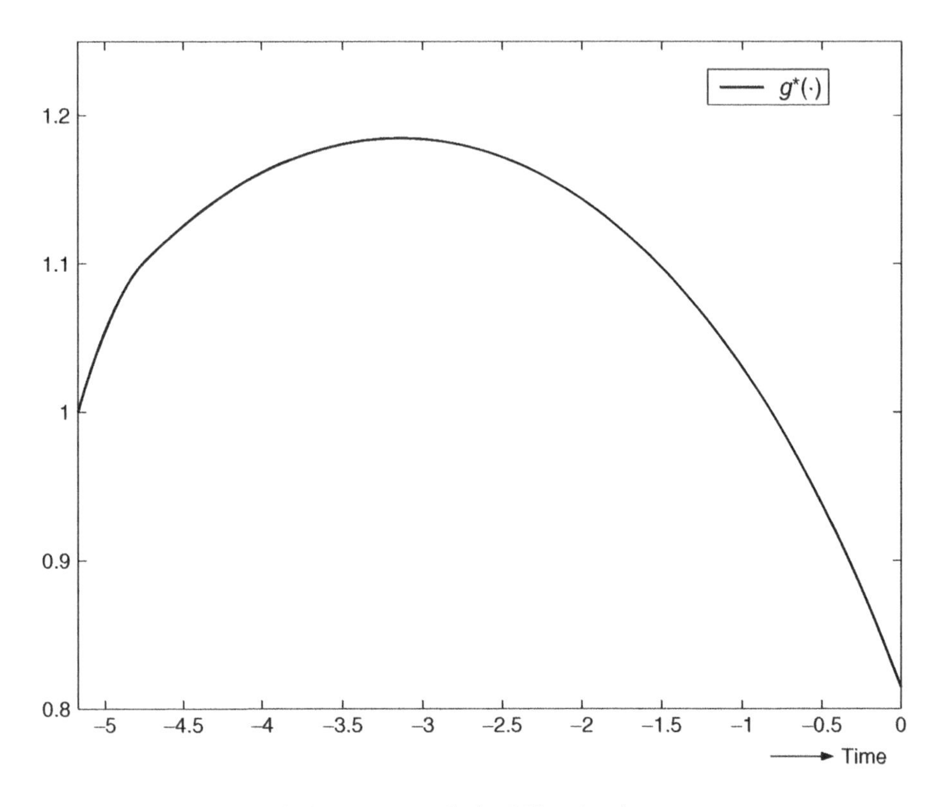

Figure 9.2: Input rate path for M/exp/∞ input process.

with $\nu_2 = p_2/(\delta - p_1/\nu_1)$. Like in the exponential case, $v(\cdot)$ is asymptotically linear. For $p_1 = 0.25$ and $\nu_1 = 5$, we find $c_1^\star = 1.173$, and $s^\star = 4.700$, $t^\star = 5.210$, when using $c_1 = 1.1$. Also for this example $m(\cdot) \leq \rho(\cdot)$, and hence there is tightness; the graph looks similar to Figure 9.3.

- *Pareto.* If D has a Pareto distribution, then $\mathbb{P}(D > t) = (1/(1+t))^\alpha$. The variance function is given by Equation (2.6). Notice that we have $\alpha = \frac{3}{2}$, yielding $v(t) \sim t\sqrt{t}$, which corresponds to lrd traffic. Numerical calculations show that $c_1^\star = 1.115$, and for $c_1 = 1.1$ we obtain $s^\star = 4.373$, $t^\star = 5.432$. Again it turns out that $m(\cdot)$ is majorized by $\rho(\cdot)$. We empirically found, however, that there is *not* always tightness in the M/Par/∞ case. Interestingly, if b is larger, for instance $b = 1$, then Condition (9.14) is not met.

9.6 Priority queues

In Sections 9.2 and 9.3 we analyzed overflow in the second queue of a tandem system. This analysis was enabled by the fact that we had explicit knowledge of

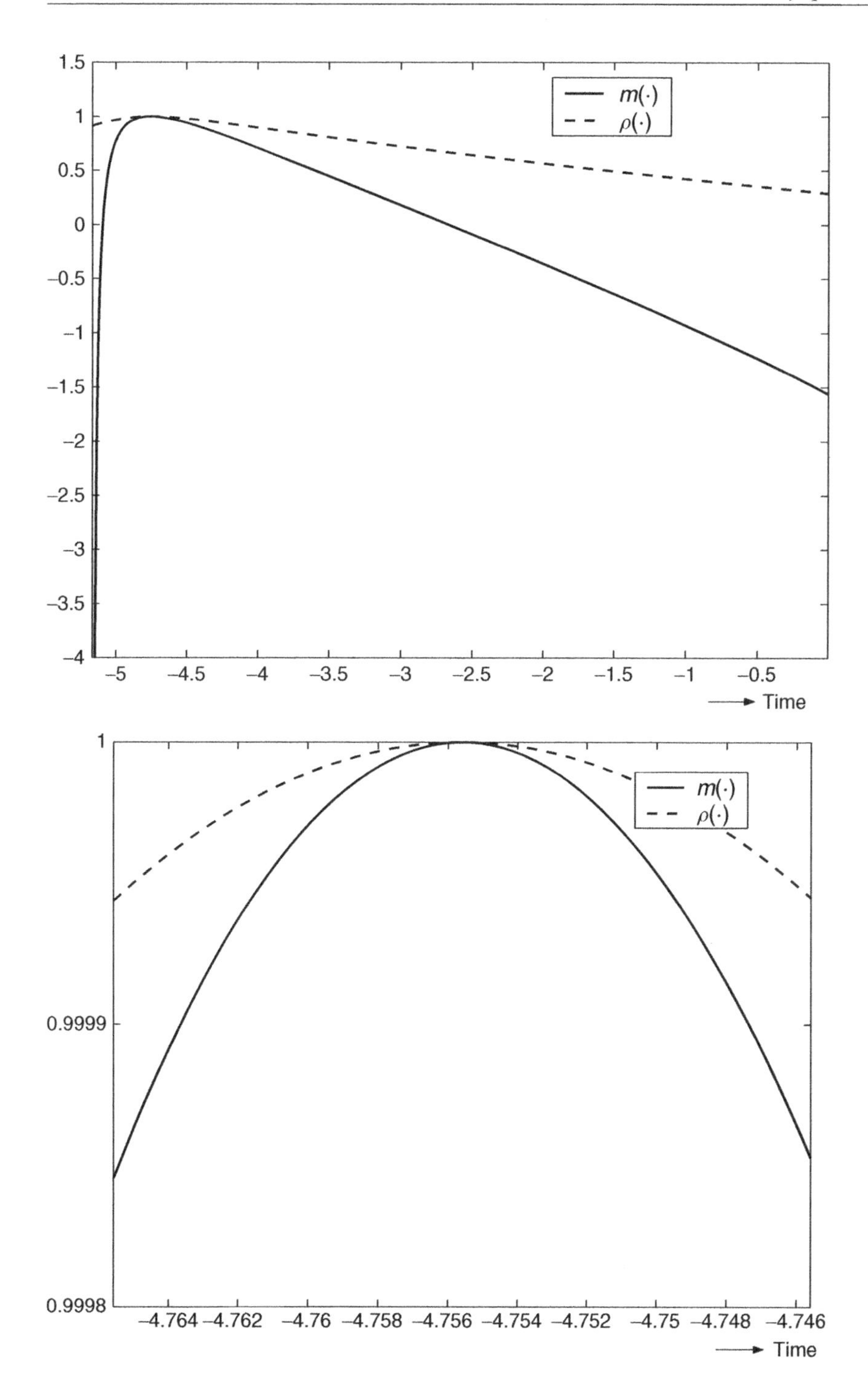

Figure 9.3: M/exp/∞ input process.

both the *first* queue and the *total* queue. In the present section, we use the same type of arguments to solve the (two-queue) priority system.

Analysis. We consider a priority system with a link of capacity nc, fed by traffic of two classes, each with its own queue. Traffic of class 1 does not 'see' class 2 at all, and consequently we know how the *high-priority* queue $Q_{h,n}$ behaves. Also, due to the work-conserving property of the system, the *total* queue length $Q_{h,n} + Q_{\ell,n}$ can be characterized. Now we are able, applying the same arguments as for the tandem queue, to analyze the decay rate of the probability of exceeding some buffer threshold in the lp queue. This similarity between tandem and priority systems has been observed before, see for instance [93].

We let the system be fed by n i.i.d. hp sources, and an equal number of i.i.d. lp sources; both classes are independent. We assume that both hp and lp sources are Gaussian, and satisfy the requirements imposed in Section 9.1. Define the means by μ_h and μ_ℓ, and the variance functions by $v_h(\cdot)$ and $v_\ell(\cdot)$, respectively; also $\mu := \mu_h + \mu_\ell$ (where $\mu < c$) and $v(\cdot) := v_h(\cdot) + v_\ell(\cdot)$. We note that in this priority setting we cannot restrict ourselves to centered processes. We denote the amount of traffic from the ith hp source in $(s, t]$, with $s < t$, by $A_{h,i}(s, t)$; we define $A_{\ell,i}(s, t)$ analogously. Also $\Gamma_h(s, t)$, $\Gamma_\ell(s, t)$, and R_h, R_ℓ are defined as before.

Remark 9.6.1 Notice that this setting also covers the case that the number of sources of both classes are *not* equal. Assume for instance that there are $n\alpha$ lp sources. Multiplying μ_ℓ and $v_\ell(\cdot)$ by α and applying the fact that the normal distribution is infinitely divisible, we arrive at n i.i.d. sources. $\diamond$

In the tandem situation we could, without loss of generality, center the Gaussian sources. It can be checked easily that such a reduction property does not hold in the priority setting, since there is no counterpart of Remark 9.1.3. Hence we cannot assume, without loss of generality, that $\mu_h = \mu_\ell = 0$.

Analogously to Lemma 9.1.1, we obtain that $\mathbb{P}(Q_{\ell,n} > nb)$ equals

$$\mathbb{P}\left(\exists t > 0 : \forall s > 0 : \frac{1}{n}\sum_{i=1}^{n} A_{h,i}(-t, -s) + \frac{1}{n}\sum_{i=1}^{n} A_{\ell,i}(-t, 0) > b + c(t - s)\right).$$

Let $I_c^{(\mathrm{p})}(b)$ be the exponential decay rate of $\mathbb{P}(Q_{\ell,n} > nb)$; analogously to Theorem 9.2.1 it can be shown that this decay rate exists. Similar to the tandem case, with $f(\cdot) \equiv (f_h(\cdot), f_\ell(\cdot))$,

$$\mathcal{S}_{s,t}^{(\mathrm{p})} := \{f \in \Omega \times \Omega : f_h(-s) - f_h(-t) - f_\ell(-t) > b + c(t - s)\};$$

$$\mathcal{U}_{s,t}^{(\mathrm{p})} := \left\{ f \in \Omega \times \Omega : \begin{array}{l} -f_h(-t) - f_\ell(-t) \geq b + ct; \\ f_h(-s) - f_h(-t) - f_\ell(-t) \geq b + c(t - s) \end{array} \right\}; \tag{9.18}$$

$$\mathbb{P}(Q_{\ell,n} > nb) = \mathbb{P}\left(\left(\frac{1}{n}\sum_{i=1}^{n} A_{h,i}(\cdot); \frac{1}{n}\sum_{i=1}^{n} A_{\ell,i}(\cdot)\right) \in \bigcup_{t>0}\bigcap_{s>0} \mathcal{S}_{s,t}^{(\mathrm{p})}\right).$$

Theorem 9.6.2 *The following lower bound applies:*

$$I_c^{(p)}(b) \geq \inf_{t>0} \sup_{s>0} \inf_{f \in \mathcal{U}_{s,t}^{(p)}} \mathbb{I}(f), \tag{9.19}$$

with $\overline{f}_h(t) := f_h(t) - \mu_h t$, $\overline{f}_\ell(t) := f_\ell(t) - \mu_\ell t$, *and*

$$\mathbb{I}(f) := \frac{1}{2}||\overline{f}_h||^2_{R_h} + \frac{1}{2}||\overline{f}_\ell||^2_{R\ hl}.$$

The infimum over $f \in \mathcal{U}_{s,t}^{(p)}$ can be computed explicitly, as in Lemma 9.2.3. As the analysis is analogous to the tandem case, but the expressions are more complicated, we only sketch the procedure. Again there is a regime in which one of the two constraints is redundant. Define

$$k_p(s,t) := \mathbb{E}(A_h(s) \mid A_h(t) + A_\ell(t) = b + ct).$$

Using the convexity of the large-deviations rate function, it can be shown that, if

$$\mathbb{E}(A_h(t-s) + A_\ell(t) \mid A_h(t) + A_\ell(t) = b + ct) > b + c(t-s),$$

only the first constraint in the set (9.18) is tightly met; it is equivalent to require that $k_p(s,t) < cs$. (If $k_p(s,t) \geq cs$ either both constraints in the set (9.18) are met with equality, or only the second constraint is met with equality; exact conditions for these two cases are easy to derive, but these are not relevant in this discussion). As earlier, under $k_p(s,t) < cs$, we obtain the decay rate

$$\inf_{f \in \mathcal{U}_{s,t}^{(p)}} \mathbb{I}(f) = \frac{(b + (c-\mu)t)^2}{2v(t)}, \tag{9.20}$$

cf. the single queue with link rate nc; in the other cases, the expressions are somewhat more involved. Denote the value of $t > 0$ (in this section) by t_c that minimizes the right-hand side of Equation (9.20).

Similar to the tandem case, there is a regime (i.e., a set of values of the link rate c) in which $I_c^{(p)}(b)$ coincides with the decay rate of a single queue. In this regime, which we call regime (A), conditional on a large value of the total queue length, it is likely that the hp queue is empty, such that all traffic that is still in the system is in the lp queue. Hence, for all c in

$$\{c \mid \forall s > 0 : k_p(s, t_c) < cs\} \tag{9.21}$$

we conclude

$$I_c^{(p)}(b) = \frac{(b + (c-\mu)t_c)^2}{2v(t_c)}.$$

If c is not in the set (9.21), we can use the methodology of Sections 9.2 and 9.3 to find a condition under which the lower bound of Theorem 9.6.2 is tight; we call this regime (B).

Two regimes. In the tandem case, we found that the single-queue result holds for $c_1 \geq c_1^\star$, whereas it does not hold for $c_1 < c_1^\star$; the threshold value $c_1^\star$ was found explicitly in Section 9.2. In the priority setting, there is not such a clear dichotomy. Consider, for instance, the situation in which both types of sources correspond to Brownian motions; $v_h(t) \equiv \lambda_h t$, $v_\ell(t) \equiv \lambda_\ell t$, and $\lambda := \lambda_h + \lambda_\ell$. Define

$$\Xi := \sqrt{\mu_\ell^2 + \frac{\lambda_\ell}{\lambda_h}(c - \mu_h)^2}.$$

Then straightforward calculus yields that for $(\lambda_h - \lambda_\ell)c \leq \lambda_h(\mu_h + 2\mu_\ell) - \lambda_\ell \mu_h$, regime (A) applies (i.e., the single-queue result holds):

$$I_c^{(\mathrm{p})}(b) = \frac{2b(c-\mu)}{\lambda},$$

whereas otherwise we are in regime (B):

$$I_c^{(\mathrm{p})}(b) = \frac{b(\Xi - \mu_\ell)}{\lambda_\ell};$$

this is shown by verifying that the lower bound of Theorem 9.6.2 is tight for the specific case of Brownian motion input. Using $\mu_h + \mu_\ell < c$, it can be verified easily that this implies that for $\lambda_h \leq \lambda_\ell$ the single-queue solution applies, whereas for $\lambda_h > \lambda_\ell$ only for

$$c \leq \frac{\lambda_h(\mu_h + 2\mu_\ell) - \lambda_\ell \mu_h}{\lambda_h - \lambda_\ell},$$

the single-queue solution applies. We return to this model in Section 11.3.

Discussion. Large deviations for priority queues have been studied in several papers. We mention here the works [204, 292]. We briefly review these results, and compare them with our analysis. Our lower bound reads

$$I_{c,1}^{(\mathrm{p})}(b) := \inf_{t>0} \sup_{s>0} \Upsilon_p(s,t), \quad \text{with} \quad \Upsilon_p(s,t) := \inf_{f \in \mathcal{U}_{s,t}^{(\mathrm{p})}} \mathbb{I}(f).$$

Just as we did, [204] identifies two cases. They get the same solution for our regime (A), i.e., the situation in which, given a long total queue length, the hp queue is relatively short, cf. also the empty buffer approximation in [29].

In regime (B) the hp queue tends to be large, given that the total queue is long. To prevent this from happening, [204] proposes a heuristic that minimizes $\mathbb{I}(f)$ over

$$\{f \in \Omega \times \Omega : \exists t > 0 : -f_h(-t) - f_\ell(-t) \geq b + ct; -f_h(-t) \leq ct\}. \tag{9.22}$$

Because regime (B) applies, the optimum paths in the set (9.22) are such that the constraints on f are tightly met; consequently, set (9.22) is a subset of $\mathcal{U}_{t,t}^{(\mathrm{p})}$. Hence the resulting decay rate, which we denote by $I^{(\mathrm{p})}_{c,\mathrm{II}}(b)$, yields a lower bound, but our lower bound will be closer to the real decay rate:

$$I^{(\mathrm{p})}_{c,\mathrm{II}}(b) := \inf_{t>0} \Upsilon_p(t,t) \leq \inf_{t>0}\sup_{s>0} \Upsilon_p(s,t) = I^{(\mathrm{p})}_{c,\mathrm{I}}(b).$$

Remark 9.6.3 In the simulation experiments performed in [204], it was found that the lower bound $I^{(\mathrm{p})}_{c,\mathrm{II}}(b)$ is usually close to the exact value. Our numerical experiments (cf. the examples on the tandem queue in Section 9.5) show that the hp buffer usually starts to fill shortly after the total queue starts its busy period. This means that, in many cases, the error made by taking $s = t$ is relatively small. It explains why the heuristic based on set (9.22) performs well. ◇

Wischik [292] focuses on discrete time, and allows more general traffic than just Gaussian sources. Translated into continuous time, in regime (B), his lower bound on the decay rate $I^{(\mathrm{p})}_{c,\mathrm{III}}(b)$ (Theorem 14) minimizes $\mathbb{I}(f)$ over

$$\{f \in \Omega \times \Omega : \exists t > 0 : \exists s > 0 : -f_h(-t) - f_\ell(-t) \geq b + ct; -f_h(-s) \leq cs\}; \tag{9.23}$$

again a straightforward comparison gives that our lower bound $I^{(\mathrm{p})}_{c,\mathrm{I}}(b)$ is closer to the actual decay rate:

$$I^{(\mathrm{p})}_{c,\mathrm{III}}(b) := \inf_{t>0}\inf_{s>0} \Upsilon_p(s,t) \leq \inf_{t>0}\sup_{s>0} \Upsilon_p(s,t) = I^{(\mathrm{p})}_{c,\mathrm{I}}(b).$$

Bibliographical notes

This chapter is largely based on Mandjes and Van Uitert [201], see also [199]. Interestingly, along the lines of Mannersalo and Norros [204, 206, 225, 226] also the following approximation can be proposed:

$$I^{(\mathrm{t})}_c(b) \approx \inf_{t \geq t^0} \frac{(b + c_2 t)^2}{2v(t)}; \tag{9.24}$$

in [204, 206, 225, 226] this is called a *(rough) full link approximation*. The idea behind this approximation is the following. If $t_{c_2} \geq t^0$, and traffic has been generated at a more or less constant rate in $[-t_{c_2}, 0]$, then no (or hardly any) traffic is left in the first queue at time 0, and the approximation seems reasonably accurate. If, on the other hand, $t_{c_2} < t^0$, then there will be traffic left in the first queue at time 0, so the input rate needs to be pushed down; therefore $t = t^0$ has to be chosen, such that the sources are forced to transmit at about rate c_1, and the first queue remains (nearly) empty. Numerical experiments have indicated that this approximation is quite accurate, see Mandjes, Mannersalo, and Norros [192].

In this chapter, we have seen that the lower bound presented in Section 9.2 is tight for iOU, but not for fBm. The difficulty when looking for the most likely path is that, for fixed t, we have to deal with an infinite intersection of events, indexed by $s \in (0, t)$. We found a lower bound on the decay rate of the intersection, which corresponded to the least likely event in the intersection. As remarked earlier, the lower bound is tight if this least likely event essentially implies the other events in the intersection. Apparently, for iOU this is the case (and is the most likely path equivalent to the weighted sum of two covariance functions), but for fBm it is not.

The question remained what the most likely path should be for fBm. To investigate this issue, Mandjes, Mannersalo, Norros, and Van Uitert [193] have been looking first at a more elementary case. Consider the decay rate of $p_n[\mathcal{V}]$, with

$$\mathcal{V} := \bigcap_{t\in(0,1)} \mathcal{V}_t \quad \text{with } \mathcal{V}_t := \{f \in \Omega : f(t) \geq t\},$$

which could be interpreted as the event of having a busy period of length at least 1, in a queue drained at unit rate. Norros [223] already provided several bounds on this decay rate:

- as $\mathcal{V}_t \subseteq \mathcal{V}$, we have

$$-\lim_{n\to\infty} \frac{1}{n} \log p_n[\mathcal{V}] \geq \sup_{t\in(0,1)} \frac{t^2}{2t^{2H}} = \frac{1}{2};$$

- as the norm of any feasible path is an upper bound, we have

$$-\lim_{n\to\infty} \frac{1}{n} \log p_n[\mathcal{V}] \leq \mathbb{I}(\chi) =: \vartheta(H),$$

  where $\chi(t) = t$, for $t \in (0, 1)$.

The function $\vartheta(\cdot)$ could be evaluated explicitly, and numerical investigations indicated that there was still a modest gap between the lower bound (i.e., $\frac{1}{2}$) and the upper bound. In [193], the exact value for the decay rate was found. Notably, the most likely path $f^\star$ is for $H \in (\frac{1}{2}, 1)$ such that $f^\star(t) = t$ for $t \in [0, \tau] \cup \{1\}$ (where $\tau < \frac{1}{2}$), and that $f^\star(t) > t$ for $t \in (\tau, 1)$; similarly, in the regime $H \in (0, \frac{1}{2})$, we have that $f^\star(t) = t$ for $t \in \{0\} \cup [\tau, 1]$. Interestingly, the most likely path is now a linear combination of uncountably many covariance functions. The analysis is substantially more involved than that of this chapter, but a number of concepts could still be used; more specifically, the concept of least likely events turned out to be very useful. [193] also shows that the 'smoothness' of the Gaussian process under consideration plays an important role here, which also explains why for iOU the lower bound of this chapter was tight, but for fBm it was not.

The techniques to identify the most likely path in $\mathcal{V}$ can be adapted to the tandem case. In Mandjes, Mannersalo, and Norros [192] (a program to compute) the corresponding decay rate is presented: a function ζ is found such that

$$\lim_{n\to\infty} \frac{1}{n} \log \mathbb{P}(Q_{2,n} \geq nb) = \zeta(b, c_1, c_2, H).$$

In case $c_1 < c_1^\star$, we again have that a part of the most likely path is linear. It is noted that these also provide the large-buffer asymptotics: reasoning as in Exercise 5.6.3 leads to

$$\lim_{\mathrm{B}\to\infty} \frac{1}{\mathrm{B}^{2-2H}} \log \mathbb{P}(Q_2 \geq \mathrm{B}) = \zeta(1, \mathrm{C}_1, \mathrm{C}_2, H). \tag{9.25}$$

The fact that there is, in the tandem setting, a critical service rate $c_1^\star$ has been observed several times earlier. In this respect, we mention the pioneering study of Chang, Heidelberger, Juneja, and Shahabuddin [47]. They focus on srd sources in a discrete-time setting for a class of queuing networks that is slightly more general than tandems (intree networks). In [47] also a fast simulation procedure is proposed. The dichotomy, with a regime in which the first queue is 'transparent' and a regime in which the first queue really 'shapes' the traffic, was also found in several other studies: see for instance Chang [45], Mandjes [183], and Dębicki, Mandjes and Van Uitert [66]. Also notice the similarity with the priority queue, see [3].

When analyzing a tandem system under the many-sources scaling, we have implicitly characterized the (aggregate) output stream of the first queue. Wischik [291] concentrates, also in the many-sources setting, on the statistical characteristics of the *individual* output streams; he shows that these are essentially unchanged by passage through the first queue. Wischik and Ganesh [293] argue that the (aggregate) output stream is 'as long-range dependent' as the input stream (in terms of its so-called *Hurstiness*).

We already explained in Section 9.6 how our results on the priority system compare to those found by Mannersalo and Norros [204] (which was essentially a rough full link approximation, comparable to Equation (9.24) described above) and by Wischik [292]. It turns out that the case with fBm input allows an explicit solution, see Mandjes, Mannersalo, and Norros [191]; this analysis again relies on the findings of [193].

We conclude this chapter by remarking that Shakkottai and Srikant [263] focus on *delay* asymptotics (rather than the asymptotics of the buffer content distribution) under the many-sources scaling, see also Sections 10.7 and 11.3. A related study on priority queues is by Delas, Mazumdar, and Rosenberg [71].

Chapter 10

Generalized processor sharing

Traditionally, every type of allocation was supported by a different network: there were logically separated voice networks, data networks, etc. The current trend, however, is to integrate these networks, such that a growing range of traffic types is supported over a common network infrastructure. As described in the introduction of this book, these traffic types are highly heterogeneous, with respect to both their performance requirements and their statistical properties.

It is clear that FIFO queues lack the capability of offering multiple performance levels. Hence, if a FIFO queue is used to support traffic classes with heterogeneous performance requirements, all classes should be offered the most stringent of these requirements. This approach inevitably leads to inefficient use of network resources: some of the classes get a considerably better performance than requested. This explains why performance differentiation is desirable.

The need for efficient performance-differentiating mechanisms motivates the development of discriminatory scheduling disciplines. These are designed such that they actively distinguish between the streams of the various traffic types. Packet versions of the ideal fluid discipline *generalized processor sharing* (abbreviated as GPS in this monograph) (see, e.g., [73, 235, 236]) are considered to be suitable candidates. In GPS, each class is guaranteed a certain minimum service rate; if one of the classes does not fully use this guaranteed rate, the residual capacity is redistributed among the other classes (in proportion to their guaranteed rates). Note that this makes GPS a work-conserving discipline. GPS is considered as an attractive compromise between isolation and sharing: each traffic class is protected against 'misbehavior' of other classes, whereas at the same time substantial multiplexing gains between classes can be achieved.

In this chapter, we focus on two classes sharing the total service capacity c according to GPS. We assign guaranteed rate $\phi_i c$ to class i, which can be claimed by class i at any time – the ϕ_i are referred to as *weights*, $i = 1, 2$. Without loss of

Large deviations for Gaussian queues M. Mandjes

generality, it can be assumed that the weights sum to 1. Both classes are assigned a queue that fills when the input rate temporarily exceeds the capacity available. When both classes are backlogged, i.e., have nonempty queues, both are served at their guaranteed rate. If one of the classes does not fully use its guaranteed rate, then the unused capacity is made available to the other class.

It is clear that, in order to fully benefit from GPS, the weights should be chosen appropriately. This weight setting is not a straightforward task that usually relies on expressions (or approximations) for the buffer-content distributions of the queues. Weight setting procedures available from the literature are often restricted to special classes of input traffic; see, e.g., [94, 165] for the case of leaky-bucket regulated traffic.

We remark that the priority queue, as analyzed in Section 9.6, is a special case of GPS: one of the weights is chosen to be 1 and the other as 0. Clearly, the priority queue does not offer the flexibility of GPS. More particularly, under priority scheduling the low-priority class can suffer from starvation effects when, during a substantial period of time, the high-priority class claims the full service capacity. Under GPS such a situation can be prevented, by giving the 'low-priority' class a sufficiently high weight.

This chapter can be considered as the 'GPS analog' of the previous chapter: we develop sample-path large deviations for a GPS system fed by Gaussian traffic. As in the previous chapters, we again consider the many-sources scaling: the capacity of the queue is nc, and there are n sources of both type 1 and type 2 (but this is by no means a real restriction, as Remark 9.6.1 applies again). The purpose of this chapter is to find the counterpart of Theorem 9.6.2: we wish to characterize the decay rate of the overflow probabilities (of both queues).

The structure of this chapter is slightly different from that of Chapter 9. For the models dealt with there, i.e., the tandem and priority queue, we knew how to analyze one of the two individual queues of the network (the first queue in the tandem case, and the hp queue in the priority case), as well as the total queue (defined as the sum of the buffer contents of the individual queues). This knowledge enabled us to analyze the other individual queue (i.e., the downstream queue in the tandem case, and the lp queue in the priority case). Unfortunately, this approach breaks down in the GPS case, as the individual queues are so intimately related that it is impossible to analyze one without the other. In other words, the only queue that is straightforward to analyze is the total queue. This makes GPS considerably harder to analyze.

A few words on the literature may be added here to put this chapter in the right perspective. In [205], the same setting was considered (logarithmic many sources large deviations for GPS with Gaussian input); useful intuition and heuristics were developed, along the lines of the rough full link approximation mentioned in the bibliographical notes of Chapter 9.

This approximation was a generalization of that found in [204] for priority queues. In the previous chapter, we enhanced this approximation; notably a lower bound on the decay rate of overflow in the low-priority queue was found, as well as conditions under which this lower bound coincides with the exact value. In more specific terms, the main goal of the present chapter is to obtain similar rigorous many-sources large-deviations results for the two-class GPS system.

A second asymptotic regime is the so-called *large-buffer* regime; see Chapter 5. It is, however, not clear to what extent the assumption of large buffers applies in practice – particularly, real-time applications do not tolerate large delays, and hence in this situation large buffers are not appropriate. We come back to results on this regime in the bibliographical notes at the end of this chapter.

The organization of the remainder of this chapter is as follows. Section 10.1 formally introduces GPS. In Section 10.2 we derive upper and lower bounds for the overflow probabilities in the two-queue GPS system. These are generic in that they apply not only to Gaussian inputs but in fact to any input traffic model. Then we evaluate these bounds in the many-sources framework. First we consider the regime in which the mean rate of the type-2 sources, $n\mu_2$, is below their guaranteed rate $n\phi_2 c$; lower and upper bounds on the decay rate are addressed in Sections 10.3 and 10.4. Section 10.5 deals with the (somewhat easier) case $n\mu_2 \geq n\phi_2 c$. We then prove tightness of the derived bounds under certain conditions, and present an intuitive motivation why tightness can be expected more generally; see Section 10.6. Section 10.7 considers the many-sources asymptotics of long delays in GPS.

10.1 Preliminaries on GPS

We consider a system where traffic is served according to the GPS mechanism, consisting of two queues sharing a link of capacity nc. We assume the system to be fed by traffic from two classes, where class i uses queue i ($i = 1, 2$). Without loss of generality, it is assumed that both classes consist of n flows (see Remark 9.6.1); as usual, we also assume that these flows are i.i.d. We assign a weight $\phi_i \geq 0$ to class i and, again without loss of generality, assume that these add up to 1, i.e., $\phi_1 + \phi_2 = 1$. The GPS mechanism then works as follows. Class i receives service at rate $n\phi_i c$ when both classes are backlogged. Because class i gets at least service at rate $n\phi_i c$ when it has backlog, we will refer to it as the *guaranteed rate* of class i. If one of the classes has no backlog and is transmitting at a rate less than or equal to its guaranteed rate, then this class is served at its transmission rate, while the other class receives the remaining service capacity. If both classes are sending at rates less than their guaranteed rates, then they are both served at their sending rate, and some service capacity is left unused. We assume that the buffer sizes of both queues are infinitely large.

Without loss of generality, we focus on the workload of the first queue. The goal of this chapter is to derive the logarithmic asymptotics for the probability that the stationary workload exceeds a threshold nb. Denoting the stationary workload in the ith GPS queue at time 0 by $Q_{i,n} \equiv Q_{i,n}(0)$, the probability of our interest reads

$$\mathbb{P}(Q_{1,n} \geq nb),$$

and we wish to determine its decay rate, i.e.,

$$I_c^{(g)}(b) := -\lim_{n\to\infty} \frac{1}{n} \log \mathbb{P}(Q_{1,n} \geq nb).$$

We denote the amount of traffic generated by the jth flow of class i in the interval $(s,t]$, $j = 1,\ldots,n$, $i = 1,2$ by $A_{j,i}(s,t)$. Defining $B_{i,n}(s,t)$ as the total service that was available for class i in the interval $(s,t]$, we have the following identity:

$$Q_{i,n}(t) = Q_{i,n}(s) + \sum_{j=1}^{n} A_{j,i}(s,t) - B_{i,n}(s,t), \quad \forall s < t, \text{ with } s,t \in \mathbb{R}. \tag{10.1}$$

The stationary queue can be represented by

$$Q_{i,n}(0) = \sup_{t>0} \left\{ \sum_{j=1}^{n} A_{j,i}(-t,0) - B_{i,n}(-t,0) \right\}, \tag{10.2}$$

where the negative of the optimizing t corresponds to the beginning of the busy period that includes time 0, as argued in [249]. In the next section, we rewrite our problem in terms of the empirical mean processes $n^{-1}\sum_{j=1}^{n} A_{j,i}(\cdot,\cdot)$, $i = 1,2$.

10.2 Generic upper and lower bound on the overflow probability

In a GPS framework, the workloads of the queues are intimately related: it is not possible to write down an explicit expression for $Q_{i,n}(0)$, for $i = 1,2$, without using the evolution of the workload in the other queue. This makes the analysis of GPS systems hard. In this section, we derive explicit upper and lower bounds for $Q_{1,n}(0)$ in terms of the processes $\sum_{j=1}^{n} A_{j,i}(\cdot,\cdot)$, $i = 1,2$.

In the remainder of this chapter, we have to distinguish between two regimes. The most involved regime is $\mu_2 < \phi_2 c$, which we refer to as *underload* for class 2. In this regime, class 2 is stable regardless of the behavior of the other class. The other regime is $\mu_2 \geq \phi_2 c$; here class 2 is said to be in *overload*. Although the bounds that are derived in this section hold for both regimes, they are only useful in the regime with underload for class 2 – they will be exploited in Sections 10.3

and 10.4. The analysis for the regime with class 2 in overload does not require any sophisticated bound, and is presented in Section 10.5.

Note that the results in this section hold regardless of the distribution of the inputs; they are also valid for non-Gaussian traffic. We also remark that, in order to justify the use of the large deviations results (in particular 'Schilder', i.e., Theorem 4.2.3) we have to formally verify whether the sets under consideration are indeed open or closed. This (technical) issue can be dealt with in precisely the same fashion as in [201, Theorem 3.1], and is left out here.

Trivially, we can rewrite the overflow probability as

$$\mathbb{P}\left(Q_{1,n}(0) \geq nb\right) = \mathbb{P}\left(\bigcup_{x\geq 0}\{Q_{1,n}(0) + Q_{2,n}(0) \geq nx + nb,\ Q_{2,n}(0) \leq nx\}\right). \tag{10.3}$$

Because of the work-conserving nature of GPS, it is easily seen that the following relation holds for the total queue:

$$Q_{1,n}(0) + Q_{2,n}(0) = \sup_{t>0}\left\{\sum_{j=1}^{n}(A_{j,1}(-t,0) + A_{j,2}(-t,0)) - nct\right\}. \tag{10.4}$$

Substituting this relation for $Q_{1,n}(0) + Q_{2,n}(0)$ in the right-hand side of Equation (10.3), we find

$$\mathbb{P}\left(\bigcup_{x\geq 0}\left\{\begin{array}{l}\sup_{t>0}\left\{\sum_{j=1}^{n}(A_{j,1}(-t,0) + A_{j,2}(-t,0)) - nct\right\} \geq nx + nb, \\ Q_{2,n}(0) \leq nx\end{array}\right\}\right). \tag{10.5}$$

We denote the optimizing t in the above supremum by $t^\star$. Following [249], $-t^\star$ can be interpreted as the beginning of the busy period of the total queue containing time 0.

Next we consider $Q_{2,n}(0)$. Let us denote the beginning of the busy period of queue 2 containing time 0 by $-s^\star$. Then, clearly $s^\star \in [0, t^\star]$, since the busy period of the total queue cannot start after the start of the busy period of queue 2. Now using the supremum representation (10.2), we obtain

$$Q_{2,n}(0) = \sup_{s\in(0,t^\star]}\left\{\sum_{j=1}^{n} A_{j,2}(-s,0) - B_{2,n}(-s,0)\right\}. \tag{10.6}$$

In order to find bounds for $\mathbb{P}(Q_{1,n}(0) \geq nb)$, it follows from Equation (10.5) that we need to bound the class-2 workload at time 0, $Q_{2,n}(0)$. Given its representation in Equation (10.6), this means that we have to find bounds on the service that was available for class 2 during the busy period containing time 0.

We introduce the following additional notations:

$$\mathcal{E}_n := \left\{ \begin{array}{l} \exists x \geq 0, t > 0 : \forall s \in (0, t] : \\ (1/n) \sum_{j=1}^{n} \left(A_{j,1}(-t, 0) + A_{j,2}(-t, 0) \right) \geq x + b + ct, \\ (1/n) \sum_{j=1}^{n} A_{j,2}(-s, 0) \leq x + \phi_2 cs \end{array} \right\};$$

$$\mathcal{F}_n := \left\{ \begin{array}{l} \exists x \geq 0, t > 0 : \forall s \in (0, t] : \exists u \in [0, s) : \\ (1/n) \sum_{j=1}^{n} \left(A_{j,1}(-t, 0) + A_{j,2}(-t, 0) \right) \geq x + b + ct, \\ (1/n) \sum_{j=1}^{n} \left(A_{j,2}(-s, 0) + A_{j,1}(-s, -u) \right) \leq x + cs - \phi_1 cu \end{array} \right\}.$$

In the following lemmas, we derive the lower and upper bound for the overflow probability of class 1.

Lemma 10.2.1 *The following lower bound applies:*

$$\mathbb{P}\left(Q_{1,n}(0) \geq nb\right) \geq \mathbb{P}\left(\mathcal{E}_n\right).$$

Proof. Recall that $-s^\star$ denotes the beginning of the busy period of queue 2 that contains time 0. Hence, the workload of class 2 is positive in the interval $(-s^\star, 0]$, indicating that class 2 claims at least its guaranteed rate in this interval: $B_{2,n}(-s^\star, 0) \geq n\phi_2 cs^\star$. Using this lower bound in Equation (10.6), we derive

$$Q_{2,n}(0) \leq \sup_{s \in (0, t^\star]} \left\{ \sum_{j=1}^{n} A_{j,2}(-s, 0) - \phi_2 ncs \right\}. \tag{10.7}$$

The lower bound for $\mathbb{P}(Q_{1,n}(0) \geq nb)$ is now found upon substituting Equation (10.7) for $Q_{2,n}(0)$ in Equation (10.5). □

Lemma 10.2.2 *The following upper bound applies:*

$$\mathbb{P}\left(Q_{1,n}(0) \geq nb\right) \leq \mathbb{P}\left(\mathcal{F}_n\right).$$

Proof. From Equation (10.6), we see that we need an upper bound on $B_{2,n}(-s^\star, 0)$. We distinguish between two scenarios: (a) queue 1 is strictly positive during $(-s^\star, 0]$ and (b) queue 1 has been empty at some time in $(-s^\star, 0]$.

(a) Since both queues are strictly positive during $(-s^\star, 0]$, both classes claim their guaranteed rate, i.e., $B_{2,n}(-s^\star, 0) = n\phi_2 cs^\star$.

(b) Trivially, $B_{2,n}(-s^\star, 0) \leq ncs^\star - B_{1,n}(-s^\star, 0)$. Bearing in mind that queue 1 has been empty in $(-s^\star, 0]$, we define

$$u^\star := \inf\{u \in [0, s^\star) : Q_{1,n}(-u) = 0\}.$$

Hence, both queues were strictly positive during $(-u^\star, 0]$, and consequently both classes are assigned their guaranteed rates. Together with Equation (10.1), this yields

$$\begin{aligned} B_{1,n}(-s^\star, 0) &= B_{1,n}(-s^\star, -u^\star) + B_{1,n}(-u^\star, 0) \\ &= Q_{1,n}(-s^\star) + \sum_{j=1}^{n} A_{j,1}(-s^\star, -u^\star) + n\phi_1 c u^\star \\ &\geq \inf_{u\in[0,s^\star)} \left\{ \sum_{j=1}^{n} A_{j,1}(-s^\star, -u) + n\phi_1 c u \right\}. \end{aligned}$$

This implies

$$B_{2,n}(-s^\star, 0) \leq ncs^\star - \inf_{u\in[0,s^\star)} \left\{ \sum_{j=1}^{n} A_{j,1}(-s^\star, -u) + n\phi_1 c u \right\}. \quad (10.8)$$

As the right-hand side of Equation (10.8) is larger than $n\phi_2 cs^\star$, items (a) and (b) imply that using Equation (10.8) in Equation (10.6) yields

$$\begin{aligned} Q_{2,n}(0) \geq \sup_{s\in(0,t]} &\left\{ \sum_{j=1}^{n} A_{j,2}(-s, 0) - ncs \right. \\ &\left. + \inf_{u\in[0,s)} \left\{ \sum_{j=1}^{n} A_{j,1}(-s, -u) + n\phi_1 c u \right\} \right\}. \end{aligned}$$

Substituting this for $Q_{2,n}(0)$ in Equation (10.5) then yields the desired upper bound. □

Remark 10.2.3 Compare the sets $\mathcal{E}_n$ and $\mathcal{F}_n$; evidently, $\mathcal{E}_n \subseteq \mathcal{F}_n$. Any path f of the sample-mean process in $\mathcal{F}_n$ defines epochs $u^\star$ and $s^\star$ (as identified in the proof of Lemma 10.2.2). It is not hard to see that if $u^\star = s^\star$, f is also in $\mathcal{E}_n$. From the proof of Lemma 10.2.2, taking $u^\star = s^\star$ means that scenario (a) applies, where queue 1 is strictly positive during the busy period of queue 2 containing time 0. These simple observations turn out to play a crucial role in the discussion presented in Section 10.6. ◇

Remark 10.2.4 Bounds similar to those used in the proofs of Lemmas 10.2.1 and 10.2.2 have been applied in [296]. A crucial novelty of our approach, compared to that of [296], is that it explicitly indicates when the bounds match, relying on the interpretation of $t^\star$, $s^\star$, and $u^\star$; see also Remark 10.2.3.

There is, however, an important other difference compared with [296].

- In [296], the 'trivial GPS upper bound' (i.e., $B_{i,n}(s,t) \geq n\phi_i c(t-s)$) is used in the upper bound, whereas the lower bound is more involved (and relies on a bound closely related to Equation (10.8)).

- In *our* approach, however, it is crucial that $Q_{1,n}(0)$ is large, which happens when (i) the total queue $Q_{1,n}(0) + Q_{2,n}(0)$ is *large*, and (ii) at the same time $Q_{2,n}(0)$ is relatively *small*; see Equation (10.5). This explains that our *lower bound* uses the trivial GPS upper bound, but now applied to queue 2. Similarly, our *upper bound* uses the lower bound of [296], but now applied to queue 2.

As a result, our bounds on $\mathbb{P}(Q_{1,n}(0) \geq nb)$ are tighter than those in [296]. In the large-buffer setting, the weaker bounds of [296] were sufficiently sharp, but in our many-sources setting, this is not the case. ◇

10.3 Lower bound on the decay rate: class 2 in underload

From now, on we impose the restriction that the inputs are Gaussian. Inputs of type i have mean rate μ_i and variance curve $v_i(\cdot)$; $\Gamma_i(s,t)$ is defined as the covariance between $A_i(s)$ and $A_i(t)$, where $A_i(\cdot)$ is the traffic generated by a source of type i. For ease, we also introduce $\mu := \mu_1 + \mu_2$ (which is assumed to be smaller than c), and $v(\cdot) := v_1(\cdot) + v_2(\cdot)$.

Throughout this chapter, we suppose that Assumption 6.2.1 holds for both classes (besides Assumption 4.2.2, of course).

In this section and in the following sections, we analyze the decay rate $I_c^{(g)}(b)$. Sections 10.3 and 10.4 concern the regime in which class 2 is in underload, i.e., $\mu_2 < \phi_2 c$. In Section 10.3, we determine the decay rate of the upper bound on $\mathbb{P}(Q_{1,n}(0) \geq nb)$ as presented in Lemma 10.2.2. Then, in Section 10.4, we calculate the decay rate of the lower bound on $\mathbb{P}(Q_{1,n}(0) \geq nb)$ as presented in Lemma 10.2.1. Our analysis is based on the same methodology as before: use the large deviations results of Cramér (multivariate version) and Schilder. However, to apply 'Schilder', first a couple of remarks need to be made.

A few notes on 'Schilder'. To make the notation somewhat more compact, we introduce the functionals A_i, $i = 1, 2$. By $A_i[f](s,t)$ we denote the value of $n^{-1}\sum_{j=1}^{n} A_{j,i}(s,t)$ for the (given) path $f(\cdot)$, i.e.,

$$A_i[f](s,t) := f_i(t) - f_i(s);$$

for notational convenience, we use $f(\cdot)$ to denote the two-dimensional path $(f_1(\cdot), f_2(\cdot))^{\mathrm{T}}$.

Recalling that 'Schilder' (as presented in Theorem 4.2.3) assumes *centered* Gaussian processes, we find (in self-evident notation) the following large-deviations rate function of a given path f: with $\overline{f}_1(t) := f_1(t) - \mu_1 t$, $\overline{f}_2(t) := f_2(t) - \mu_2 t$, we have

$$\mathbb{I}(f) := \frac{1}{2}||\overline{f}_2||^2_{R_1} + \frac{1}{2}||\overline{f}_2||^2_{R_2},$$

similar to the priority case.

The lower bound. Because of Lemma 10.2.2,

$$I_c^{(g)}(b) \equiv -\limsup_{n\to\infty} \frac{1}{n} \log \mathbb{P}\left(Q_{1,n}(0) \geq nb\right) \geq -\limsup_{n\to\infty} \frac{1}{n} \log \mathbb{P}\left(\mathcal{F}_n\right).$$

We now investigate the decay rate in the right-hand side of the previous display. Defining the set of paths

$${}_x\mathcal{S}^{(g)}_{s,t,u} := \left\{ f \in \Omega \times \Omega : \begin{array}{l} A_1[f](-t,0) + A_2[f](-t,0) \geq x + b + ct, \\ A_1[f](-s,-u) + A_2[f](-s,0) \leq x - \phi_1 cu + cs \end{array} \right\}$$

and

$${}_x\mathcal{S}^{(g)}_t := \bigcap_{s\in(0,t]} \bigcup_{u\in[0,s)} {}_x\mathcal{S}^{(g)}_{s,t,u},$$

Schilder's sample-path ldp (Theorem 4.2.3) yields

$$-\lim_{n\to\infty} \frac{1}{n} \log \mathbb{P}\left(\mathcal{F}_n\right) = \inf_{x\geq 0} J^L(x), \quad \text{where } J^L(x) := \inf_{t>0} \inf_{f\in {}_x\mathcal{S}^{(g)}_t} \mathbb{I}(f). \tag{10.9}$$

Notice that we used that the decay rate of a *union* of events is just the infimum over the individual decay rates. Although we do not have such a relation for an *intersection* of events, it is possible to find an explicit lower bound, as presented in the next proposition.

Proposition 10.3.1 *The following lower bound applies:*

$$I_c^{(g)}(b) \equiv -\limsup_{n\to\infty} \frac{1}{n} \log \mathbb{P}\left(Q_{1,n}(0) \geq nb\right) \geq -\inf_{x\geq 0} J^L(x)$$

where

$$J^L(x) \geq \inf_{t>0} \sup_{s\in(0,t]} \inf_{u\in[0,s)} \inf_{f\in {}_x\mathcal{S}^{(g)}_{s,t,u}} \mathbb{I}(f). \tag{10.10}$$

Proof. The first claim follows directly from the above. We now prove the second claim. Because for all $s \in (0, t]$, for given t,

$$ {}_x\mathcal{S}_t^{(g)} \subseteq \bigcup_{u \in [0,s)} {}_x\mathcal{S}_{s,t,u}^{(g)}, $$

we have for all $s \in (0, t]$,

$$ \inf_{f \in {}_x\mathcal{S}_t^{(g)}} \mathbb{I}(f) \geq \inf_{f \in \bigcup_{u \in [0,s)} {}_x\mathcal{S}_{s,t,u}^{(g)}} \mathbb{I}(f). $$

Hence, it also holds for the maximizing s,

$$ \inf_{f \in {}_x\mathcal{S}_t^{(g)}} \mathbb{I}(f) \geq \sup_{s \in (0,t]} \inf_{f \in \bigcup_{u \in [0,s)} {}_x\mathcal{S}_{s,t,u}^{(g)}} \mathbb{I}(f). $$

This proves the second claim. □

10.4 Upper bound on the decay rate: class 2 in underload

This section concentrates on the decay rate of the lower bound on $\mathbb{P}(Q_{1,n}(0) \geq nb)$ as given in Lemma 10.2.1. The procedure turns out to be more involved than that of Section 10.3.

Because of Lemma 10.2.1,

$$ I_c^{(g)}(b) \equiv -\liminf_{n\to\infty} \frac{1}{n} \log \mathbb{P}\left(Q_{1,n}(0) \geq nb\right) \leq -\liminf_{n\to\infty} \frac{1}{n} \log \mathbb{P}\left(\mathcal{E}_n\right). $$

We now investigate the decay rate in the right-hand side of the previous display. Define the set of paths

$$ {}_x\mathcal{S}_{s,t}^{(g)} := \left\{ f \in \Omega \times \Omega : \begin{array}{l} A_1[f](-t,0) + A_2[f](-t,0) \geq x + b + ct, \\ A_2[f](-s,0) \leq x + \phi_2 cs \end{array} \right\}. $$

Similar to the first claim in Proposition 10.3.1, Schilder's sample-path LDP yields the following upper bound.

Proposition 10.4.1 *The following upper bound applies:*

$$ I_c^{(g)}(b) \equiv -\liminf_{n\to\infty} \frac{1}{n} \log \mathbb{P}\left(Q_{1,n}(0) \geq nb\right) \leq \inf_{x \geq 0} J^U(x), $$

where

$$ J^U(x) := \inf_{t>0} \inf_{f \in \bigcap_{s \in (0,t]} {}_x\mathcal{S}_{s,t}^{(g)}} \mathbb{I}(f). $$

Again, because of the fact that an intersection is involved, no explicit expression for $J^U(x)$ is available. We therefore take the following approach: we first derive a lower bound for $J^U(x)$, and then explain when this lower bound matches the exact value of $J^U(x)$ ('tightness'). More precisely, the objective of this section is to prove that, under some assumptions,

$$J^U(x) = \inf_{t>0} \sup_{s\in(0,t]} \inf_{f\in {}_x\mathcal{S}^{(g)}_{s,t}} \mathbb{I}(f). \tag{10.11}$$

Remark 10.4.2 Notice the similarity between the right-hand sides of Equations (10.10) and (10.11), in particular if the optimizing s and u in Equation (10.10) coincide; see also Remark 10.2.3. ◇

Lower bound on $J^U(x)$. The following lemma gives a lower bound for $J^U(x)$. Its proof is analogous to that of the second claim in Proposition 10.3.1 and hence omitted.

Lemma 10.4.3

$$J^U(x) \geq \inf_{t>0} \sup_{s\in(0,t]} \inf_{f\in {}_x\mathcal{S}^{(g)}_{s,t}} \mathbb{I}(f).$$

The lower bound in Lemma 10.4.3 can be expressed more explicitly. To this end, we first concentrate on calculating the minimum of $\mathbb{I}(f)$ over $f \in {}_x\mathcal{S}^{(g)}_{s,t}$, for fixed s and t. The result, as stated in Exercise 10.4.4, requires the introduction of two functions. First, recall the large-deviations rate function $\Lambda(\cdot,\cdot)$ of the bivariate normal random variable $(A_1(-t,0) + A_2(-t,0), A_2(-s,0))$,

$$\Lambda(y,z) := \frac{1}{2}(y-\mu t,\ z-\mu_2 s)^{\mathrm{T}}\Sigma(s,t)^{-1}\begin{pmatrix} y-\mu t \\ z-\mu_2 s \end{pmatrix},$$

where the covariance matrix $\Sigma(s,t)$ is defined as

$$\Sigma(s,t) := \begin{pmatrix} v(t) & \Gamma_2(s,t) \\ \Gamma_2(s,t) & v_2(s) \end{pmatrix}.$$

We also define, for $i = 1, 2$,

$$k_i(x,s,t) := \mu_i s + \frac{(x+b+(c-\mu)t)}{v(t)}\Gamma_i(s,t).$$

Exercise 10.4.4 Show that, for $s \in (0,t]$,

$$\inf_{f\in {}_x\mathcal{S}^{(g)}_{s,t}} \mathbb{I}(f) = \Upsilon_x(s,t)$$

$$:= \begin{cases} \Lambda(x+b+ct,\ x+\phi_2 cs), & \text{if } k_2(x,s,t) > x+\phi_2 cs; \\ (x+b+(c-\mu)t)^2/2v(t), & \text{if } k_2(x,s,t) \leq x+\phi_2 cs. \end{cases}$$

Solution. The proof is analogous to that of Lemma 9.2.3. Using Theorem 4.1.8,

$$\inf_{f\in {}_x\mathcal{S}^{(g)}_{s,t}} \mathbb{I}(f) = \inf \Lambda(y,z),$$

where the infimum is over y and z such that $y \geq x+b+ct$ and $z \leq x+\phi_2 cs$.

Because $\Lambda(\cdot,\cdot)$ is convex in y and z, we can use the Lagrangian to find the infimum over y and z:

$$\mathcal{L}(y,z,\xi_1,\xi_2) = \Lambda(y,z) - \xi_1(y-x-b-ct) + \xi_2(z-x-\phi_2 cs),$$

with $\xi_1,\xi_2 \geq 0$. Two cases may occur, depending on the specific values of x, s, and t. (i) If x, s, and t are such that $k_2(x,s,t) > x+\phi_2 cs$, then both constraints are binding, i.e., $y^\star = x+b+ct$ and $z^\star = x+\phi_2 cs$. (ii) If x, s, and t are such that $k_2(x,s,t) \leq x+\phi_2 cs$, then only the first constraint is binding, i.e., $y^\star = x+b+ct$, and $z^\star = k_2(x,s,t)$. ◇

Now Lemma 10.4.3 and Exercise 10.4.4 yield the final lower bound for $J^U(x)$, as stated in the next corollary.

Corollary 10.4.5

$$J^U(x) \geq \inf_{t>0} \sup_{s\in(0,t]} \Upsilon_x(s,t).$$

Interpretation of $\Upsilon_x(s,t)$. We now intuitively explain the form of $\Upsilon_x(s,t)$; cf. the arguments mentioned in Section 9.2. The decay rate $\mathbb{I}(f)$ can be interpreted as the cost of having a path f, and, likewise, $\Upsilon_x(s,t)$ as the cost of generating a traffic according to a path in the set ${}_x\mathcal{S}^{(g)}_{s,t}$.

The proof of Exercise 10.4.4 shows that the *first* constraint, i.e., $y \geq x+b+ct$, is always binding, whereas the *second* constraint, i.e., $z \leq x+\phi_2 cs$, is sometimes binding, depending on the value of $k_2(x,s,t)$ compared to $x+\phi_2 cs$. Observe that $k_2(x,s,t)$ is in fact a conditional expectation:

$$k_2(x,s,t) \equiv \mathbb{E}[A_2(-s,0) \mid A_1(-t,0)+A_2(-t,0) = x+b+ct].$$

The two cases of Exercise 10.4.4 can now be interpreted as follows. (i) The optimal value for z is $x+\phi_2 cs$. In this case, $k_2(x,s,t)$, which is the expected value of the amount of traffic sent by class 2 in $(-s,0]$ given that in total $x+b+ct$ is sent during $(-t,0]$, is larger than $x+\phi_2 cs$: with high probability the second constraint is *not met* just by imposing the first constraint. In terms of cost, this means that in this regime additional cost is incurred by imposing the second constraint. (ii) The optimal value for z is precisely $k_2(x,s,t)$, and is smaller than $x+\phi_2 cs$: $A_1(-t,0)+A_2(-t,0) = x+b+ct$ implies $A_2(-s,0) \leq x+\phi_2 cs$ with high probability. Intuitively, this means that, given that the first constraint is satisfied, the second constraint is already met with high probability.

Using this reasoning, it follows after some calculations that we can rewrite $\Upsilon_x(s,t)$ in a helpful way, as shown in the next corollary (which is the immediate counterpart of Corollary 9.2.5). The first term accounts for the cost of satisfying the first constraint in ${}_x\mathcal{S}^{(g)}_{s,t}$, and the second term (which is possibly 0) for the second constraint.

Corollary 10.4.6

$$\Upsilon_x(s,t) = \frac{(x+b+ct-\mathbb{E}[A_1(-t,0)+A_2(-t,0)])^2}{2\mathbb{V}\mathrm{ar}[A_1(-t,0)+A_2(-t,0)]} + \frac{\max^2\left\{\begin{array}{c}\mathbb{E}[A_2(-s,0)\mid A_1(-t,0)+A_2(-t,0)= \\ x+b+ct]-x-\phi_2 cs, 0\end{array}\right\}}{2\mathbb{V}\mathrm{ar}[A_2(-s,0)\mid A_1(-t,0)+A_2(-t,0)=x+b+ct]}.$$

Two regimes for ϕ_2. Corollary 10.4.6 implies that

$$\inf_{t>0}\sup_{s\in(0,t]}\Upsilon_x(s,t)\geq \inf_{t>0}\frac{(x+b+(c-\mu)t)^2}{2v(t)}. \tag{10.12}$$

Let the optimum in the right-hand side for given c be attained in t_c (which is, in fact, a function of x, but we suppress x here, as x is held fixed in the remainder of this section). Suppose that for all $s\in(0,t_c]$ it holds that $k_2(x,s,t_c)\leq x+\phi_2 cs$, then obviously the inequality in Equation (10.12) is tight. This corresponds to a critical weight $\phi_2^{c,U}(x)$ above which there is tightness. This critical value is given by

$$\begin{aligned}\phi_2^{c,U}(x) &:= \inf\left\{\phi_2 : \sup_{s\in(0,t_c]}\{k_2(x,s,t_c)-x-\phi_2 cs\}\leq 0\right\}\\ &\equiv \sup_{s\in(0,t_c]}\frac{k_2(x,s,t_c)-x}{cs}.\end{aligned} \tag{10.13}$$

The resulting two regimes can be intuitively explained as follows.

(A) *Large* ϕ_2. If $\phi_2 > \phi_2^{c,U}(x)$, using the interpretation in terms of conditional expectations, the buffer content of queue 2 at time 0 is likely to be below nx. Hence, if in total $n(x+b+ct_c)$ is sent during $(-t_c,0]$, it is likely that at time 0, the buffer content of class 1 has reached level nb.

(B) *Small* ϕ_2. If $\phi_2 < \phi_2^{c,U}(x)$, then the guaranteed rate for class 2 is relatively small, implying that its buffer content may easily grow. Again, in total (at least) $n(x+b+ct_c)$ has been sent during the interval $(-t_c,0]$, but now it is *not* obvious that most of it goes to the buffer of class 1. Class 2 has to be 'forced' to take *at most* its guaranteed rate during this interval.

Conditions for tightness. As the overflow behavior in case of (A) $\phi_2 \geq \phi_2^{c,U}(x)$ is essentially different from that in case of (B) $\phi_2 < \phi_2^{c,U}(x)$, we will consider the two regimes separately in this section.

The procedure followed will be the same for both regimes. Let us denote the optimizing s and t in Corollary 10.4.5 by $s^\star$ and $t^\star$, respectively. (Notice that $s^\star$ and $t^\star$ are functions of x, but, for conciseness, we again suppress the argument x.) First, we use Schilder's theorem to determine the most probable path in ${}_x\mathcal{S}^{(g)}_{s^\star,t^\star}$ for the regime of ϕ_2 under consideration. Denoting this optimal path by $f^\star$, we then check whether

$$f^\star \in \left(\bigcup_{t \geq 0} \bigcap_{s \in (0,t]} {}_x\mathcal{S}^{(g)}_{s,t} \right). \tag{10.14}$$

If so, the optimal path giving rise to the lower bound of Corollary 10.4.5 is, in fact, the optimal path for $J^U(x)$. Consequently, under condition (10.14), $J^U(x)$ and its lower bound coincide.

Regime (A): ϕ_2 larger than critical weight. Because of the definition of $\phi_2^{c,U}(x)$, it holds for all $\phi_2 \geq \phi_2^{c,U}(x)$ that

$$\inf_{t>0} \sup_{s \in (0,t]} \Upsilon_x(s,t) = \frac{(x+b+(c-\mu)t_c)^2}{2v(t_c)},$$

as identified before. The next theorem states that, for these ϕ_2, the lower bound on $J^U(x)$ (see Corollary 10.4.5) actually *equals* $J^U(x)$. We omit its proof because it essentially follows from the proof of Theorem 3.8 in [201].

Theorem 10.4.7 *If $\phi_2 \geq \phi_2^{c,U}(x)$, then*

$$J^U(x) = \inf_{t>0} \sup_{s \in (0,t]} \Upsilon_x(s,t) = \frac{(x+b+(c-\mu)t_c)^2}{2v(t_c)},$$

and the most probable paths are, for $r \in [-t_c, 0)$,

$$\begin{aligned} f_1^\star(r) &= -\mathbb{E}[A_1(r,0) \mid A_1(-t_c,0) + A_2(-t_c,0) = x+b+ct_c] \\ &= -k_1(x,-r,t_c); \\ f_2^\star(r) &= -\mathbb{E}[A_2(r,0) \mid A_1(-t_c,0) + A_2(-t_c,0) = x+b+ct_c] \\ &= -k_2(x,-r,t_c). \end{aligned}$$

Regime (B): ϕ_2 smaller than critical weight. The analysis of this regime is considerably more involved than that of the situation of a large ϕ_2, and we refer to

[200] for the detailed proof. It is noted that the proof requires Assumptions 4.2.2 and 6.2.1.

The arguments used are similar to those in Section 9.3, and in fact check whether Equation (10.14) holds; they enable us to find a technical condition under which the lower bound of Lemma 10.4.3 is tight. For further reference, we denote this condition by $\mathcal{C}_x$ (note that the condition depends on x).

10.5 Analysis of the decay rate: class 2 in overload

In this section, the decay rate of $\mathbb{P}(Q_{1,n}(0) \geq nb)$ is calculated for the regime $\phi_2 c \leq \mu_2$.

Theorem 10.5.1 *If $\phi_2 \leq \mu_2/c$, then*

$$I_c^{(\mathrm{g})}(b) \equiv -\lim_{n\to\infty} \frac{1}{n} \log \mathbb{P}(Q_{1,n}(0) \geq nb) = \inf_{t\geq 0} \frac{(b + (\phi_1 c - \mu_1)t)^2}{2v_1(t)}. \tag{10.15}$$

Proof. We first show that the desired expression is a lower bound. Denote the stationary workload of queue i by $Q_{i,n}^{nc}(0)$ if it is served (in isolation) at a constant rate nc. Then the lower bound follows from

$$\begin{aligned}\mathbb{P}(Q_{1,n}(0) \geq nb) &\leq \mathbb{P}(Q_{1,n}^{n\phi_1 c}(0) \geq nb)\\ &= \mathbb{P}\left(\exists t > 0 : \frac{1}{n}\sum_{j=1}^{n} A_{j,1}(-t, 0) \geq b + \phi_1 ct\right),\end{aligned}$$

due to $Q_{1,n}(0) \leq Q_{1,n}^{n\phi_1 c}(0)$.

The upper bound is a matter of computing the rate function of a feasible path. Let $t^\star$ be the optimizer in the right-hand side of Equation (10.15). For $r \in [-t^\star, 0)$ define

$$\begin{aligned}f_1^\star(r) &:= -\mathbb{E}[A_1(r, 0) \mid A_1(-t^\star, 0) = b + \phi_1 ct^\star]\\ &= \mu_1 r - \frac{(b + (\phi_1 c - \mu_1)t^\star)}{v_1(t^\star)}\Gamma_1(-r, t^\star);\\ f_2^\star(r) &:= -\mathbb{E}[A_2(r, 0) \mid A_1(-t^\star, 0) = b + \phi_1 ct^\star] = \mu_2 r.\end{aligned}$$

This path clearly leads to overflow in queue 1 of the GPS system (as the type-2 sources claim their weight, such that exactly service rate $n\phi_1 c$ is left for the type-1 sources). The norm of $f_2^\star(\cdot)$ is obviously 0, as these sources are transmitting at mean rate; the rate function corresponding to $f_1^\star(\cdot)$ equals the desired expression. □

10.6 Discussion of the results

In this section, we discuss the results obtained in the previous sections. We identify three regimes for the value of ϕ_2, corresponding to three generic overflow scenarios. Case (i) directly relates to the overload regime of Section 10.5 and Cases (ii) and (iii) to the underload regime of Sections 10.3 and 10.4.

For Case (i) our analysis immediately yields the exact decay rate; see Theorem 10.5.1. For Cases (ii) and (iii), however, we saw that the situation is more complicated, in that we could only find *bounds* on the decay rate. We strongly believe, however, that under fairly general conditions these bounds coincide. This claim can be backed up by convincing numerical evidence, as reported in [200]. Besides, it turns out that in the case of Brownian motion inputs the results turn out to be exact; see Section 11.4. In this section we give a further intuitive explanation why we think that the bounds are tight under broad conditions.

Case (i): class 2 in overload. First, consider the situation $\phi_2 \leq \mu_2/c =: \phi_2^o$. In this scenario, the type-2 sources claim their guaranteed rate $n\phi_2 c$ with overwhelming probability, so that overflow in queue 1 resembles overflow in a FIFO queue with link rate $n\phi_1 c$; this principle plays a crucial role in the proof of Theorem 10.5.1. We repeat it here for a comparison with Cases (ii) and (iii).

For $\phi_2 \in [0, \phi_2^o]$:

$$I_c^{(g)}(b) = \inf_{t>0} \frac{(b + (\phi_1 c - \mu_1)t)^2}{2v_1(t)}.$$

Case (ii): Class 2 in underload, with ϕ_2 small. As argued in Sections 10.3 and 10.4, in this regime it is not sufficient to require that $n(x + b + ct)$ traffic is generated in t units of time, since, with high probability, a considerable amount of traffic will be left in queue 2. Hence, additional effort is required to ensure that the buffer content of queue 2 stays below nx.

Based on heuristic arguments, we present two claims.

A. *Regarding the optimal values of u and s.* Recall the probabilistic upper bound in Lemma 10.2.2. In the proof of that lemma, $-s^\star$ denotes the beginning of the busy period of the second queue, which contains time 0. Hence, the second queue remains backlogged during the interval $(-s^\star, 0]$ and claims at least its guaranteed rate $n\phi_2 c$, leaving at most rate $n\phi_1 c$ to the first queue. Paralleling the proof of Lemma 10.2.2, two scenarios are possible: in scenario (a) queue 1 was continuously backlogged during $(-s^\star, 0]$, whereas in scenario (b), queue 1 has been empty after time $-s^\star$, i.e., queue 1 was empty at some time $-u^\star$ during the busy period of queue 2.

Scenario (b) is not likely to be optimal, for the following reason. As queue 1 was empty at $-u^\star$, it does not benefit from any effort before $-u^\star$; queue 1 has to build up its entire buffer in the interval $(-u^\star, 0]$. Now recall that queue 2 already started to show deviant behavior from time $-s^\star < -u^\star$, claiming its guaranteed rate. However this additional effort of queue 2 before time $-u^\star$ is of no 'benefit' for queue 1. In order for queue 1 to fully exploit that queue 2 takes its guaranteed rate during $(-s^\star, 0]$, it should be continuously backlogged during this interval, as in scenario (a). We therefore expect that in the most likely scenario $u^\star = s^\star$.

B. *Regarding the optimal value of x.* We introduced x in the right-hand side of Equation (10.3). From this representation, it follows immediately that nx can be interpreted as the amount of traffic left in queue 2 (at the epoch when the total queue size reaches $n(x+b)$).

We have argued earlier that queue 2 has to claim its guaranteed rate during $(-s^\star, 0]$. If a positive amount of traffic is left in queue 2 at time 0, the type-2 sources apparently 'generated too much traffic'; the guaranteed rate could have been claimed with less effort. We therefore expect that in the most likely scenario $x^\star = 0$. Notice that an essential condition here is that $\phi_2 > \mu_2/c$, as otherwise a build-up of traffic in queue 2 would not be 'wasted effort'.

Because of Claims A and B, we expect that this regime applies to $\phi_2 \in [\phi_2^o, \phi_2^c]$, with

$$\phi_2^c := \sup_{s\in(0,t_c(0)]} \frac{k_2(0,s,t_c(0))}{cs};$$

cf. Equation (10.13). Defining

$$\begin{pmatrix} z_1(t) \\ z_2(s) \end{pmatrix} := \begin{pmatrix} b+(c-\mu)t \\ (\phi_2 c - \mu_2)s \end{pmatrix},$$

we expect the following relation to hold:

For $\phi_2 \in [\phi_2^o, \phi_2^c]$:

$$I_c^{(g)}(b) = \frac{1}{2}\inf_{t\geq 0}\sup_{s\in[0,t]} \begin{pmatrix} z_1(t) \\ z_2(s) \end{pmatrix}^{\mathrm{T}} \begin{pmatrix} v_1(t)+v_2(t) & \Gamma_2(s,t) \\ \Gamma_2(s,t) & v_2(s) \end{pmatrix}^{-1} \begin{pmatrix} z_1(t) \\ z_2(s) \end{pmatrix},$$

under the technical condition $\mathcal{C}_0$ (recall that the condition $\mathcal{C}_x$ was introduced in Section 10.4).

Case (iii): Class 2 in underload, with ϕ_2 large. Here, overflow of the total queue implies overflow of queue 1. Consequently, we expect the following relation.

For $\phi_2 \in [\phi_2^c, 1]$:

$$I_c^{(g)}(b) = \inf_{t>0} \frac{(b + (c - \mu)t)^2}{2v(t)}.$$

10.7 Delay asymptotics

So far, we have concentrated on asymptotics of the buffer content distribution. Often, however, performance requirements are expressed in terms of (packet) delay rather than (packet) loss. In FIFO queues with constant service speed, the buffer content distribution can be directly translated into the delay distribution. In queues operating under GPS, however, such a procedure is far from straightforward, as the service speed is variable; in fact, the service speed allocated to one queue depends on the buffer content of the other queue. Hence, the analysis of delay in GPS is hard.

This section has two goals. First, we derive, for general traffic processes, bounds on the delay distribution. Then, we specialize with respect to Gaussian inputs and analyze the associated decay rates under the many-sources regime.

Bounds on the delay distribution. We consider the many-sources setting of the previous chapters. We focus on the delay experienced by a packet ('fluid molecule') of class 1, having arrived at an arbitrary point in time, say time 0; we denote this delay by $D_{1,n} \equiv D_{1,n}(0)$. It is clear that

$$\mathbb{P}(D_{1,n} > d) = \mathbb{P}\left(Q_{1,n} > B_{1,n}(0, d)\right),$$

where, as before, $Q_{i,n} \equiv Q_{i,n}(0)$ is the steady-state queue length of class i and $B_{i,n}(s, t)$ is the amount of service available to class i in $(s, t]$.

Therefore, to derive a lower bound on $\mathbb{P}(D_{1,n} > d)$, we need to find an upper bound on $B_{1,n}(0, d)$. As $B_{1,n}(0, d) = ncd - B_{2,n}(0, d)$, this amounts to finding a lower bound on $B_{2,n}(0, d)$. Analogous to Inequality (10.8)

$$B_{1,n}(0, d) = ncd - B_{2,n}(0, d) \leq ncd - \inf_{u\in[0,d]} \left\{ \sum_{j=1}^{n} A_{2,j}(0, u) + n\phi_2 c(d - u) \right\}.$$

This yields the lower bound

$$\mathbb{P}(D_{1,n} > d) \geq \mathbb{P}\left(Q_{1,n} > ncd - \inf_{u\in[0,d]} \left\{ \sum_{j=1}^{n} A_{2,j}(0, u) + n\phi_2 c(d - u) \right\} \right)$$

$$= \mathbb{P}\left(\bigcup_{x\geq 0} \left\{ \begin{array}{l} Q_{1,n}(0) + Q_{2,n}(0) > nx + ncd - \\ \qquad \inf_{u\in[0,d]} \left\{ \sum_{j=1}^{n} A_{2,j}(0, u) + n\phi_2 c(d - u) \right\}, \\ Q_{2,n}(0) \leq nx \end{array} \right\} \right);$$

cf. Expression (10.5). Now, due to 'Reich',

$$Q_{1,n}(0) + Q_{2,n}(0) = \sup_{t \geq 0} \left(\sum_{j=1}^{n} \left(A_{1,j}(-t, 0) + A_{2,j}(-t, 0) \right) - nct \right);$$

$$Q_{2,n}(0) = \sup_{s \geq 0} \left(\sum_{j=1}^{n} A_{2,j}(-s, 0) - B_{2,n}(-s, 0) \right).$$

The optimizing t (s), say $t^\star$ ($s^\star$), has the interpretation of the beginning of the busy period of the total (second) queue containing time 0. Hence, we have that $s^\star \leq t^\star$. Also notice that trivially, for any s_1, s_2 such that $s_1 < s_2$, $B_{2,n}(s_1, s_2) \geq n\phi_2 c(s_2 - s_1)$.

Now, define

$${}_x\mathcal{V}^{(g)}_{s,t,u} := \left\{ f \in \Omega \times \Omega : \begin{array}{l} A_1[f](-t, 0) + A_2[f](-t, u) \geq x + ct + \phi_1 cd + \phi_2 cu, \\ A_2[f](-s, 0) \leq x + \phi_2 cs \end{array} \right\},$$

and

$$\mathcal{V}^{(g)} := \bigcup_{x,t \geq 0} \bigcap_{s \in [0,t]} \bigcup_{u \in [0,d]} {}_x\mathcal{V}^{(g)}_{s,t,u}.$$

Then, the above inequalities yield the following lemma.

Lemma 10.7.1 *The following lower bound applies:*

$$\mathbb{P}(D_{1,n} > d) \geq p_n[\mathcal{V}^{(g)}].$$

We now turn to upper bounds. The following upper bound is a consequence of

$$B_{1,n}(0, d) \geq ncd - Q_{2,n}(0) - \sum_{j=1}^{n} A_{2,j}(0, d);$$

this is a direct implication of the fact that, in an interval $[0, d]$, a queue never claims more than the queue length at time 0, increased by the amount of traffic arriving in $[0, d]$. Define

$$\mathcal{W}_1^{(g)} := \{ f \in \Omega \times \Omega : \exists t \geq 0 : A_1[f](-t, 0) + A_2[f](-t, d) \geq c(t + d) \}.$$

Lemma 10.7.2 *The following upper bound applies:*

$$\mathbb{P}(D_{1,n} > d) \leq p_n[\mathcal{W}_1^{(g)}] = \mathbb{P}\left(Q_{1,n}(0) + Q_{2,n}(0) \geq ncd - \sum_{j=1}^{n} A_{j,2}(0, d) \right).$$

A second upper bound follows from $B_{1,n}(0,d) \geq n\phi_1 cd$, which entails that $\mathbb{P}(D_{1,n} > d) \leq \mathbb{P}(Q_{1,n}(0) \geq n\phi_1 cd)$. This can be further bounded by applying Lemma 10.2.2. To this end, we define

$$ {}_x\mathcal{W}^{(g)}_{s,t,u} := \left\{ f \in \Omega \times \Omega : \begin{array}{l} A_1[f](-t,0) + A_2[f](-t,0) \geq x + \phi_1 cd + ct, \\ A_1[f](-s,-u) + A_2[f](-s,0) \leq x + cs - \phi_1 cu \end{array} \right\}, $$

and

$$ \mathcal{W}^{(g)}_2 := \bigcup_{x,t\geq 0} \bigcap_{s\in[0,t]} \bigcup_{u\in[0,s]} {}_x\mathcal{W}^{(g)}_{s,t,u}. $$

Lemma 10.7.3 *The following upper bound applies:*

$$ \mathbb{P}(D_{1,n} > d) \leq p_n[\mathcal{W}^{(g)}_2]. $$

We stress that the bounds derived in Lemmas 10.7.1, 10.7.2, and 10.7.3 apply to any general input process with stationary increments, i.e., not to just Gaussian processes.

Decay rates of the delay distribution. The goal is now to characterize

$$ L^{(g)}_c(d) := -\lim_{n\to\infty} \frac{1}{n} \log \mathbb{P}(D_{1,n} > d). $$

We first consider the situation of queue 2 in overload: $\phi_2 \leq \phi^o_2 = \mu_2/c$. In the notation of Section 10.5, we have that $B_{1,n}(0,d) \geq n\phi_1 cd$ entails that

$$ \mathbb{P}(D_{1,n} > d) \leq \mathbb{P}(Q_{1,n}(0) \geq n\phi_1 cd) \leq \mathbb{P}(Q^{n\phi_1 c}_{1,n}(0) \geq n\phi_1 cd). $$

It leads to the following result.

Proposition 10.7.4 *If $\phi_2 \leq \phi^o_2$, then*

$$ L^{(g)}_c(d) = -\inf_{t\geq 0} \frac{(\phi_1 cd + (\phi_1 c - \mu_1)t)^2}{2v_1(t)}. $$

Proof. Theorem 10.5.1 implies that

$$ \begin{aligned} \lim_{n\to\infty} \frac{1}{n} \log \mathbb{P}(Q_{1,n}(0) \geq n\phi_1 cd) &= \lim_{n\to\infty} \frac{1}{n} \log \mathbb{P}(Q^{n\phi_1 c}_{1,n}(0) \geq n\phi_1 cd) \\ &= -\inf_{t\geq 0} \frac{(\phi_1 cd + (\phi_1 c - \mu_1)t)^2}{2v_1(t)}. \end{aligned} $$

Hence, we only need to prove that the most likely path that corresponds to the event $\{Q^{n\phi_1 c}_{1,n}(0) \geq n\phi_1 cd\}$ is indeed such that also $\{D_{1,n} > d\}$. Take the path from the proof of Theorem 10.5.1, with b replaced by $\phi_1 cd$. This path is such that, at time 0, queue 1 has buffer content $\phi_1 cd$. Also, queue 2 continuously claims service rate

$\phi_2 c$ in the interval $[0, d]$ (as $\mu_2 \geq \phi_2 c$). In other words, the service rate available to queue 1 in this interval was $\phi_1 c$. We conclude that the delay of a 'fluid molecule' arriving at time 0 is indeed d. □

We now turn to the harder case in which class 2 is in underload, i.e., $\phi_2 > \mu_2/c$. First, we check what we can learn from Lemma 10.7.2. It is clear that it implies, for all ϕ_2, that

$$L_c^{(g)}(d) \geq \inf_{t \geq 0} \frac{(c(t+d) - \mu_1 t - \mu_2(t+d))^2}{2v_1(t) + 2v_2(t+d)}, \tag{10.16}$$

but we may wonder for which ϕ_2 there is actually equality.

To this end, first observe that the most likely path of type 2 traffic in $\mathcal{W}_1^{(g)}$ reads

$$f_2^\star(r) = \mu_2 r - \frac{((c-\mu_2)(t^\star + d) - \mu_1 t^\star)}{v_1(t^\star) + v_2(t^\star + d)} \cdot (\Gamma_2(t^\star, t^\star + d) - \Gamma_2(t^\star + r, t^\star + d)),$$

where $t^\star$ is the optimizer in Equation (10.16) and $r \in [-t^\star, d]$. The corresponding path of the input rate is

$$g_2^\star(r) := (f_2^\star)'(r) = \mu_2 + \frac{((c-\mu_2)(t^* + d) - \mu_1 t^*)}{2v_1(t^*) + 2v_2(t^* + d)}(v_2'(d - r) + v_2'(t^* + r)).$$

From Lemma 10.7.2, we conclude that there is tightness in Equation (10.16) if, along the most likely path in $\mathcal{W}_1^{(g)}$,

$$B_{1,n}(0, d) = ncd - \sum_{j=1}^{n} A_{2,j}(0, d) - Q_{2,n}(0).$$

This is obviously true when queue 2 never becomes positive, and a sufficient condition for this is

$$\phi_2 \geq \phi_2^d := \frac{1}{c} \cdot \sup_r g_2^\star(r).$$

We obtain the following result.

Proposition 10.7.5 *For all ϕ_2,*

$$L_c^{(g)}(d) \geq \inf_{t \geq 0} \frac{(c(t+d) - \mu_1 t - \mu_2(t+d))^2}{2v_1(t) + 2v_2(t+d)}.$$

This equation holds with equality for all $\phi_2 \geq \phi_2^d$.

Now we see what can be concluded from Lemma 10.7.2. The following result follows analogously to Proposition 10.3.1. We first define the large-deviations rate function of the bivariate random variable

$$(A_1(-t,0) + A_2(-t,0), A_1(-s,-u) + A_2(-s,0))$$

through

$$\Lambda(y,z) := \frac{1}{2}\begin{pmatrix} y - (\mu_1+\mu_2)t \\ z - (\mu_1+\mu_2)s + \mu_2 u \end{pmatrix}^{\mathrm{T}} \Sigma(s,t,u)^{-1} \begin{pmatrix} y - (\mu_1+\mu_2)t \\ z - (\mu_1+\mu_2)s + \mu_2 u \end{pmatrix};$$

here $\Sigma(s,t,u)$ denotes the covariance matrix:

$$\Sigma(s,t,u) := \begin{pmatrix} v_1(t)+v_2(t) & \Gamma_1(s,t,u)+\Gamma_2(s,t) \\ \Gamma_1(s,t,u)+\Gamma_2(s,t) & v_1(s-u)+v_2(s) \end{pmatrix};$$

$$\Gamma_1(s,t,u) := \mathbb{C}\mathrm{ov}(A_1(-t,0), A_1(-s,-u)) = \frac{(v(t-u) - v(u) + v(s) - v(t-s))}{2}.$$

Again, observe that the probability of an intersection of events is smaller than the probability of any event in the intersection, and hence also smaller than the least likely event. This idea yields the following lower bound on the decay rate.

Proposition 10.7.6 *For all* ϕ_2,

$$L_c^{(\mathrm{g})}(d) \geq \inf_{x\geq 0}\inf_{t\geq 0}\sup_{s\in[0,t]}\inf_{u\in[0,s]}\inf_{f\in_x \mathcal{W}_{s,t,u}^{(\mathrm{g})}} \mathbb{I}(f).$$

Here,

$$\inf_{f\in_x \mathcal{W}_{s,t,u}^{(\mathrm{g})}} \mathbb{I}(f) = \inf \Lambda(y,z),$$

where the infimum in the right-hand side is over all $y \geq x + ct + \phi_1 cd$ *and* $z \leq x + cs - \phi_1 cu$.

Tightness conditions (i.e., conditions under which the lower bound of Proposition 10.7.6 is actually attained) can be found as in Section 9.3.

Bibliographical notes

Most of the material presented in this chapter has been taken from Mandjes and Van Uitert [200]. For the same setting (GPS in many-sources regime, with Gaussian inputs), Mannersalo and Norros [205] developed approximations (of the same spirit as the rough full link approximations mentioned in the bibliographical notes of Chapter 9).

It is noted that GPS in the large-buffer regime (rather than the many-sources regime) is well understood. A significant contribution was made by Zhang [296] – see also [297] for the multiple-queue case – but an essential assumption in his work is that the input traffic is supposed to be short-range dependent (to ensure the existence of the asymptotic cumulant function); long-range dependent input (such as fBm) was not covered. A related setting was considered by Bertsimas, Paschalidis, and Tsitsiklis [33]. Other interesting papers on GPS are [99, 123, 124].

Other papers on the large-buffer regime are, e.g., Borst, Mandjes, and Van Uitert [38, 39], where the focus is on a single node fed by a mixture of heavy-tailed and light-tailed sources, and Massoulié [207]. Van Uitert and Borst [276, 277] analyze a GPS network with heavy-tailed inputs.

Large-buffer asymptotics with Gaussian input. Dȩbicki and Van Uitert [70] can be viewed as the large-buffer counterpart of the present chapter; see also [65]. In the large-buffer setting, the regimes are determined not only by the weights but also by the 'dominant' class; we say that class i is dominant over class j if

$$\lim_{t\to\infty}\frac{v_j(t)}{v_i(t)}=0.$$

For ease, we specialize to the situation where type i input is fBm with $v_i(t)=t^{2H_i}$; then class i is dominant over class j if $H_i > H_j$. The cases considered in [70] are as follows:

(i) $\mu_1 < \phi_1 \mathrm{C}$ and $\mu_2 > \phi_2 \mathrm{C}$. In this situation queue 2 claims $\phi_2 \mathrm{C}$ all the time, such that

$$\lim_{\mathrm{B}\to\infty}\frac{1}{\mathrm{B}^{2-2H_1}}\log\mathbb{P}(Q_1\geq \mathrm{B}) = -\frac{1}{2}\left(\frac{1}{1-H_1}\right)^{2-2H_1}\left(\frac{\phi_1\mathrm{C}-\mu_1}{H_1}\right)^{2H_1}; \tag{10.17}$$

cf. Proposition 5.6.1.

(ii) $\mu_1 < \phi_1 \mathrm{C}$ and $\mu_2 < \phi_2 \mathrm{C}$. Now, two cases should be distinguished:

(ii^a) Class 1 dominant ($H_1 > H_2$). If the total queue is large, this is most likely because of queue 1 being large. Along these lines it can be argued that class 2 essentially claims just μ_2. The service rate left over to class 1 is therefore $\mathrm{C}-\mu_2$. Hence,

$$\lim_{\mathrm{B}\to\infty}\frac{1}{\mathrm{B}^{2-2H_1}}\log\mathbb{P}(Q_1\geq \mathrm{B}) = -\frac{1}{2}\left(\frac{1}{1-H_1}\right)^{2-2H_1}\left(\frac{\mathrm{C}-\mu}{H_1}\right)^{2H_1}. \tag{10.18}$$

(ii^b) Class 2 is dominant ($H_2 > H_1$). In this case, we have that, if the total queue is large, this is most likely because of queue 2 being large. In other words, one could say that queue 2 claims $\phi_2\mathrm{C}$. We therefore obtain again asymptotics of the form (10.17).

(iii) $\mu_1 > \phi_1\mathrm{C}$ and $\mu_2 < \phi_2\mathrm{C}$. Again, two cases should be distinguished:

(iii^a) Class 1 dominant ($H_1 > H_2$). Also in this situation class 2 essentially claims just μ_2, so that Equation (10.18) applies again.

(iii^b) Class 2 is dominant ($H_2 > H_1$). This case turns out to be the most complicated. It is clear that in this scenario class 1 should transmit at rate μ_1 (which is no rare event; it corresponds to its average behavior). The question that remains is, how much capacity (from the $\phi_2\mathrm{C}$ allocated to queue 2) does class 2 claim? It should claim more than μ_2, as otherwise queue 1 does not build up, but how much more should it claim?

To answer this question, suppose that class 2 generates traffic at rate r. It is 'useless' to choose r larger than $\phi_2\mathrm{C}$, as class 2 can never claim more than $\phi_2\mathrm{C}$. If $r \geq \phi_2\mathrm{C}$, then queue 1 grows at rate $\mu_1 - \phi_1\mathrm{C}$. If r is smaller than $\phi_2\mathrm{C}$, then queue 1 grows at rate $\mu_1 - \mathrm{C} + r$. Summarizing, queue 1 builds up essentially linearly, with slope

$$\mu_1 - (\mathrm{C} - \min\{r, \phi_2\mathrm{C}\}) = \mu_1 - \max\{\mathrm{C} - r, \phi_1\mathrm{C}\}.$$

Stated differently, queue 1 grows as the downstream queue in a tandem system fed at rate r and with service capacities $\phi_2\mathrm{C}$ (first queue) and $\mathrm{C} - \mu_1$ (second queue). This explains why $\mathbb{P}(Q_1 \geq \mathrm{B})$ looks like

$$\mathbb{P}\left(\sup_{t\geq 0}(A_2(-t,0) - (\mathrm{C} - \mu_1)t) - \sup_{s\in[0,t]}(A_2(-s,0) - \phi_2\mathrm{C}s) \geq \mathrm{B}\right),$$

which is interpreted as the buffer content of the downstream queue of a tandem network; cf. Lemma 9.1.1. It is noted that arguments in the same vein can be found in [296]: the formulae for the decay rates given there show a similar relation between queues operating under GPS and tandem systems.

We therefore obtain that, with the notation as in Equation (9.25),

$$\lim_{\mathrm{B}\to\infty} \frac{1}{\mathrm{B}^{2-2H_2}} \log \mathbb{P}(Q_1 \geq \mathrm{B}) = \zeta(1, \phi_2\mathrm{C} - \mu_2, \mathrm{C} - \mu_1 - \mu_2, H_2). \tag{10.19}$$

Notice that, remarkably, Equations (10.17) and (10.18) both depend on H_1, whereas Equation (10.19) depends on H_2. In none of the expressions *both* H_1 and H_2 appear. Also, H_2 appears only when class 2 is dominant, but in addition class 1 should be in overload ($\mu_1 > \phi_1\mathrm{C}$).

Interestingly, the cases identified above map on the cases of Section 10.6. If the asymptotics are as in Equation (10.17), class 2 essentially just takes away $\phi_2 c$, as in Case (i) of Section 10.6. If Equation (10.18) applies, then overflow of the total queue essentially corresponds to overflow of queue 1, comparable to Case (iii) of Section 10.6. Finally, if Equation (10.19) applies, queue 2 takes away a rate between its average rate and its guaranteed rate, similar to Case (ii) of Section 10.6.

Delay. Delay analysis of GPS was considered in detail by Paschalidis [237]; in this reference large delay asymptotics are given, but with a focus on short-range dependent input. Section 10.7 is based on Lieshout and Mandjes [174].

Chapter 11

Explicit results for short-range dependent inputs

The previous chapters concentrated on sample-path large deviations in the many-sources regime. Various queuing models were considered: the single-node FIFO queue, and also tandem, priority, and GPS systems. The expressions for the decay rate were often only implicitly given, as the solution of an optimization problem.

In this chapter, we focus on the special case of short-range dependent (srd) input. Under this assumption, it turns out that these optimization problems can actually be solved, and insightful, explicit expressions are derived.

This chapter is organized as follows. Section 11.1 defines what we mean by srd in the context of this chapter: we introduce the concept *asymptotically linear variance*. In Section 11.2, the tandem queue is analyzed, whereas Section 11.3 deals with the priority queue, and Section 11.4 with the GPS system. Section 11.5 gives a discussion of the results and some concluding remarks.

11.1 Asymptotically linear variance; some preliminaries

In this chapter, we consider two types of input. We have already seen the first, namely, Brownian motion (not necessarily standard Brownian motion), several times; the second, namely, Gaussian sources with *asymptotically linear variance* is new. For convenience, we define the *centered process* $\overline{A}(\cdot)$ by putting $\overline{A}(t) := A(0,t) - \mu t$, where $\mu := \mathbb{E}A(0,1)$. As in the previous chapters, we assume throughout this chapter that Assumptions 4.2.2 and 11.1.1 are fulfilled.

Large deviations for Gaussian queues M. Mandjes

Assumption 11.1.1 *We assume that*
(i) $v(\cdot)$ *is continuous, differentiable on* $(0, \infty)$*;*
(ii) $v(\cdot)$ *is strictly increasing.*

Definition 11.1.2 *Consider a Gaussian source with stationary increments.*

(i) Brownian motion: *The source* $\mathrm{BM}(\lambda, \mu)$ *has mean input rate* μ *and variance function* $v(t) = \lambda t$*, for* $t \geq 0$.

(ii) Asymptotically linear variance: *The source* $\mathrm{ALV}(\kappa, \lambda, \mu)$ *has mean input rate* μ *and a variance function* $v(\cdot)$ *satisfying*

$$\lim_{t \to \infty} v(t) - \lambda t = \kappa.$$

Exercise 11.1.3 *Ornstein–Uhlenbeck input.* As before, we represent an (integrated) Ornstein–Uhlenbeck input process by choosing $v(t) = t - 1 + e^{-t}$. Prove that this source corresponds to an ALV process.

Solution. It is immediately seen that this source corresponds to an ALV process with $\kappa = -1$ and $\lambda = 1$. ◇

Exercise 11.1.4 *M/G/∞ inputs with Pareto job sizes.* A single M/G/∞ source consists of jobs that arrive according to a Poisson process of rate $\bar{\lambda}$. They stay in the system during some holding time that is distributed as a random variable D (with $\mathbb{E}D < \infty$). During this holding time, any job generates a constant traffic stream at a rate of, say, 1. We assume $\mathbb{P}(D > t) = (t+1)^{-\alpha}$, i.e., D has a Pareto tail. Take $\alpha > 1$; then $\mathbb{E}D = (\alpha - 1)^{-1} < \infty$. We now consider the Gaussian counterpart of this input process. Is it ALV?

Solution. The mean input rate is trivially $\mu := \bar{\lambda}\,\mathbb{E}D$ per unit time, whereas the variance function reads (assume for ease that $\alpha \notin \{2, 3\}$)

$$v(t) = \nu \cdot \left(1 - (t+1)^{3-\alpha} + (3-\alpha)t\right) \quad \text{with } \nu := \frac{2\bar{\lambda}}{(3-\alpha)(2-\alpha)(\alpha-1)},$$

as we have seen in Section 2.5. Importantly, if $\alpha \in (1, 2)$, then the traffic process has essentially long-range dependent properties, as $v(t)$ is superlinear; if $\alpha > 3$, then the process is ALV with $\kappa = \nu$ and $\lambda = \nu(3 - \alpha)$. The intermediate case $\alpha \in (2, 3)$ will be commented on in Section 11.5. ◇

The following lemma can be proved analogously to Exercise 4.1.9.

Lemma 11.1.5 *Let* $(X_i, Y_i) \in \mathbb{R}^2$ *i.i.d. bivariate normal random variables, with mean vector* $(\mu_X, \mu_Y)^{\mathrm{T}}$ *and two-dimensional covariance matrix*

$$\Sigma = \begin{pmatrix} \sigma_X^2 & \rho(X, Y) \\ \rho(X, Y) & \sigma_Y^2 \end{pmatrix}.$$

Fix $a > \mu_X$ and $b > \mu_Y$. Then

$$\lim_{n\to\infty}\frac{1}{n}\log\mathbb{P}\left(\frac{1}{n}\sum_{i=1}^{n}X_i \geq a, \frac{1}{n}\sum_{i=1}^{n}Y_i \leq b\right) = -\frac{1}{2}\frac{(a-\mu_X)^2}{\sigma_X^2},$$

if

$$\mathbb{E}(Y \mid X = a) = \mu_Y + \frac{\rho(X,Y)}{\sigma_X^2}(a-\mu_X) < b;$$

otherwise

$$\lim_{n\to\infty}\frac{1}{n}\log\mathbb{P}\left(\frac{1}{n}\sum_{i=1}^{n}X_i \geq a, \frac{1}{n}\sum_{i=1}^{n}Y_i \leq b\right)$$

$$= -\frac{1}{2}(a-\mu_X, b-\mu_Y)^{\mathrm{T}}\Sigma^{-1}\begin{pmatrix} a-\mu_X \\ b-\mu_Y \end{pmatrix}.$$

The following application of 'Schilder' follows analogously to Exercise 4.2.4.

Lemma 11.1.6 *Let a, t be positive, and define $F_{a,t} := \{f \mid f(t) \geq a\}$. Then,*

$$\lim_{n\to\infty}\frac{1}{n}\log\mathbb{P}\left(\frac{1}{n}\sum_{i=1}^{n}\overline{A}_i(\cdot) \in F_{a,t}\right) = -\frac{1}{2}\frac{a^2}{v(t)};$$

the optimizing path $f^\star(\cdot)$ is, for $r \in \mathbb{R}$, given by

$$f^\star(r) = \frac{\Gamma(r,t)}{v(t)} \cdot a.$$

Remark 11.1.7 For centered BM, the optimizing path $f^\star(\cdot) \in F_{a,t}$ from Lemma 11.1.6, equals 0 for negative r, grows linearly with slope a/t for $r \in [0,t]$, and remains at level a for $r \geq t$. ◇

Results for single FIFO queue. We first consider the single FIFO queue, i.e., n i.i.d. Gaussian sources feeding into a queue with link rate nc. The paths leading to overflow in the FIFO setting are

$$\mathcal{S}^{(\mathrm{f})}(b) := \{f \in \Omega : \exists t > 0 : A[f](-t,0) > b + ct\};$$

see Section 6.1. To find the corresponding decay rate, as we have seen earlier, 'Schilder' implies that $\mathbb{I}(f)$ needs to be minimized over all $f \in \mathcal{S}^{(\mathrm{f})}(b)$:

$$I_c^{(\mathrm{f})}(b) := -\lim_{n\to\infty}\frac{1}{n}\log\mathbb{P}(Q_n \geq nb) = \inf_{f\in\mathcal{S}^{(\mathrm{f})}(b)}\mathbb{I}(f)$$

$$= \inf_{t\geq 0}\frac{(b+(c-\mu)t)^2}{2v(t)}. \tag{11.1}$$

The following lemma evaluates $I_c^{(f)}(b)$ in Equation (11.1) for the specific situations of BM and ALV input.

Lemma 11.1.8 *Consider a single FIFO queue with link rate nc, fed by n i.i.d. sources.*

(i) *If the sources are* $\mathrm{BM}(\lambda, \mu)$, *then, for* $b \geq 0$,

$$I_c^{(f)}(b) = 2 \cdot \frac{c - \mu}{\lambda} \cdot b.$$

(ii) *If the sources are* $\mathrm{ALV}(\kappa, \lambda, \mu)$, *then*

$$\left(I_c^{(f)}(b) - 2 \cdot \frac{c - \mu}{\lambda} \cdot b \right) \longrightarrow -2\kappa \left(\frac{c - \mu}{\lambda} \right)^2, \qquad b \to \infty.$$

Proof. Part (i) directly follows from computing the infimum over $t \geq 0$ in Equation (11.1). Part (ii) is a consequence of [40, Theorem 3]. This theorem states the following. Let $\theta^\star$ denote the unique positive solution to

$$\lim_{t \to \infty} \frac{1}{t} \log \mathbb{E} \exp(\theta \overline{A}(t)) = (c - \mu)\theta.$$

Furthermore, suppose that $-\lim_{t \to \infty} \log \mathbb{E} \exp(\theta^\star \overline{A}(t) - \theta^\star (c - \mu)t) =: \nu$ exists. Then it holds that the shape function is asymptotically linear: $I_c^{(f)}(b) - \theta^\star b \to \nu$, for $b \to \infty$.

It is not hard to check that, in our Gaussian setting, for $\mathrm{ALV}(\kappa, \lambda, \mu)$ sources,

$$\lim_{t \to \infty} \frac{1}{t} \log \mathbb{E} \exp(\theta \overline{A}(t)) = \lim_{t \to \infty} \frac{1}{2t} \cdot \theta^2 v(t) = \frac{\lambda}{2} \cdot \theta^2,$$

yielding $\theta^\star = 2(c - \mu)/\lambda$. Also,

$$\begin{aligned}
\lim_{t \to \infty} \log \mathbb{E} \exp(\theta^\star \overline{A}(t) - \theta^\star (c - \mu)t) &= \lim_{t \to \infty} \frac{1}{2} (\theta^\star)^2 v(t) - \theta^\star (c - \mu)t \\
&= \lim_{t \to \infty} 2 \left(\frac{c - \mu}{\lambda} \right)^2 \cdot (v(t) - \lambda t) \\
&= 2\kappa \left(\frac{c - \mu}{\lambda} \right)^2.
\end{aligned}$$

This proves the statement. □

In the following alternative proof of part (ii), we explicitly use the ALV properties of the sources. We include the proof, because several proofs in the sequel of this chapter are along the same lines.

Alternative proof of part (ii) of Lemma 11.1.8. Use Equation (11.1), and choose $\overline{\epsilon} > 0$ arbitrarily. It is clear that, invoking the fact that $v(\cdot)$ is strictly increasing (Assumption 11.1.1.(ii)) and the ALV characterization, for b large enough and arbitrary M,

$$\inf_{t\leq Mb} \frac{(b+(c-\mu)t)^2}{2v(t)} \geq \frac{b^2}{2v(Mb)} \geq \frac{b^2}{2M\lambda b+\overline{\epsilon}} = O\left(\frac{1}{2M\lambda}\cdot b\right).$$

Also,

$$\left.\frac{(b+(c-\mu)t)^2}{2v(t)}\right|_{t:=b/(c-\mu)} = O\left(2\cdot\frac{c-\mu}{\lambda}\cdot b\right).$$

When M is chosen to be sufficiently small, this implies that we can restrict ourselves to the infimum over t in $[Mb,\infty)$. For any $\epsilon > 0$, we can select a b sufficiently large, such that for t in this range $|v(t)-\lambda t-\kappa| < \epsilon$. Hence,

$$\inf_{t\geq Mb} \frac{(b+(c-\mu)t)^2}{2v(t)} \leq \inf_{t\geq Mb} \frac{(b+(c-\mu)t)^2}{2(\lambda t+\kappa-\epsilon)}.$$

The latter optimum equals

$$2\cdot\frac{c-\mu}{\lambda}\cdot b - 2(\kappa-\epsilon)\left(\frac{c-\mu}{\lambda}\right)^2, \quad \text{achieved for } t = \frac{b}{c-\mu} - 2\cdot\frac{\kappa-\epsilon}{\lambda}.$$

The upper bound follows after $\epsilon \downarrow 0$ and $b\to\infty$. The lower bound is analogous, with ϵ replaced by $-\epsilon$. □

In concrete terms, the goal of the present chapter is to find the counterparts of Lemma 11.1.8 for the 'complex buffer architectures' tandem, priority, and GPS. Our analysis in the next sections shows that this is possible, partly relying on the bounds derived in the previous two chapters.

11.2 Tandem queue with srd input

Consider a tandem system of queues, fed by n i.i.d. Gaussian sources; the output of the first queue feeds into the second queue. The queues have link speeds nc_1 and nc_2, respectively; to avoid a trivial system, we assume $c_2 < c_1$. The system is stable: a source's mean rate μ is smaller than c_2. In this section, we analyze the probability that the stationary buffer content of the second queue, $Q_{2,n}$, exceeds level nb, for BM and ALV input.

From Chapter 9, we have that

$$\lim_{n\to\infty}\frac{1}{n}\log\mathbb{P}(Q_{2,n} > nb) = -I_c^{(\mathrm{t})}(b), \quad \text{where } I_c^{(\mathrm{t})}(b) := \inf_{f\in\mathcal{S}^{(\mathrm{t})}(b)} \mathbb{I}(f);$$

the 'overflow set' $\mathcal{S}^{(t)}(b)$ is defined as

$$\mathcal{S}^{(t)}(b) := \{f \in \Omega : \exists t > t^0 : \forall s \in (0, t) : A[f](-t, -s) > b + c_2 t - c_1 s\},$$

with $t^0 := db$ being the 'minimal' time till overflow, starting from an empty system; $d := (c_1 - c_2)^{-1}$. The main results of this section are Theorems 11.2.3 and 11.2.4–the former treats the BM case, whereas the latter focuses on ALV input. We start by providing two lemmas, Lemmas 11.2.1 and 11.2.2; Lemma 11.2.1 is valid for any type of Gaussian inputs.

Lemma 11.2.1 *For any* $b \geq 0$,

$$I_c^{(t)}(b) \geq \inf_{t>t^0} \frac{(b + (c_2 - \mu)t)^2}{2v(t)}.$$

Proof. Notice that evidently

$$\mathcal{S}^{(t)}(b) \subseteq \{f \in \Omega : \exists t > t^0 : A[f](-t, 0) > b + c_2 t\}.$$

This immediately proves the statement. □

Lemma 11.2.2 *Suppose the sources are* BM(λ, μ). *If* $c_1 \geq 2c_2 - \mu$, *then*

$$I_c^{(t)}(b) \leq 2 \cdot \frac{c_2 - \mu}{\lambda} \cdot b.$$

If $c_1 \in (c_2, 2c_2 - \mu)$, *then*

$$I_c^{(t)}(b) \leq \frac{1}{2} \cdot \frac{(c_1 - \mu)^2}{(c_1 - c_2)\lambda} \cdot b.$$

Proof. (i) Suppose $c_1 \geq 2c_2 - \mu$. Then the path that generates traffic at a rate of $2(c_2 - \mu)$, between $-b/(c_2 - \mu)$ and 0, is in $\mathcal{S}^{(t)}(b)$. (ii) Suppose $c_1 \in (c_2, 2c_2 - \mu)$. Then the path that generates traffic at rate c_1, between $-db$ and 0, is in $\mathcal{S}^{(t)}(b)$. Theorem 4.2.3 ('Schilder') implies that the norm of any feasible path constitutes an upper bound on the decay rate. Now applying Remark 11.1.7, the statement follows immediately. □

The following theorems give our main results for short-range dependent traffic. It states that its precise shape depends on whether $c_1 \geq 2c_2 - \mu$, or not.

Theorem 11.2.3 *Suppose the sources are* BM(λ, μ). *If* $c_1 \geq 2c_2 - \mu$, *then*

$$I_c^{(t)}(b) = 2 \cdot \frac{c_2 - \mu}{\lambda} \cdot b.$$

If $c_1 \in (c_2, 2c_2 - \mu)$, *then*

$$I_c^{(t)}(b) = \frac{1}{2} \cdot \frac{(c_1 - \mu)^2}{(c_1 - c_2)\lambda} \cdot b.$$

Proof. The result is a direct application of Lemmas 11.2.1 and 11.2.2; by performing the minimization over $t > t^0$ in Lemma 11.2.1 we obtain the desired expression. □

We now concentrate on the situation in which the tandem queue is fed by ALV sources. We obtain the following counterpart of Lemma 11.1.8.(ii).

Theorem 11.2.4 *Suppose the sources are* ALV(κ, λ, μ). *If* $c_1 \geq 2c_2 - \mu$, *then*

$$\left(I_c^{(t)}(b) - 2 \cdot \frac{c_2 - \mu}{\lambda} \cdot b\right) \longrightarrow -2\kappa\left(\frac{c_2 - \mu}{\lambda}\right)^2 \quad \text{as } b \to \infty.$$

If $c_1 \in (c_2, 2c_2 - \mu)$, *then*

$$\left(I_c^{(t)}(b) - \frac{1}{2} \cdot \frac{(c_1 - \mu)^2}{(c_1 - c_2)\lambda} \cdot b\right) \longrightarrow -\frac{\kappa}{2}\left(\frac{c_1 - \mu}{\lambda}\right)^2 \quad \text{as } b \to \infty.$$

Proof. Our proof consists of a lower bound and an upper bound. The lower bound follows directly from Lemma 11.2.1, whereas the upper bound is a matter of finding the norm of a feasible path.

Lower bound. Choose $\epsilon > 0$ arbitrarily. Take b sufficiently large, such that for all t larger than db we have that $|v(t) - \lambda t - \kappa| \leq \epsilon$. Applying the generic lower bound of Lemma 11.2.1, we find

$$I_c^{(t)}(b) \geq \inf_{t \geq db} \frac{(b + (c_2 - \mu)t)^2}{2v(t)} \geq \inf_{t \geq db} \frac{(b + (c_2 - \mu)t)^2}{2(\lambda t + \kappa + \epsilon)}.$$

If the latter infimum would be over all positive t–rather than all t larger than $t_0 = db$–it would be attained at

$$t^\star(b) := \frac{b}{c_2 - \mu} - 2\left(\frac{\kappa + \epsilon}{\lambda}\right);$$

it is also observed that, for $b \to \infty$, the condition $t^\star(b) > db$ reads $c_1 > 2c_2 - \mu$.

From the above, we conclude that for case $t^\star(b) > db$ the lower bound

$$I_c^{(t)}(b) \geq \frac{(b + (c_2 - \mu)t^\star(b))^2}{2(\lambda t^\star(b) + \kappa + \epsilon)} = 2 \cdot \frac{c_2 - \mu}{\lambda} \cdot \left(b - (c_2 - \mu) \cdot \frac{\kappa + \epsilon}{\lambda}\right)$$

applies, whereas for $t^\star(b) < db$ we have

$$\begin{aligned} I_c^{(t)}(b) &\geq \frac{(b + (c_2 - \mu)db)^2}{2(\lambda db + \kappa + \epsilon)} \\ &= \frac{b^2(1 + (c_2 - \mu)d)^2}{2(\lambda db + \kappa + \epsilon)} = \frac{1}{2} \cdot \frac{(c_1 - \mu)^2}{(c_1 - c_2)\lambda} \cdot b - \frac{\kappa + \epsilon}{2}\left(\frac{c_1 - \mu}{\lambda}\right)^2 + o\left(\frac{1}{b}\right). \end{aligned}$$

Now, let $\epsilon \downarrow 0$ and $b \to \infty$, and we get the desired lower bound.

Upper bound. Choose $\epsilon > 0$ arbitrarily. Define

$$s_\epsilon^\star(b) = \max\left\{d_\epsilon b,\ \frac{b}{c_2-\mu} - \frac{2\kappa}{\lambda}\right\}, \quad \text{with } d_\epsilon := \frac{1+\epsilon}{(c_1-\mu)-(1+\epsilon)(c_2-\mu)}. \tag{11.2}$$

Observe that, in particular,

$$(1+\epsilon)(b+(c_2-\mu)s_\epsilon^\star(b)) \leq (c_1-\mu)s_\epsilon^\star(b). \tag{11.3}$$

Define the path

$$f(r) := -\frac{\Gamma(-s_\epsilon^\star(b), r)}{v(s_\epsilon^\star(b))} \cdot (b+(c_2-\mu)s_\epsilon^\star(b)) - \mu r.$$

First, we show that this path is feasible (for b large), i.e., $f \in \mathcal{S}^{(\mathrm{t})}(b)$. Obviously, we have that $A[f](-s_\epsilon^\star(b), 0) = b + c_2 s_\epsilon^\star(b)$. What is remaining to be proved is that $A[f](-s, 0) < c_1 s$ for all $s \in (0, s_\epsilon^\star(b))$. With straightforward calculations and using the standard relations between $\Gamma(\cdot,\cdot)$ and $v(\cdot)$, it turns out that this is equivalent to requiring, for all $\gamma \in (0,1)$,

$$\left(1 + \frac{v(\gamma s_\epsilon^\star(b))}{v(s_\epsilon^\star(b))} - \frac{v((1-\gamma)s_\epsilon^\star(b))}{v(s_\epsilon^\star(b))}\right)\left(b+(c_2-\mu)s_\epsilon^\star(b)\right) < 2(c_1-\mu)\gamma s_\epsilon^\star(b). \tag{11.4}$$

Fixing $\gamma \in (0,1)$, the definition of ALV (in conjunction with $s_\epsilon^\star(b) \to \infty$ as $b \to \infty$) implies that there exists a $b_0(\gamma) \geq 0$, such that for any $b \geq b_0(\gamma)$, we have

$$\left(1 + \frac{v(\gamma s_\epsilon^\star(b))}{v(s_\epsilon^\star(b))} - \frac{v((1-\gamma)s_\epsilon^\star(b))}{v(s_\epsilon^\star(b))}\right) \leq 2(1+\epsilon)\gamma.$$

Observe that $b_0(0)$ and $b_0(1)$ could be chosen as finite numbers, and that we can choose a function $b_0(\cdot)$ that is continuous on $[0,1]$; cf. Assumption 11.1.1.(i). But then requirement (11.4) is true for all $b \geq \max_{\gamma\in[0,1]} b_0(\gamma)$ (which is finite, as any continuous function attains a maximum on a finite interval), due to Equation (11.3). We thus conclude that our path is feasible.

Recall that, because of Theorem 4.2.3, the norm of any feasible path constitutes an upper bound on the decay rate. We now compute (an upper bound to) the norm of our path $f(\cdot) \in \mathcal{S}^{(\mathrm{t})}(b)$. Take $\delta > 0$ arbitrarily. Choose b sufficiently large so that

$$|v(s_\epsilon^\star(b)) - \lambda s_\epsilon^\star(b) - \kappa| \leq \delta$$

(which is possible due to the definition of ALV). Because of the feasibility of the path, and applying Lemma 11.1.6, we obviously have

$$I_c^{(\mathrm{t})}(b) \leq \frac{(b+(c_2-\mu)s_\epsilon^\star(b))^2}{2v(s_\epsilon^\star(b))} \leq \frac{(b+(c_2-\mu)s_\epsilon^\star(b))^2}{2(\lambda s_\epsilon^\star(b)+\kappa-\delta)}. \tag{11.5}$$

For b large, it is clear that for $c_1 - \mu \leq 2(c_2 - \mu)(1+\epsilon)$ the maximum in Equation (11.2) is attained by the first argument between the brackets, i.e., $d_\epsilon b$. In this case, the upper bound (11.5) becomes

$$I_c^{(\mathrm{t})}(b) \leq \frac{1}{2} \cdot \frac{(c_1-\mu)^2}{((c_1-\mu)-(1+\epsilon)(c_2-\mu))\lambda} \cdot b - \frac{\kappa-\delta}{2}\left(\frac{c_1-\mu}{\lambda(1+\epsilon)}\right)^2 + O\left(\frac{1}{b}\right).$$

On the other hand, for $c_1 - \mu > 2(c_2 - \mu)(1+\epsilon)$,

$$I_c^{(\mathrm{t})}(b) \leq 2 \cdot \frac{c_2-\mu}{\lambda} \cdot \left(b - (c_2-\mu)\cdot\frac{\kappa-\delta}{\lambda}\right).$$

Now, let $\epsilon, \delta \downarrow 0$ and $b \to \infty$, and we have established the statement. □

11.3 Priority queue with srd input

In a priority queue, a link of capacity nc is considered, fed by traffic of two classes, each with its own queue. Traffic of class 1 does not 'see' class 2 at all, and consequently we know how the *high-priority queue* $Q_{h,n}$ behaves; see Lemma 11.1.8. A more challenging task is the characterization of overflow in the low priority queue. Let us consider the setting of Section 9.6. Our analysis relies on the following expression for the decay rate of overflow in the lp queue; notice that in Section 9.6 an alternative representation was used.

Lemma 11.3.1 *The decay rate of overflow in the lp queue is given by*

$$\lim_{n\to\infty}\frac{1}{n}\log\mathbb{P}(Q_{\ell,n} > nb) = -I_c^{(\mathrm{p})}(b), \quad \textit{where} \quad I_c^{(\mathrm{p})}(b) := \inf_{f\in\mathcal{S}^{(\mathrm{p})}(b)} \mathbb{I}(f).$$

Here the 'overflow set' is defined as $\mathcal{S}^{(\mathrm{p})}(b) :=$

$$\left\{ f \in \Omega\times\Omega : \begin{array}{c} \exists t, x > 0 : \forall s > 0 : \\ A_h[f_h](-t,0) + A_\ell[f_\ell](-t,0) > b + ct + x, \\ A_h[f_h](-s,0) \leq cs + x \end{array} \right\}.$$

Proof. To make sure that the lp queue exceeds nb, there must be an $x > 0$ such that the total queue exceeds $nb + nx$, whereas the hp queue remains below nx. Now the stated follows from

$$Q_{h,n} + Q_{\ell,n} = \sup_{t>0}\left(\sum_{i=1}^{n}(A_{h,i}(-t,0) + A_{\ell,i}(-t,0)) - nct\right);$$

$$Q_{h,n} = \sup_{s>0}\left(\sum_{i=1}^{n} A_{h,i}(-s,0) - ncs\right),$$

where $A_{h,i}(\cdot)$ ($A_{\ell,i}(\cdot)$, respectively) corresponds to the traffic stream generated by the ith hp (lp) source. □

Notice that the 'overflow set' of the above lemma is slightly different from the one we used in Section 9.6, but it can be verified easily that both sets are equal. The next lemma applies the above characterization to derive a lower bound on the decay rate.

Lemma 11.3.2 *Let* $S^{(p)}(b,t,x) := \{(a_h, a_\ell) \mid a_h + a_\ell > b + ct + x, a_h \leq ct + x\}$. *Then*

$$I_c^{(p)}(b) \geq \inf_{t,x>0} k^{(p)}(b,t,x),$$

with

$$k^{(p)}(b,t,x) := \frac{1}{2} \inf_{(a_h,a_\ell)\in S^{(p)}(b,t,x)} \left(\frac{(a_h - \mu_h t)^2}{v_h(t)} + \frac{(a_\ell - \mu_\ell t)^2}{v_\ell(t)} \right).$$

Also, if $(b - \mu_\ell t)v_h(t) \leq ((c - \mu_h)t + x)v_\ell(t)$, *then*

$$k^{(p)}(b,t,x) = k_1^{(p)}(b,t,x) := \frac{1}{2} \cdot \frac{(b + (c-\mu)t + x)^2}{v(t)},$$

whereas otherwise

$$k^{(p)}(b,t,x) = k_2^{(p)}(b,t,x) := \frac{1}{2} \cdot \left(\frac{((c-\mu_h)t + x)^2}{v_h(t)} + \frac{(b - \mu_\ell t)^2}{v_\ell(t)} \right).$$

Proof. First, it is noted that the following trivial inclusion holds: $\mathcal{S}^{(p)}(b) \subseteq$

$$\left\{ f \in \Omega \times \Omega : \exists t, x > 0 : \begin{array}{l} A_h[f_h](-t,0) + A_\ell[f_\ell](-t,0) > b + ct + x, \\ A_h[f_h](-t,0) \leq ct + x \end{array} \right\}.$$

This immediately yields the lower bound $\inf_{x,t>0} k^{(p)}(b,t,x)$.

The explicit expression for $k^{(p)}(b,t,x)$ is derived as follows. Applying Lemma 11.1.5, it is readily derived that the first constraint in $S^{(p)}(b,t,x)$, i.e., $a_h + a_\ell > b + ct + x$ is always tight. The second constraint is tight if

$$\mu_h t + \frac{v_h(t)}{v(t)}(b + (c-\mu)t + x) > ct + x.$$

If it is tight, the infimum is achieved at $(a_h, a_\ell) = (ct + x, b)$, otherwise it is achieved at

$$(a_h, a_\ell) = \left(\mu_h t + \frac{v_h(t)}{v(t)}(b + (c-\mu)t + x), \mu_\ell t + \frac{v_\ell(t)}{v(t)}(b + (c-\mu)t + x) \right).$$

Direct calculations yield the stated result. □

Introduce the following notation: $\mu := \mu_h + \mu_\ell$, $\kappa := \kappa_h + \kappa_\ell$, and $\lambda := \lambda_h + \lambda_\ell$.

Lemma 11.3.3 *Suppose that the high-priority sources are* BM(λ_h, μ_h) *and that the low-priority sources are* BM(λ_ℓ, μ_ℓ). *Then,*

$$\inf_{t,x>0} k_1^{(\mathrm{p})}(b,t,x) = 2 \cdot \frac{c-\mu}{\lambda} \cdot b;$$

$$\inf_{t,x>0} k_2^{(\mathrm{p})}(b,t,x) = \frac{\Xi - \mu_\ell}{\lambda_\ell} \cdot b,$$

where

$$\Xi \equiv \Xi(\lambda_h, \mu_h, \lambda_\ell, \mu_\ell) := \sqrt{\mu_\ell^2 + \frac{\lambda_\ell}{\lambda_h}(c-\mu_h)^2}.$$

Proof. This is a matter of standard computations. In the first minimization, it turns out that $x_1^\star = 0$, $t_1^\star = b/(c-\mu)$, whereas in the second

$$x_2^\star = 0, \quad t_2^\star = \sqrt{\frac{\lambda_h}{\lambda_h \mu_\ell^2 + \lambda_\ell (c-\mu_h)^2}} \cdot b.$$

The desired result is obtained by inserting these into the objective function. □

Theorem 11.3.4 *Suppose that the high-priority sources are* BM(λ_h, μ_h) *and that the low-priority sources are* BM(λ_ℓ, μ_ℓ). *If* $\lambda_h(c-\mu_h-2\mu_\ell) \leq \lambda_\ell(c-\mu_h)$, *then*

$$I_c^{(\mathrm{p})}(b) = 2 \cdot \frac{c-\mu}{\lambda} \cdot b.$$

If $\lambda_h(c-\mu_h-2\mu_\ell) > \lambda_\ell(c-\mu_h)$, *then*

$$I_c^{(\mathrm{p})}(b) = \frac{\Xi - \mu_\ell}{\lambda_\ell} \cdot b.$$

Proof. *Lower bound.* We use Lemma 11.3.2. Define three sets:

$$T_1 := \{(t,x) \in \mathbb{R}_+^2 \mid (b-\mu_\ell t)v_h(t) \leq ((c-\mu_h)t + x)v_\ell(t)\};$$

T_2 with the '$\leq$'-sign replaced by '$\geq$', and $\overline{T}$ with the '$\leq$'-sign replaced by '$=$'. Notice that $k_1^{(\mathrm{p})}(b,\cdot,\cdot)$ and $k_2^{(\mathrm{p})}(b,\cdot,\cdot)$ coincide for (t,x) in $\overline{T}$. Let $t_i^\star, x_i^\star$ $(i=1,2)$ be defined as in the proof of Lemma 11.3.3.

First, consider the infimum of $k^{(\mathrm{p})}(b,t,x)$ over $(t,x) \in T_1$. Clearly, the optimum is in $T_1 \setminus \overline{T}$ iff $(b-\mu_\ell t_1^\star)v_h(t_1^\star) < ((c-\mu_h)t_1^\star + x_1^\star)v_\ell(t_1^\star)$, or, equivalently,

$$\lambda_h(c-\mu_h-2\mu_\ell) < \lambda_\ell(c-\mu_h); \tag{11.6}$$

otherwise the optimum over T_1 is attained in $\overline{T}$.

Then, consider the infimum of $k^{(p)}(b, t, x)$ over $(t, x) \in T_2$. Now, the optimum is in $T_2 \setminus \overline{T}$ iff

$$(b - \mu_\ell t_2^\star)v_h(t_2^\star) > ((c - \mu_h)t_2^\star + x_2^\star)v_\ell(t_2^\star),$$

and otherwise at the boundary $\overline{T}$. More tedious calculations yield that this condition is equivalent to

$$\lambda_h(c - \mu_h - 2\mu_\ell) > \lambda_\ell(c - \mu_h). \tag{11.7}$$

Because both conditions (11.6) and (11.7) are mutually exclusive, this proves the lower bound.

Upper bound. The upper bound is just a matter of computing the norms of paths in $\mathcal{S}^{(p)}(b)$, just as in the tandem case. □

Remark 11.3.5 It is straightforward, but tedious, to check that both expressions for $I_c^{(p)}(b)$ (from Theorem 11.3.4) coincide if $\lambda_h(c - \mu_h - 2\mu_\ell) = \lambda_\ell(c - \mu_h)$. During these computations, we also find that then $\Xi(\lambda_h - \lambda_\ell) = \mu_\ell\lambda$. ◇

We now turn to the situation of ALV sources.

Lemma 11.3.6 *Suppose that the high-priority sources are* $\text{ALV}(\kappa_h, \lambda_h, \mu_h)$ *and that the low-priority sources are* $\text{ALV}(\kappa_\ell, \lambda_\ell, \mu_\ell)$. *Then,*

$$\lim_{b\to\infty}\left(\inf_{x,t>0} k_1^{(p)}(b, t, x) - 2 \cdot \frac{c-\mu}{\lambda} \cdot b\right) = -2\kappa\left(\frac{c-\mu}{\lambda}\right)^2.$$

Proof. This is equivalent to Lemma 11.1.8. □

Lemma 11.3.7 *Suppose that the high-priority sources are* $\text{ALV}(\kappa_h, \lambda_h, \mu_h)$ *and that the low-priority sources are* $\text{ALV}(\kappa_\ell, \lambda_\ell, \mu_\ell)$. *Then,*

$$\lim_{b\to\infty}\left(\inf_{x,t>0} k_2^{(p)}(b, t, x) - \frac{\Xi - \mu_\ell}{\lambda_\ell} \cdot b\right) = -\frac{(c-\mu_h)^2}{\lambda_h}\left(\frac{\kappa_h}{2\lambda_h} - \Phi\right) + \Phi^2\Xi \cdot \frac{\mu_\ell}{2\kappa_\ell},$$

with

$$\Phi := -\frac{\kappa_\ell}{\lambda_\ell}\left(1 - \frac{\mu_\ell}{\Xi}\right).$$

Proof. First, observe that, for any given value of b, t, the infimum of $k_2^{(p)}(b, t, x)$ over x is attained in 0. Therefore, consider $\inf_{t\geq 0} k_2^{(p)}(b, t, 0)$. We borrow the argument of the alternative proof of part (ii) of Lemma 11.1.8.

- We first show that we can restrict ourselves to $t \geq Mb$. Choosing $M < \mu_\ell^{-1}$, and $\overline{\epsilon} > 0$ arbitrarily, and using Assumption 11.1.1.(ii) and the ALV properties, we notice that for b large

$$\inf_{t<Mb} k_2^{(\mathrm{p})}(b,t,0) \geq \frac{(1-\mu_\ell M)^2}{2} \cdot \frac{b^2}{M\lambda_\ell b + \kappa_\ell + \overline{\epsilon}} = O\left(\frac{(1-\mu_\ell M)^2}{2M\lambda_\ell} \cdot b\right). \tag{11.8}$$

Also, it is not hard to verify that for $b \to \infty$

$$k_2^{(\mathrm{p})}\left(b, b\cdot\sqrt{\frac{\lambda_h}{\lambda_\ell(c-\mu_h)^2 + \lambda_h\mu_\ell^2}}, 0\right) = O\left(\frac{\Xi - \mu_\ell}{\lambda_\ell} \cdot b\right). \tag{11.9}$$

Choosing $M > 0$ sufficiently small, Equation (11.8) majorizes Equation (11.9), and hence we can restrict ourselves in $\inf_t k_2^{(\mathrm{p})}(b,t,0)$ to $t \geq Mb$.

- Choose b large enough, such that, for all $t \geq Mb$, both $|v_h(t) - \lambda_h t - \kappa_h| < \epsilon$ and $|v_\ell(t) - \lambda_\ell t - \kappa_\ell| < \epsilon$. Then, applying Lemma A.1 (see the appendix of this chapter) and allowing $\epsilon \downarrow 0$ and $b \to \infty$ we obtain the stated result. □

Theorem 11.3.8 *If $\lambda_h(c-\mu_h-2\mu_\ell) \leq \lambda_\ell(c-\mu_h)$, then*

$$\lim_{b\to\infty}\left(I_c^{(\mathrm{p})}(b) - 2\cdot\frac{c-\mu}{\lambda}\cdot b\right) = -2\kappa\left(\frac{c-\mu}{\lambda}\right)^2.$$

If $\lambda_h(c-\mu_h-2\mu_\ell) > \lambda_\ell(c-\mu_h)$, then

$$\lim_{b\to\infty}\left(I_c^{(\mathrm{p})}(b) - \frac{\Xi-\mu_\ell}{\lambda_\ell}\cdot b\right) = -\frac{(c-\mu_h)^2}{\lambda_h}\left(\frac{\kappa_h}{2\lambda_h} - \Phi\right) + \Phi^2\Xi\cdot\frac{\mu_\ell}{2\kappa_\ell}.$$

Proof. The lower bound is analogous to the lower bound in Theorem 11.3.4, but with Lemmas 11.3.6 and 11.3.7 replacing Lemma 11.3.3. The upper bound is analogous to the upper bound in the tandem case, i.e., in the proof of Theorem 11.2.4. □

Waiting time asymptotics. The first part of this section was devoted to buffer overflow in the lp queue, or, more precisely, the probability that the buffer content of the lp queue exceeds some predefined level. We now focus on the probability of a long delay in the lp queue. To this end, following [224], consider the notion of *virtual* waiting time. The random variable $D_{\ell,n}(0)$ is defined as the time it takes to transmit a 'fluid molecule' that enters the lp queue at time 0. We define

$$L_c^{(\mathrm{p})}(d) := -\lim_{n\to\infty}\frac{1}{n}\log\mathbb{P}(D_{\ell,n}(0) > d).$$

Lemma 11.3.9

$$L_c^{(\mathrm{p})}(d) \geq \frac{1}{2}\sup_{u\in(0,d)}\inf_{s>0}\frac{((c-\mu_h-\mu_\ell)s + (c-\mu_h)u)^2}{v_h(s+u) + v_\ell(s)}.$$

Proof. The set of paths such that the virtual delay exceeds d equals

$$\mathcal{T}_c^{(\mathrm{p})}(d) := \left\{ \begin{array}{l} f \in \Omega : \forall u \in (0,d) : \exists s > 0 : \ A_h[f_h](-s,u) \\ \hfill + A_\ell[f_\ell](-s,0) > c(u+s) \end{array} \right\};$$

see [224, Section 4]. The statement follows immediately by applying Lemma 11.1.6; as before, the fact that the probability of an intersection is smaller than the probability of each of the individual events is used. □

When computing the decay rate of the waiting time, a crucial role is played by the parameter ϱ, defined by

$$\varrho := \frac{c-\mu_h}{c-\mu} - 2 \cdot \frac{\lambda_h}{\lambda}.$$

Theorem 11.3.10 *Suppose that the high-priority sources are* $\mathrm{BM}(\lambda_h, \mu_h)$ *and that the low-priority sources are* $\mathrm{BM}(\lambda_\ell, \mu_\ell)$. *Then, for* $\varrho > 0$,

$$L_c^{(\mathrm{p})}(d) = 2\left(\frac{c-\mu}{\lambda}\right)\left(\frac{\lambda_h\mu_\ell + \lambda_\ell(c-\mu_h)}{\lambda}\right)d;$$

for $\varrho \le 0$,

$$L_c^{(\mathrm{p})}(d) = \frac{1}{2}\cdot\left(\frac{(c-\mu_h)^2}{\lambda_h}\right)d.$$

Suppose that the high-priority sources are $\mathrm{ALV}(\kappa_h, \lambda_h, \mu_h)$, *and that the low-priority sources are* $\mathrm{ALV}(\kappa_\ell, \lambda_\ell, \mu_\ell)$. *Then, for* $\varrho > 0$,

$$\left(L_c^{(\mathrm{p})}(d) - 2\left(\frac{c-\mu}{\lambda}\right)\left(\frac{\lambda_h\mu_\ell + \lambda_\ell(c-\mu_h)}{\lambda}\right)d\right) \longrightarrow -2\kappa\left(\frac{c-\mu}{\lambda}\right)^2;$$

for $\varrho \le 0$,

$$\left(L_c^{(\mathrm{p})}(d) - \frac{1}{2}\cdot\left(\frac{(c-\mu_h)^2}{\lambda_h}\right)d\right) \longrightarrow -\frac{\kappa_2}{2}\left(\frac{c-\mu_2}{\lambda_2}\right)^2.$$

Proof. The lower bounds follow directly from Lemma 11.3.9, as follows. For BM traffic the optimizing $s^\star(u)$, for given u, equals

$$s^\star(u) = \left(\frac{c-\mu_h}{c-\mu_h-\mu_\ell} - 2\cdot\frac{\lambda_h}{\lambda}\right)\cdot u = \varrho u,$$

provided $\varrho > 0$. The resulting function of u increases on $[0,d]$, so the optimum is attained at $u^\star = d$. For ALV traffic we get, if $\varrho > 0$,

$$s^\star(u) - \left(\frac{c-\mu_h}{c-\mu_h-\mu_\ell} - 2\cdot\frac{\lambda_h}{\lambda}\right)\cdot u \longrightarrow -2\cdot\frac{\kappa}{\lambda},$$

also leading to $u^\star = d$. Now straightforward algebra (use Lemma A.1!) gives the lower bound. If $\varrho \leq 0$, then the optimizing $s^\star(u)$ is 0.

The upper bound is a matter of computing the norms of feasible paths, as before. □

Remark 11.3.11 We remark that, in case of BM, the above proofs indicate that, for $\varrho > 0$, the most likely path is such that (i) between time epochs $-s^\star(d)$ and d any hp source generates traffic at a constant rate

$$\varsigma_\ell := (1+\varrho)(c-\mu) = 2(c-\mu)\frac{\lambda_\ell}{\lambda} + \mu_\ell,$$

whereas (ii) between $-s^\star(d)$ and 0 any lp source transmits at rate

$$\varsigma_h := c - \varrho(c-\mu) = 2(c-\mu)\frac{\lambda_h}{\lambda} + \mu_h.$$

Observe that the behavior of the hp sources between time 0 and d does play a role (as this may interfere with the lp traffic); the behavior of the lp sources between 0 and d is irrelevant, as this cannot further delay the 'fluid molecule' entering the lp queue at time 0.

Notice that we have that $\varsigma_\ell > \mu_\ell$ and $\varsigma_h > \mu_h$; in other words, both types of sources generate traffic at a rate higher than their mean rate. Outside the intervals indicated above, the sources obey their mean rates. Consequently, between $-s^\star(d)$ and 0, the total queue builds up at a rate $\varsigma_h + \varsigma_\ell = 2c - \mu$ and it can be checked that it has the size $(c-\mu)\varrho \cdot d$ at time 0.

In case $\varrho \leq 0$, any hp source just takes away a rate c in the time interval between 0 and d, thus causing overflow; both queues were empty at time 0. ◇

11.4 GPS queue with srd input

In this section, we consider a system where traffic is served according to a generalized processor sharing (GPS) mechanism, consisting of two queues sharing a link of capacity nc. We assume the system to be fed by traffic from two classes, where class i uses queue i, for $i = 1, 2$. It is assumed that both classes consist of n flows (but, again, due to the infinite divisibility of the normal distribution, this is not a restriction).

As discussed in the previous chapter, class i receives service at rate $n\phi_i c$ when both classes are backlogged. Because class i gets at least service at rate $n\phi_i c$ when it has backlog, we will refer to it as the *guaranteed rate* of class i. Without loss of generality, we focus on the workload of the first queue. The goal here is to analyze the decay rate of the probability that the stationary workload exceeds a threshold nb. Hence, denoting the stationary workload in the ith GPS queue at time 0 by $Q_{i,n} \equiv Q_{i,n}(0)$, the probability of our interest is $\mathbb{P}(Q_{1,n} \geq nb)$.

In Lemma 10.2.2 it was shown that the 'overflow set' is bounded from above by

$$\overline{S^{(g)}}(b) := \left\{ f \in \Omega \times \Omega : \begin{array}{c} \exists t, x > 0 : \forall s \in (0,t) : \exists u \in (0,s) : \\ A_1[f_1](-t,0) + A_2[f_2](-t,0) > b + ct + x, \\ A_1[f_1](-s,-u) + A_2[f_2](-s,0) \leq cs - \phi_1 cu + x \end{array} \right\},$$

we have that

$$I_c^{(g)}(b) := -\lim_{n\to\infty} \frac{1}{n} \log \mathbb{P}(Q_{1,n} \geq nb) \geq \inf_{f \in \overline{S^{(g)}}(b)} \mathbb{I}(f).$$

In this section, we concentrate on BM input–the case of ALV is harder, and we comment on it later (Section 11.5). The input of class i consists of n sources of the type BM(λ_i, μ_i). The main result is that the expressions intuitively derived in Section 10.6 are correct for BM input.

Define also $\mu := \mu_1 + \mu_2$ and $\lambda := \lambda_1 + \lambda_2$, and the 'reduced rates': $\overline{c} := c - \mu$ and $\overline{c}_i = c\phi_i - \mu_i$ (for $i = 1, 2$). The system is stable: $\mu < c$.

We start by presenting an introductory lemma.

Lemma 11.4.1 *Suppose that the class-1 sources are* BM(λ_1, μ_1) *and that the class-2 sources are* BM(λ_2, μ_2). *Let us define the set* $S^{(g)}(b, x, u, t) := \{(a_1, a_2) \mid a_1 > b + ct + x, a_2 \leq ct - \phi_1 cu + x\}$. *Then,*

$$I_c^{(g)}(b) \geq \inf_{x,t>0, u\in(0,t)} k^{(g)}(b, x, u, t),$$

where $k^{(g)}(b, x, u, t) := \frac{1}{2} \inf_{(a_1,a_2)\in S^{(g)}(b,x,u,t)} H(a_1, a_2)$, *with*

$$H(a_1, a_2) := \left(\frac{(a_1 - \mu t)^2}{\lambda_1 u} - \frac{2(a_1 - \mu t)(a_2 - \mu t + \mu_1 u)}{\lambda_1 u} + \frac{(a_2 - \mu t + \mu_1 u)^2}{\lambda_1 u} \frac{\lambda t}{\lambda t - \lambda_1 u} \right).$$

Also, if $\lambda t(b + \overline{c}_1 u) \leq \lambda_1 u(b + \overline{c}t + x)$, *then*

$$k^{(g)}(b, x, u, t) = k_1^{(g)}(b, x, u, t) := \frac{1}{2} \cdot \frac{(b + \overline{c}t + x)^2}{\lambda t},$$

whereas otherwise

$$k^{(g)}(b, x, u, t) = k_2^{(g)}(b, x, u, t) := \frac{1}{2} \cdot \left(\frac{(\overline{c}t - \overline{c}_1 u + x)^2}{\lambda t - \lambda_1 u} + \frac{(b + \overline{c}_1 u)^2}{\lambda_1 u} \right).$$

Proof. First, it is noted that

$$\overline{S^{(g)}}(b) \subseteq \left\{ f \in \Omega \times \Omega : \begin{array}{c} \exists t, x > 0 : \exists u \in (0,t) : \\ A_1[f_1](-t,0) + A_2[f_2](-t,0) > b + ct + x, \\ A_1[f_1](-t,-u) + A_2[f_2](-t,0) \leq ct - \phi_1 cu + x \end{array} \right\}.$$

Identifying a_1 with $A_1(-t,0)+A_2(-t,0)$ and a_2 with $A_1(-t,-u)+A_2(-t,0)$, the lower bound is derived as follows. Notice that these random variables have variances λt and $\lambda_1(t-u)+\lambda_2 t=\lambda t-\lambda_1 u$, respectively, while their covariance reads $\lambda t-\lambda_1 u$. Also,

$$(a_1-\mu t, a_2-\mu t+\mu_1 u)^{\mathrm{T}}\begin{pmatrix}\lambda t & \lambda t-\lambda_1 u\\ \lambda t-\lambda_1 u & \lambda t-\lambda_1 u\end{pmatrix}^{-1}\begin{pmatrix}a_1-\mu t\\ a_2-\mu t+\mu_1 u\end{pmatrix}$$

equals $H(a_1,a_2)$. This proves the first part of the lemma. The evaluation of the maximum over $(a_1,a_2)\in S^{(g)}(b,x,u,t)$ is analogous to the priority case (use Lemma 11.1.5). □

In the following lemma we explicitly derive a closed-form expression for $k_1^{(g)}(b,x,u,t)$ (minimized over t,u,x). It can be proved in a standard manner, and therefore omit the proof.

Lemma 11.4.2 *Suppose that the class-1 sources are* $\mathrm{BM}(\lambda_1,\mu_1)$ *and that the class-2 sources are* $\mathrm{BM}(\lambda_2,\mu_2)$. *Then,*

$$\inf_{t,x>0,u\in(0,t)} k_1^{(g)}(b,x,u,t)=2\cdot\frac{c-\mu}{\lambda}\cdot b.$$

The derivation of the infimum of $k_2^{(g)}(b,x,u,t)$ (minimized over t,u,x) is considerably harder. We first define

$$T(\gamma):=\bar{c}_1^2\gamma^2+(\bar{c}-\bar{c}_1\gamma)^2\cdot\frac{\lambda_1\gamma}{\lambda-\lambda_1\gamma};\quad t(\gamma):=\frac{1}{\sqrt{T(\gamma)}}.$$

Lemma 11.4.3 *Suppose that the class-1 sources are* $\mathrm{BM}(\lambda_1,\mu_1)$ *and that the class-2 sources are* $\mathrm{BM}(\lambda_2,\mu_2)$. *Then*

$$\inf_{t,x>0,u\in(0,t)} k_2^{(g)}(b,x,u,t)=\frac{1}{2}\inf_{\gamma\in(0,1)}U(\gamma)\cdot b,$$

with

$$U(\gamma):=\frac{(1+\bar{c}_1\gamma t(\gamma))^2}{\lambda_1\gamma t(\gamma)}+\frac{(\bar{c}-\bar{c}_1\gamma)^2 t(\gamma)}{\lambda-\lambda_1\gamma}.$$

Proof. First, observe that, for any given value of b,t,u, the infimum over x is attained in 0. Therefore, consider $k_2^{(g)}(b,0,u,t)$ to be optimized over positive t and $u\in(0,t)$. Now, write $u=\gamma t$, with $\gamma\in(0,1)$. Perform the optimization over t. Straightforward calculus yields that the minimum is attained at $t=bt(\gamma)$. By inserting this, the statement is proved. □

The following two lemmas determine the infimum of $U(\gamma)$ over $\gamma\in(0,1)$.

Lemma 11.4.4 *Suppose*

$$\phi_1 \geq \phi_1^c := \frac{\lambda_1 - \lambda_2}{\lambda_1 + \lambda_2} \cdot \left(1 - \frac{\mu}{c}\right) + \frac{\mu_1}{c}.$$

For all $\gamma \in (0, 1)$, it holds that $\gamma t(\gamma) \leq t(1)$.

Proof. It is equivalent to checking that $T(\gamma) \geq \gamma^2 T(1)$ for $\gamma \in (0, 1)$, or

$$E_\ell(\gamma) := \left(\frac{\overline{c} - \overline{c}_1\gamma}{\overline{c}_2}\right)^2 \geq \frac{(\lambda - \lambda_1\gamma)\gamma}{\lambda_2} =: E_r(\gamma).$$

Due to the fact that $E_\ell(\cdot)$ and $E_r(\cdot)$ correspond to parabolas (where the former is convex and the latter is concave), it is enough to verify whether $E_\ell'(1) \leq E_r'(1)$. This yields the condition

$$-2 \cdot \frac{\overline{c}_1}{\overline{c}_2} \leq 1 - \frac{\lambda_1}{\lambda_2},$$

which is in turn equivalent to $\phi_1 \geq \phi_1^c$. □

Lemma 11.4.5 *If $\phi_1 \geq \phi_1^c$, then $\inf_{\gamma \in (0,1)} U(\gamma) = U(1)$.*

Proof. We need to check if, for all $\gamma \in (0, 1)$,

$$\frac{(1 + \overline{c}_1\gamma t(\gamma))^2}{\lambda_1\gamma t(\gamma)} + \frac{(\overline{c} - \overline{c}_1\gamma)^2 t(\gamma)}{\lambda - \lambda_1\gamma} \geq \frac{(1 + \overline{c}_1 t(1))^2}{\lambda_1 t(1)} + \frac{\overline{c}_2^2 t(1)}{\lambda_2}.$$

Straightforward algebraic manipulations yield that this is equivalent to

$$\lambda_2(\lambda - \lambda_1\gamma)t(1) - (\lambda - \lambda_1\gamma)\gamma V_1(\gamma)t(\gamma) + \lambda_2 t(1)V_2(\gamma)t^2(\gamma) \geq 0, \quad \text{with}$$

$$V_1(\gamma) := \lambda_1\overline{c}_2^2 t^2(1) + \lambda_2(1 + \overline{c}_1^2 t^2(1));$$

$$V_2(\gamma) := (\lambda - \lambda_1\gamma)\overline{c}_1^2\gamma^2 + \lambda_1\gamma(\overline{c} - \overline{c}_1\gamma)^2.$$

Now, it is not hard to see that $V_1(\gamma) = 2\lambda_2$ and $V_2(\gamma) = (\lambda - \lambda_1\gamma)/t^2(\gamma)$, so it remains to verify that

$$2\lambda_2 \cdot (\lambda - \lambda_1\gamma) \cdot (t(1) - \gamma t(\gamma)) \geq 0,$$

but this holds (for $\phi_1 \geq \phi_1^c$) because of Lemma 11.4.4. □

With the above results, we can prove our main theorem for the two-queue GPS system with Brownian inputs.

Theorem 11.4.6 *Suppose that the class-1 sources are* $\mathrm{BM}(\lambda_1, \mu_1)$ *and that the class-2 sources are* $\mathrm{BM}(\lambda_2, \mu_2)$. *Then, with*

$$\phi_2^c = 1 - \phi_1^c; \quad \phi_2^o = \frac{\mu_2}{c},$$

it holds that (i) for $\phi_2 \in [0, \phi_2^o]$,

$$I_c^{(g)}(b) = 2 \cdot \frac{\phi_1 c - \mu_1}{\lambda_1} \cdot b;$$

(ii) for $\phi_2 \in [\phi_2^o, \phi_2^c]$,

$$I_c^{(g)}(b) = \frac{1}{2} \cdot U(1) \cdot b;$$

(iii) for $\phi_2 \in [\phi_2^c, 1]$,

$$I_c^{(g)}(b) = 2 \cdot \frac{c - \mu}{\lambda_1 + \lambda_2} \cdot b.$$

Proof. Case (i) follows directly from Section 10.5. Class 2 is in overload, and 'takes away it weight' without any effort. As a consequence, in essence, class 1 sees a queue with service rate $n\phi_1 c$.

The proof of cases (ii) and (iii) mimics the proof of Theorem 11.3.4. The *lower bound* uses Lemma 11.4.1. Then, the proof is as in the lower bound of Theorem 11.3.4, with the sets

$$T_1 := \{(t, u, x) \in \mathbb{R}_+^3 \mid \lambda t(b + \bar{c}_1 u) \leq \lambda_1 u(b + \bar{c}t + x)\};$$

T_2 with the '$\leq$'-sign replaced by '$\geq$', and $\overline{T}$ with the '$\leq$'-sign replaced by '$=$'. The infima of the $k_i^{(g)}(b, x, u, t)$ (over t, u, x) for $i = 1, 2$ follow then from Lemmas 11.4.2, 11.4.3, and 11.4.5.

The *upper bound* is just a matter of verifying that the paths of the lower bound are feasible, and computing their norm. □

Example 11.4.7 Here, we illustrate the result of the previous theorem by an example. We suppose that both types of sources correspond to Brownian motions, with $\mu_1 = 0.2$, $\mu_2 = 0.3$, $v_1(t) = 2t$, and $v_2(t) = t$. Take $c = 1$. With the buffer size of class i denoted by $B_i \equiv nb_i$, let $I_{c,i}^{(g)}(b_i)$ be the decay rate of class i, and

$$L_{c,i}^{(g)} := \frac{I_{c,i}^{(g)}(b_i)}{b_i}.$$

These $L_{c,i}^{(g)}$ are given in Figure 11.1, as a function of the weight ϕ_1.

Suppose the weight ϕ_1 (and hence implicitly also $\phi_2 = 1 - \phi_1$) has to be chosen such that the decay rate of class i is larger than δ_i (for $i = 1, 2$). Then, we need to verify whether there is a $\phi_1 \in [0, 1]$ such that both $b_1 L_{c,1}^{(g)} \geq \delta_1$ and $b_2 L_{c,2}^{(g)} \geq \delta_2$. This can easily be verified from the graph in the figure. ◇

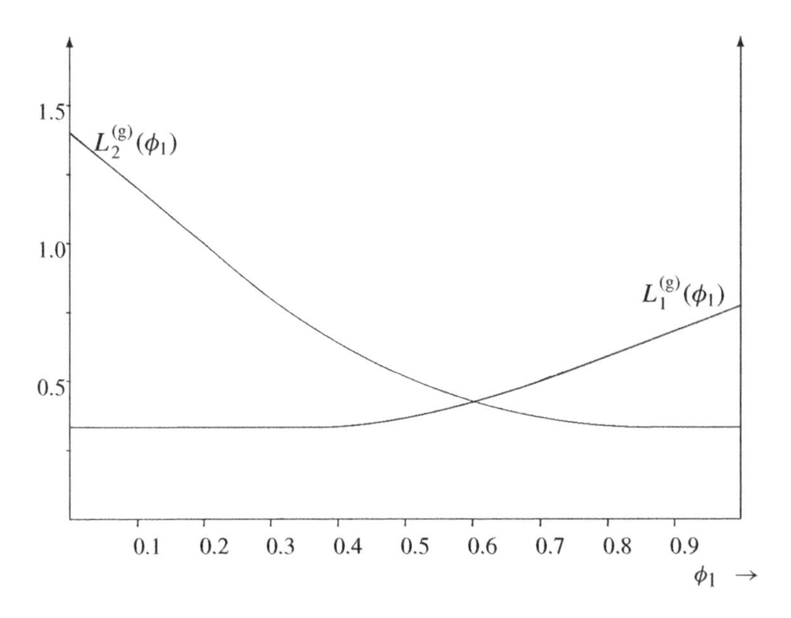

Figure 11.1: The curves $L_i^{(g)}(\phi_1)$ of Example 11.4.7.

11.5 Concluding remarks

In this chapter, we have computed, in a many-sources setting, the exponential decay rate of the overflow probability in a tandem queue, a priority system, and a system operating under a GPS scheduler. The input was assumed short-range dependent Gaussian traffic; we have distinguished between Brownian-motion input and input with an asymptotically linear variance function. A few remarks are appropriate here.

- In Exercise 11.1.4, we considered the M/G/∞ input process with Pareto jobs. It was argued that for $\alpha > 3$ the input is ALV. For $\alpha \in (2, 3)$ it is true that $v(t)/t$ tends to a constant, but $v(t) - t$ does not. Hence, the process is short-range dependent, but *not* ALV.

To get an impression of the large-buffer behavior for $\alpha \in (2, 3)$, we consider the FIFO queue fed by Gaussian sources with the (somewhat simpler) variance function $v(t) = t + t^\beta$, for $\beta \in (0, 1)$; for ease, take $\mu = 0$. It is readily verified that the optimizing $t = t(b)$ is the inverse of

$$b(t) := \frac{ct + (2 - \beta)ct^\beta}{1 + \beta t^{\beta-1}};$$

for large t, it holds that $b(t) \approx ct + 2(1 - \beta)ct^\beta$, and hence also that, for large b, the optimizing t looks like $b/c - 2(1 - \beta)(b/c)^\beta$. Now, it can be verified that

$$\left(\inf_{t>0} \frac{(b+ct)^2}{2v(t)}\right) \approx 2bc - 2c^{2-\beta}b^{\beta}.$$

We see that a variance function consisting of a linear part as well as a polynomial, sublinear part leads to a decay rate function with a linear and a polynomial, sublinear part. We expect this type of behavior to carry over to the complex buffer architectures considered in this chapter.

- In the GPS setting, we only considered the case of BM input. In the situation with ALV input, we run into technical problems. In the counterpart of Lemma 11.4.1 for ALV sources, the minimum needs to be taken over all $t \geq 0$ and $u \in (0, t)$. Because u can be chosen close to t, we expect that we have to impose regularity conditions on $v(\cdot)$ around 0, to be able to compute the minimum over t and u.

- Zhang [296] also considers behavior of GPS schedulers for short-range dependent traffic (more general than Gaussian, but discrete-time). His assumptions are in line with those in, e.g., Glynn and Whitt [119], and are of the following type.

With $A(t)$ denoting the traffic generated by a single source in an interval of length t, it is assumed that

$$\lim_{t\to\infty} t^{-1} \log \mathbb{E} \exp(\theta A(t)) < \infty \tag{11.10}$$

is finite for positive θ; for Gaussian sources, this would be equivalent to requiring that $v(t)$ is at most linear. Such a framework obviously allows for instance $v(t)/t \to \lambda$. The results obtained are of the type $I_c(b)/b \to \theta^\star$ for $b \to \infty$, where $I_c(b)$ is the decay rate of overflow in the queue under consideration and $\theta^\star$ is a positive constant.

Our requirement in the variance function, i.e., $v(t) - \lambda t \to \kappa$ for ALV sources, is more demanding (in the sense that it implies that $v(\cdot)$ is at most linear), but, in return, we get more precise results: $I_c(b) - \theta^\star b \to \nu$ as $b \to \infty$.

Bibliographical notes

This chapter is based on Mandjes [187]. Recommended reading material on the srd case is Glynn and Whitt [119], where large-buffer asymptotics are considered under what could be called *Gärtner–Ellis* type of conditions (Equation 11.10).

Appendix

Lemma A.1 *Take $A, C \in \mathbb{R}$; $B, D > 0$; $\sigma > 0$; $\tau \in \mathbb{R}$. Then,*

$$\lim_{x\to\infty} \left(\inf_{t\geq 0} \frac{(\sigma t)^2}{A+Bt} + \frac{(x-\tau t)^2}{C+Dt}\right) - \frac{2x}{D}\left(\sqrt{\tau^2 + \frac{D\sigma^2}{B}} - \tau\right)$$

$$= -\frac{\sigma^2}{B}\left(\frac{A}{B} - K_2\right) + \frac{K_2^2}{K_1} \cdot \frac{\tau}{C},$$

with

$$K_1 := \left(\sqrt{\tau^2 + \frac{D\sigma^2}{B}}\right)^{-1}, \quad K_2 := -\frac{C}{D}\left(1 - \frac{\tau}{\sqrt{\tau^2 + D\sigma^2/B}}\right).$$

Proof. First fix x, and differentiate with respect to t to find the following first-order condition:

$$\sigma^2 \cdot \frac{2At + Bt^2}{(A+Bt)^2} + 2\left(\frac{\tau t - x}{C + Dt}\right) - D\left(\frac{\tau t - x}{C + Dt}\right)^2 = 0,$$

which is solved by

$$\frac{\tau t - x}{C + Dt} = \frac{1}{D}\left(\tau + \sqrt{\tau^2 + D\sigma^2 \cdot \frac{2At + Bt^2}{(A+Bt)^2}}\right).$$

We can equivalently express x as function of t:

$$x(t) = -\frac{C}{D} \cdot \tau + \left(\frac{C}{D} + t\right)\sqrt{\tau^2 + D\sigma^2 \cdot \frac{2At + Bt^2}{(A+Bt)^2}}.$$

Now, it is readily checked that

$$\lim_{t\to\infty}\left(x(t) - t\sqrt{\tau^2 + \frac{D\sigma^2}{B}}\right) = \frac{C}{D}\left(\sqrt{\tau^2 + \frac{D\sigma^2}{B}} - \tau\right),$$

and hence $\lim_{x\to\infty} t(x) - K_1 x = K_2$. Inserting these into the objective function yields the stated result. □

Chapter 12

Brownian queues

In the previous chapter we have seen that, for the special case of short-range dependent input, a fairly explicit analysis could be performed. We could find the many-sources asymptotics for the FIFO queue, but also tandem, priority, and GPS systems (and we saw that the corresponding decay rates are linear in the buffer size).

Specializing to the case of Brownian inputs, however, it turns out that even more explicit results can be derived. The buffer content distribution of reflected Brownian motion was already given in Section 5.3, but various ramifications can be found as well. These are given in Section 12.1. In Sections 12.2 and 12.3 we consider the tandem queue with Brownian input, and we subsequently derive the buffer content distribution of the downstream queue, and the joint distribution of the buffer content of the first and second queue.

12.1 Single queue: detailed results

Maximum over a finite horizon. Here, our goal is to compute, for $\text{B}, \text{C}, T > 0$,

$$\alpha(\text{B}, \text{C}, T) = \mathbb{P}\left(\sup_{t\in[0,T]} B(t) - \text{C}t \geq \text{B}\right),$$

with $B(\cdot)$ standard Brownian motion. We first focus on $T = 1$: we give an expression for $\alpha(\text{B}, \text{C}, 1)$. Recall from Section 5.3

$$\beta(\text{B}, \text{C}, a) := \mathbb{P}\left(\sup_{t\geq 0} B(t) - \text{C}t \geq \text{B} \mid B(1) = a\right).$$

Large deviations for Gaussian queues M. Mandjes

Then it is clear that

$$1-\alpha(\text{B},\text{C},1)=\int_{-\infty}^{\text{B}+\text{C}}(1-\beta(\text{B},\text{C},y))\frac{1}{\sqrt{2\pi}}e^{-y^2/2}\,\mathrm{d}y.$$

Applying that $\beta(\text{B},\text{C},y)=\beta(\text{B},\text{C}-y,0)=\exp(-2\text{B}(\text{B}+\text{C}-y))$, direct computation gives, with $\Phi(\cdot)$ the distribution function of a standard normal random variable,

$$\alpha(\text{B},\text{C},1)=1-\Phi(\text{B}+\text{C})+e^{-2\text{BC}}\Phi(-\text{B}+\text{C}).$$

Now we consider the situation of general T. Straightforward scaling arguments yield that $\alpha(\text{B},\text{C},T)=\alpha(\text{B}/\sqrt{T},\text{C}\sqrt{T},1)$, so that

$$\alpha(\text{B},\text{C},T)=1-\Phi\left(\frac{\text{B}}{\sqrt{T}}+\text{C}\sqrt{T}\right)+e^{-2\text{BC}}\Phi\left(-\frac{\text{B}}{\sqrt{T}}+\text{C}\sqrt{T}\right). \tag{12.1}$$

Queuing dynamics. Define

$$\begin{aligned}\gamma(\text{B},\text{B}',\text{C},T)&:=\mathbb{P}(Q(0)\geq\text{B},\,Q(T)\geq\text{B}')\\&=\mathbb{P}\left(\sup_{s\geq0}\{B(-s,0)-\text{C}s\}\geq\text{B},\ \sup_{t\geq0}\{B(T-t,T)-\text{C}t\}\geq\text{B}'\right),\end{aligned}$$

with $B(s,t):=B(t)-B(s)$. It is elementary that

$$\gamma(\text{B},\text{B}',\text{C},T)=1-e^{-2\text{BC}}-e^{-2\text{B}'\text{C}}+\overline{\gamma}(\text{B},\text{B}',\text{C},T),$$

where

$$\overline{\gamma}(\text{B},\text{B}',\text{C},T):=\mathbb{P}\left(\sup_{s\geq0}\{B(-s,0)-\text{C}s\}<\text{B},\ \sup_{t\geq0}\{B(T-t,T)-\text{C}t\}<\text{B}'\right);$$

it turns out that it is slightly easier to analyze this 'complement' $\overline{\gamma}(\text{B},\text{B}',\text{C},T)$ rather than $\gamma(\text{B},\text{B}',\text{C},T)$.

A second simplification can be made. As above, we can, without loss of generality, assume that $T=1$ because

$$\overline{\gamma}(\text{B},\text{B}',\text{C},T)=\overline{\gamma}\left(\frac{\text{B}}{\sqrt{T}},\frac{\text{B}'}{\sqrt{T}},\text{C}\sqrt{T},1\right).$$

To make the expressions more compact, we write $B_{\text{C}}(s,t):=B(s,t)-\text{C}(t-s)$. By conditioning on $B(1)$ we obtain that $\overline{\gamma}(\text{B},\text{B}',\text{C},1)$ equals

$$\int_{-\infty}^{\text{B}'+\text{C}}\mathbb{P}\left(\sup_{s\geq0}B_{\text{C}}(-s,0)\leq\text{B},\sup_{t\geq0}B_{\text{C}}(1-t,1)\leq\text{B}',\,B(1)=x\right)\frac{1}{\sqrt{2\pi}}e^{-x^2/2}\,\mathrm{d}x.$$

Now we split this event into all requirements that relate to negative time, and all requirements that relate to positive time. The probability in the above equation decomposes into

$$\int_{-\infty}^{\mathrm{B}'+\mathrm{C}} f_x(\mathrm{B},\mathrm{B}',\mathrm{C}) g_x(\mathrm{B},\mathrm{B}',\mathrm{C})\,\mathrm{d}x; \tag{12.2}$$

where
$f_x(\mathrm{B},\mathrm{B}',\mathrm{C})$

$$:= \mathbb{P}\left(\sup_{s\geq 0} B_{\mathrm{C}}(-s,0) \leq \mathrm{B}, \sup_{t\geq 1} B_{\mathrm{C}}(1-t,0) \leq \mathrm{B}'+\mathrm{C}-x \mid B(1)=x\right)$$

$$= \mathbb{P}\left(\sup_{s\geq 0} B_{\mathrm{C}}(-s,0) \leq \mathrm{B}, \sup_{t\geq 1} B_{\mathrm{C}}(1-t,0) \leq \mathrm{B}'+\mathrm{C}-x\right)$$

$$= \mathbb{P}\left(\sup_{s\geq 0} B_{\mathrm{C}}(-s,0) \leq \min\{\mathrm{B},\mathrm{B}'+\mathrm{C}-x\}\right)$$

$$= 1-\alpha(\min\{\mathrm{B},\mathrm{B}'+\mathrm{C}-x\},\mathrm{C}) = 1-\exp\left(-2\min\{\mathrm{B},\mathrm{B}'+\mathrm{C}-x\}\mathrm{C}\right),$$

with $\alpha(\mathrm{B},\mathrm{C}) = \mathbb{P}(Q\geq \mathrm{B}) = \exp(-2\mathrm{BC})$, as in Section 5.3, and

$$g_x(\mathrm{B},\mathrm{B}',\mathrm{C}) := \mathbb{P}\left(\sup_{t\in(0,1)} B_{\mathrm{C}}(1-t,1) \leq \mathrm{B}' \mid B(1)=x\right)$$

$$= 1-\beta(\mathrm{B}',\mathrm{C},x) = 1-\exp\left(-2\mathrm{B}'(\mathrm{B}'+\mathrm{C}-x)\right).$$

Now the problem has reduced to inserting the expression for $f_x(\mathrm{B},\mathrm{B}',\mathrm{C})$ and $g_x(\mathrm{B},\mathrm{B}',\mathrm{C})$ into the integral (12.2); we have to distinguish between x smaller than $\mathrm{B}'-\mathrm{B}+\mathrm{C}$, and x between $\mathrm{B}'-\mathrm{B}+\mathrm{C}$ and $\mathrm{B}+\mathrm{C}$. It turns out that, in these integrals, the standard normal cumulative distribution function $\Phi(\cdot)$ can be recognized. After tedious computations, we obtain that $\gamma(\mathrm{B},\mathrm{B}',\mathrm{C},1)$ equals

$$\exp\left(-2\mathrm{B}'\mathrm{C}\right)\cdot\Phi(-\mathrm{B}+\mathrm{B}'-\mathrm{C}) + \exp\left(-2\mathrm{BC}\right)\cdot\Phi(\mathrm{B}-\mathrm{B}'-\mathrm{C}) + \exp\left(-2(\mathrm{B}+\mathrm{B}')\mathrm{C}\right)\cdot\Phi(-\mathrm{B}-\mathrm{B}'+\mathrm{C}) - \Phi(-\mathrm{B}-\mathrm{B}'-\mathrm{C}). \tag{12.3}$$

This expression enables us to compute even the transient distribution of reflected Brownian motion. We find

$$\mathbb{P}(Q(T)\geq \mathrm{B}' \mid Q(0)=\mathrm{B})$$

$$= \Phi\left(\frac{\mathrm{B}}{\sqrt{T}} - \frac{\mathrm{B}'}{\sqrt{T}} - \mathrm{C}\sqrt{T}\right) + \exp\left(-2\mathrm{B}'\mathrm{C}\right)\cdot\Phi\left(-\frac{\mathrm{B}}{\sqrt{T}} - \frac{\mathrm{B}'}{\sqrt{T}} + \mathrm{C}\sqrt{T}\right). \tag{12.4}$$

The latter expression is in line with (1.1) of [1], and p. 49 of [128]. Equation (12.4) has an appealing interpretation

$$\mathbb{P}\left(B(T)\geq -\mathrm{B}+\mathrm{B}'+\mathrm{C}T\right) + \exp\left(-2\mathrm{B}'\mathrm{C}\right)\cdot\mathbb{P}\left(B(T)\geq \mathrm{B}+\mathrm{B}'-\mathrm{C}T\right).$$

With respect to the first term, it is easily verified that $B(T) \geq -\text{B} + \text{B}' + \text{c}T$ is a sufficient condition for $Q(T) \geq \text{B}'$ given $Q(0) = \text{B}$. There are paths, however, along which less than $-\text{B} + \text{B}' + \text{c}T$ is generated (as the buffer becomes empty in between); the second term covers that effect.

Correlation function. We show below that even the correlation function of the buffer content of the Brownian queue can be derived from first principles. Define

$$\vartheta(T) := \frac{\mathbb{C}\text{ov}(Q(0), Q(T))}{\mathbb{V}\text{ar}Q(0)}.$$

Realizing that $\mathbb{C}\text{ov}(Q(0), Q(T)) = \mathbb{E}(Q(0)Q(T)) - 1/(4\text{c}^2)$ and $\mathbb{V}\text{ar}Q(0) = 1/(4\text{c}^2)$, we are left with finding $\mathbb{E}(Q(0)Q(T))$. As that $\mathbb{E}X = \int_0^\infty \mathbb{P}(X \geq x)\,\text{d}x$ for continuous, nonnegative random variables X, we observe that

$$\mathbb{E}(Q(0)Q(T)) = \int_0^\infty \int_0^\infty \gamma(\text{B}, \text{B}', \text{c}, T)\,\text{dB}\,\text{dB}'.$$

Plugging in Equation (12.3), subtracting $1/(4\text{c}^2)$, and dividing by $1/(4\text{c}^2)$, we obtain after tedious calculus,

$$\vartheta(T) = (2 - 2\text{c}^4T^2 - 4\text{c}^2T)(1 - \Phi(\text{c}\sqrt{T})) + (2\text{c}^3T\sqrt{T} + 2\text{c}\sqrt{T})\phi(\text{c}\sqrt{T}),$$

with $\phi(\cdot)$ the density function of a standard normal random variable. Notice that indeed $\vartheta(0) = 1$, as desired.

Also the asymptotics of $\vartheta(\cdot)$ for large T can be found. It is readily checked that

$$\lim_{T\to\infty} \vartheta(T) \bigg/ \left(\frac{16}{\text{c}^3T\sqrt{T}}\phi(\text{c}\sqrt{T})\right) = 1,$$

where it is used that

$$\lim_{x\to\infty} \frac{1 - \Phi(g(x))}{\phi(g(x))} \cdot \left(\frac{1}{g(x)} - \frac{1}{(g(x))^3} + \frac{3}{(g(x))^5} - \frac{15}{(g(x))^7}\right) = 1,$$

if $g(x) \to \infty$ as $x \to \infty$. In other words, $\vartheta(T)$ decays exponentially in T, with decay rate $\text{c}^2/2$. For detailed results on the transient of the Brownian queue, see [1, 2].

12.2 Tandem: distribution of the downstream queue

Consider the following setting: let a (standard) Brownian motion feed into a tandem network, with service speeds c_1 and c_2 (where obviously $\text{c}_1 > \text{c}_2$). The goal of this section is to find an expression for $\mathbb{P}(Q_2 \geq \text{B})$. The minimum time it takes to exceed level B in the second queue (starting empty) is $d\text{B}$, with $d := (\text{c}_1 - \text{c}_2)^{-1}$.

Interestingly, the following result holds for any input process with stationary independent increments, and hence in particular for Brownian motions. The proof can be found in [66].

Theorem 12.2.1 *Let $\{A(t), t \in \mathbb{R}\}$ be a stochastic process with stationary independent increments and let $\mu = \mathbb{E}A(1) < \mathrm{c}_2$. Then for each $\mathrm{B} \geq 0$, and $A_1(\cdot)$ and $A_2(\cdot)$ independent copies of the process $A(\cdot)$,*

$$\mathbb{P}(Q_2 > \mathrm{B}) = \mathbb{P}\left(\sup_{t\in[0,\infty)}\{A_1(t) - \mathrm{c}_2 t\} > \sup_{t\in[0,d\mathrm{B}]}\{-A_2(t) + \mathrm{c}_1 t\}\right). \tag{12.5}$$

We can now analyze representation (12.5) for the case where $A(\cdot) = B(\cdot)$, a standard Brownian motion (with zero drift). It appears that an explicit formula can be derived for $\mathbb{P}(Q_2 \geq \mathrm{B})$; the proof resembles the key argument given in [186]. As before, let $\Psi(\cdot)$ denote the tail distribution of a standard normal random variable.

Theorem 12.2.2 *Let $A(\cdot)$ be a standard Brownian motion. Then, for each $\mathrm{B} \geq 0$,*

$$\mathbb{P}(Q_2 \geq \mathrm{B}) = \frac{\mathrm{c}_1 - 2\mathrm{c}_2}{\mathrm{c}_1 - \mathrm{c}_2} e^{-2\mathrm{c}_2\mathrm{B}} \left(1 - \Psi\left(\frac{\mathrm{c}_1 - 2\mathrm{c}_2}{\sqrt{\mathrm{c}_1 - \mathrm{c}_2}}\sqrt{\mathrm{B}}\right)\right)$$
$$+ \frac{\mathrm{c}_1}{\mathrm{c}_1 - \mathrm{c}_2}\Psi\left(\frac{\mathrm{c}_1}{\sqrt{\mathrm{c}_1 - \mathrm{c}_2}}\sqrt{\mathrm{B}}\right).$$

Proof. Following Theorem 12.2.1 we have

$$\mathbb{P}(Q_2 > \mathrm{B}) = \mathbb{P}\left(\sup_{t\in[0,\infty)}\{B_1(t) - \mathrm{c}_2 t\} > \sup_{t\in(0,d\mathrm{B}]}\{B_2(t) + \mathrm{c}_1 t\}\right),$$

where $B_1(\cdot)$, $B_2(\cdot)$ are independent standard Brownian motions. Now use that, for each $x \geq 0$,

$$\mathbb{P}\left(\sup_{t\in[0,\infty)}\{B_1(t) - \mathrm{c}_2 t\} > x\right) = e^{-2\mathrm{c}_2 x}$$

and, see e.g., [22], for any $D > 0$,

$$\rho(x) := \frac{\mathrm{d}}{\mathrm{d}x}\mathbb{P}\left(\sup_{t\in(0,D]}\{B_2(t) + \mathrm{c}_1 t\} \leq x\right)$$
$$= \sqrt{\frac{2}{\pi D}}\exp\left(-\frac{(x - \mathrm{c}_1 D)^2}{2D}\right) - 2\mathrm{c}_1 e^{2\mathrm{c}_1 x}\Psi\left(\frac{x + \mathrm{c}_1 D}{\sqrt{D}}\right);$$

the latter formula follows also directly from Equation (12.1). We have obtained two integrals that we calculate separately:

$$\mathbb{P}(Q_2 > \text{B}) = \int_0^\infty e^{-2\text{C}_2 x} \rho(x)\,\mathrm{d}x = \mathcal{I}_1(\text{B}) - \mathcal{I}_2(\text{B}),$$

where

$$\mathcal{I}_1(\text{B}) := \int_0^\infty e^{-2\text{C}_2 x} \sqrt{\frac{2}{\pi d\text{B}}} \exp\left(-\frac{(x - \text{C}_1 d\text{B})^2}{2d\text{B}}\right) \mathrm{d}x;$$

$$\mathcal{I}_2(\text{B}) := \int_0^\infty e^{-2\text{C}_2 x} 2\text{C}_1 e^{2\text{C}_1 x} \Psi\left(\frac{x + \text{C}_1 d\text{B}}{\sqrt{d\text{B}}}\right) \mathrm{d}x.$$

Integral $\mathcal{I}_1(\text{B})$ is evaluated as follows:

$$\begin{aligned}
\mathcal{I}_1(\text{B}) &= 2\int_0^\infty \frac{1}{\sqrt{2\pi d\text{B}}} \exp\left(-\frac{x^2 - 2\text{C}_1 d\text{B}x + \text{C}_1^2 d\text{B}^2 + 4\text{C}_2 x d\text{B}}{2d\text{B}}\right) \mathrm{d}x \\
&= 2\mathrm{e}^{-\text{C}_2(\text{C}_1 - \text{C}_2)d\text{B}} \int_0^\infty \frac{1}{\sqrt{2\pi d\text{B}}} \exp\left(-\frac{(x - (\text{C}_1 - 2\text{C}_2)d\text{B})^2}{2d\text{B}}\right) \mathrm{d}x \\
&= 2\mathrm{e}^{-2\text{C}_2\text{B}} \left(1 - \Psi\left(\frac{\text{C}_1 - 2\text{C}_2}{\sqrt{\text{C}_1 - \text{C}_2}} \sqrt{\text{B}}\right)\right).
\end{aligned} \tag{12.6}$$

For $\mathcal{I}_2(\text{B})$ we find

$$\begin{aligned}
\mathcal{I}_2(\text{B}) &= \int_0^\infty 2\text{C}_1 e^{2x(\text{C}_1 - \text{C}_2)} \Psi\left(\frac{x + \text{C}_1 d\text{B}}{\sqrt{d\text{B}}}\right) \mathrm{d}x \\
&= \frac{\text{C}_1}{\text{C}_1 - \text{C}_2} \int_0^\infty 2(\text{C}_1 - \text{C}_2) e^{2x(\text{C}_1 - \text{C}_2)} \Psi\left(\frac{x + \text{C}_1 d\text{B}}{\sqrt{d\text{B}}}\right) \mathrm{d}x.
\end{aligned}$$

Now, integration by parts gives that $\mathcal{I}_2(\text{B}) = \text{C}_1 \tilde{\mathcal{I}}_2(\text{B})/(\text{C}_1 - \text{C}_2)$, with

$$\begin{aligned}
\tilde{\mathcal{I}}_2(\text{B}) &= -\Psi(\text{C}_1\sqrt{d\text{B}}) + \int_0^\infty e^{2x(\text{C}_1 - \text{C}_2)} \frac{1}{\sqrt{2\pi d\text{B}}} \exp\left(-\frac{(x + \text{C}_1 d\text{B})^2}{2d\text{B}}\right) \mathrm{d}x \\
&= -\Psi(\text{C}_1\sqrt{d\text{B}}) + \exp\left(-\frac{\text{C}_1^2 d\text{B}}{2} + \frac{(\text{C}_1 - 2\text{C}_2)^2 d\text{B}}{2}\right) \\
&\quad \times \int_0^\infty \frac{1}{\sqrt{2\pi d\text{B}}} \exp\left(-\frac{x^2 - 2xd\text{B}(\text{C}_1 - 2\text{C}_2) + (\text{C}_1 - 2\text{C}_2)^2 d\text{B}^2}{2d\text{B}}\right) \mathrm{d}x \\
&= -\Psi(\text{C}_1\sqrt{d\text{B}}) + \mathrm{e}^{-\text{C}_2(2\text{C}_1 - 2\text{C}_2)d\text{B}}
\end{aligned}$$

$$\times \int_0^\infty \frac{1}{\sqrt{2\pi d\mathrm{B}}} \exp\left(-\frac{(x-(\mathrm{C}_1-2\mathrm{C}_2)d\mathrm{B})^2}{2d\mathrm{B}}\right) \mathrm{d}x$$

$$= -\Psi(\mathrm{C}_1\sqrt{d\mathrm{B}}) + \mathrm{e}^{-2\mathrm{C}_2\mathrm{B}}\left(1-\Psi\left(\frac{\mathrm{C}_1-2\mathrm{C}_2}{\sqrt{\mathrm{C}_1-\mathrm{C}_2}}\sqrt{\mathrm{B}}\right)\right). \tag{12.7}$$

Combining Equations (12.6) and (12.7) yields the desired result. □

The next corollary finds the exact asymptotics of the buffer content distribution in the second node, by applying the general result of Theorem 12.2.2.

Corollary 12.2.3 *Let $A(\cdot)$ be a standard Brownian motion.*

(i) If $\mathrm{C}_1 > 2\mathrm{C}_2$, then

$$\mathbb{P}(Q_2 > \mathrm{B}) = \frac{\mathrm{C}_1-2\mathrm{C}_2}{\mathrm{C}_1-\mathrm{C}_2} e^{-2\mathrm{C}_2\mathrm{B}}(1+o(1)) \quad \text{as } \mathrm{B}\to\infty.$$

(ii) If $\mathrm{C}_1 = 2\mathrm{C}_2$, then

$$\mathbb{P}(Q_2 > \mathrm{B}) = \frac{1}{\sqrt{2\pi}\,\mathrm{C}_2}\frac{1}{\sqrt{\mathrm{B}}} e^{-2\mathrm{C}_2\mathrm{B}}(1+o(1)) \quad \text{as } \mathrm{B}\to\infty.$$

(iii) If $\mathrm{C}_1 < 2\mathrm{C}_2$, then

$$\mathbb{P}(Q_2 > \mathrm{B}) = \frac{1}{\sqrt{2\pi}}\left(\frac{\mathrm{C}_1-\mathrm{C}_2}{\mathrm{B}}\right)^{3/2} \frac{4\mathrm{C}_2}{\mathrm{C}_1^2(\mathrm{C}_1-2\mathrm{C}_2)^2} \exp\left(-\frac{\mathrm{C}_1^2}{2(\mathrm{C}_1-\mathrm{C}_2)}\mathrm{B}\right)(1+o(1))$$

as $\mathrm{B}\to\infty$.

Proof. Let $\mathcal{J}_1(\mathrm{B})$ and $\mathcal{J}_2(\mathrm{B})$ be defined as

$$\mathcal{J}_1(\mathrm{B}) := \frac{\mathrm{C}_1-2\mathrm{C}_2}{\mathrm{C}_1-\mathrm{C}_2} e^{-2\mathrm{C}_2\mathrm{B}}\left(1-\Psi\left(\frac{\mathrm{C}_1-2\mathrm{C}_2}{\sqrt{\mathrm{C}_1-\mathrm{C}_2}}\sqrt{\mathrm{B}}\right)\right)$$

and

$$\mathcal{J}_2(\mathrm{B}) := \frac{\mathrm{C}_1}{\mathrm{C}_1-\mathrm{C}_2}\Psi\left(\frac{\mathrm{C}_1}{\sqrt{\mathrm{C}_1-\mathrm{C}_2}}\sqrt{\mathrm{B}}\right);$$

recall from Theorem 12.2.2 that $\mathbb{P}(Q_2 > \mathrm{B}) = \mathcal{J}_1(\mathrm{B}) + \mathcal{J}_2(\mathrm{B})$. We also recall the standard 'first order' asymptotic result

$$\Psi(u) = \frac{1}{\sqrt{2\pi}}\frac{1}{u} e^{-u^2/2}(1+o(1)), \tag{12.8}$$

and 'second order' result

$$\frac{1}{\sqrt{2\pi}}\frac{1}{u}e^{-u^2/2} - \Psi(u) = \frac{1}{\sqrt{2\pi}}\frac{1}{u^3}e^{-u^2/2}(1+o(1)), \tag{12.9}$$

$u \to \infty$. We now consider the three cases separately.

(i) $c_1 > 2c_2$. Then

$$\lim_{B\to\infty}\frac{\log \mathcal{J}_1(B)}{\log \mathcal{J}_2(B)} = \frac{c_1^2}{4c_2(c_1-c_2)} = \frac{(c_1/c_2)^2}{4(c_1/c_2-1)} > 1,$$

due to $x^2 > 4(x-1) > 0$ for $x > 2$. The statement follows after applying Equation (12.8).

(ii) $c_1 = 2c_2$. Then $\mathcal{J}_1(B) = 0$, and the statement follows from Equation (12.8).

(iii) $c_1 < 2c_2$. In this case, remarkably, both $-\mathcal{J}_1(B)$ and $\mathcal{J}_2(B)$ equal (asymptotically)

$$\sqrt{\frac{1}{c_1-c_2}\frac{1}{\sqrt{2\pi}}\frac{1}{\sqrt{B}}\exp\left(-\frac{c_1}{2(c_1-c_2)}B\right)}(1+o(1)),$$

as $B \to \infty$, due to Equation (12.8). As a consequence, we have to rely on the more precise asymptotics (12.9). After tedious computations we derive the stated. □

We see that there are two regimes: with $c_1^\star := 2c_2$, there is an essentially different behavior for $c_1 > c_1^\star$ and $c_1 \le c_1^\star$; cf. the many-sources results of Sections 9.2 and 11.2.

12.3 Tandem: joint distribution

Interestingly, by another argument than the one used in the previous section, even an explicit formula of the joint distribution

$$q(B) := \mathbb{P}(Q_1 \ge B_1, Q_2 \ge B_2)$$

can be given. It turns out to be practical to first consider the probability

$$p(B) := \mathbb{P}(Q_1 \ge B_1, Q_1 + Q_2 \ge B_T).$$

For the sake of brevity, write $\chi \equiv \chi(B) := (B_T - B_1)/(c_1 - c_2)$. Furthermore, let $\phi(\cdot)$ denote the probability density function of a standard Normal random variable, as earlier.

Theorem 12.3.1 *For each* $\mathrm{B_T} \geq \mathrm{B_I} \geq 0$,

$$p(\mathrm{B}) = -\Psi(k_1(\mathrm{B})) + \Psi(k_2(\mathrm{B}))e^{-2\mathrm{B_I}} + \Psi(k_3(\mathrm{B}))e^{-2\mathrm{B_T}\mathrm{C_2}} + (1 - \Psi(k_4(\mathrm{B})))e^{-2(\mathrm{B_I}(\mathrm{C_1}-2\mathrm{C_2})+\mathrm{B_T}\mathrm{C_2})},$$

where

$$k_1(\mathrm{B}) := \frac{\mathrm{B_I} + \mathrm{C_1}\chi}{\sqrt{\chi}}; \quad \mathrm{k_2}(\mathrm{B}) := \frac{-\mathrm{B_I} + \mathrm{C_1}\chi}{\sqrt{\chi}};$$

$$k_3(\mathrm{B}) := \frac{\mathrm{B_I} + (\mathrm{C_1} - 2\mathrm{C_2})\chi}{\sqrt{\chi}}; \quad \mathrm{k_4}(\mathrm{B}) := \frac{-\mathrm{B_I} + (\mathrm{C_1} - 2\mathrm{C_2})\chi}{\sqrt{\chi}}.$$

Proof. First note that, due to time-reversibility arguments,

$$\mathbb{P}(Q_1 \leq \mathrm{B_I}, Q_1 + Q_2 \leq \mathrm{B_T}) = \mathbb{P}(\forall t \geq 0 : B(t) \leq \min\{\mathrm{B_I} + \mathrm{C_1}t, \mathrm{B_T} + \mathrm{C_2}t\}).$$

Let $y \equiv y(\mathrm{B}) := \mathrm{B_I} + \mathrm{C_1}\chi$. Hence, (χ, y) is the point where $\mathrm{B_I} + \mathrm{C_1}t$ and $\mathrm{B_T} + \mathrm{C_2}t$ intersect. For $t \in [0, \chi]$ the minimum is given by $\mathrm{B_I} + \mathrm{C_1}\chi$, whereas for $t \in [\chi, \infty)$ the minimum is $\mathrm{B_T} + \mathrm{C_2}t$. Now, conditioning on the value of $B(\chi)$, being normally distributed with mean 0 and variance χ, we get that $\mathbb{P}(Q_1 \leq \mathrm{B_I}, Q_1 + Q_2 \leq \mathrm{B_T})$ equals

$$\int_{-\infty}^{y(\mathrm{B})} \frac{1}{\sqrt{\chi(\mathrm{B})}}\phi\left(\frac{x}{\sqrt{\chi(\mathrm{B})}}\right)\eta(\mathrm{B})\vartheta(\mathrm{B})\,\mathrm{d}x.$$

with

$$\eta(\mathrm{B}) := \mathbb{P}(\forall t \in [0, \chi] : B(t) \leq \mathrm{B_I} + \mathrm{C_1}t | B(\chi) = x);$$

$$\vartheta(\mathrm{B}) := \mathbb{P}(\forall t \geq 0 : B(t) \leq y - x + \mathrm{C_2}t).$$

The first probability can be expressed (after some straightforward rescaling) in terms of the Brownian bridge: for $x < y(\mathrm{B})$,

$$\eta(\mathrm{B}) = \mathbb{P}\left(\forall t \in [0, 1] : B(t) \leq \frac{\mathrm{B_I}}{\sqrt{\chi}} + \left(\mathrm{C_1}\sqrt{\chi} - \frac{x}{\sqrt{\chi}}\right)t \mid B(1) = 0)\right)$$
$$= 1 - \exp\left(-2\mathrm{B_I}\left(\mathrm{C_1} - \frac{x}{\chi}\right)\right),$$

whereas the second translates into the supremum of a Brownian motion:

$$\vartheta(\mathrm{B}) = 1 - \exp(2(y - x)\mathrm{C_2}).$$

After substantial calculus, we obtain that $\mathbb{P}(Q_1 \leq \mathrm{B_I}, Q_1 + Q_2 \leq \mathrm{B_T})$ equals

$$\Phi(k_1(\mathrm{B})) - \Phi(k_2(\mathrm{B}))e^{-2\mathrm{B_I}\mathrm{C_1}} - \Phi(k_3(\mathrm{B}))e^{-2\mathrm{B_T}\mathrm{C_2}} + \Phi(k_4(\mathrm{B}))e^{-2(\mathrm{B_I}(\mathrm{C_1}-2\mathrm{C_2})+\mathrm{B_T}\mathrm{C_2})}.$$

The statement now follows from

$$p(\mathrm{B}) = 1 - \mathbb{P}(Q_1 \leq \mathrm{B}_1) - \mathbb{P}(Q_1 + Q_2 \leq \mathrm{B_T}) + \mathbb{P}(Q_1 \leq \mathrm{B}_1, Q_1 + Q_2 \leq \mathrm{B_T}),$$

also using that $\mathbb{P}(Q_1 \geq \mathrm{B}) = e^{-2\mathrm{BC}_1}$ and $\mathbb{P}(Q_1 + Q_2 \geq \mathrm{B}) = e^{-2\mathrm{BC}_2}$. □

It is clear that we can translate this result on $p(\mathrm{B})$ into a result on $q(\mathrm{B})$:

$$\mathbb{P}(Q_1 \leq \mathrm{B}_1, Q_2 \leq \mathrm{B}_2) = \int_0^{\mathrm{B}_1} \frac{\partial}{\partial x_1} \mathbb{P}(Q_1 \leq x_1, Q_1 + Q_2 \leq x_\mathrm{T}) \Big|_{x_1 := x;\, x_\mathrm{T} := \mathrm{B}_2 + x} \mathrm{d}x.$$

The calculations are given explicitly in [175]; here we give only the result. To this end, define $\tau \equiv \tau(\mathrm{B}) = d\mathrm{B}_2$, and

$$h_1(\mathrm{B}) := \frac{\mathrm{B}_1 + \mathrm{C}_1}{\sqrt{\tau}}, \quad h_2(\mathrm{B}) := \frac{-\mathrm{B}_1 + \mathrm{C}_1\tau}{\sqrt{\tau}}, \quad h_3(\mathrm{B}) := \frac{-\mathrm{B}_1 + (\mathrm{C}_1 - 2\mathrm{C}_2)\tau}{\sqrt{\tau}}.$$

Theorem 12.3.2 *For* $\mathrm{B}_1, \mathrm{B}_2 \geq 0$,

$$q(\mathrm{B}) = \left(\frac{\mathrm{C}_2}{\mathrm{C}_1 - \mathrm{C}_2} \Psi(h_1(\mathrm{B}))\right) + \Psi(h_2(\mathrm{B}))e^{-2\mathrm{B}_1\mathrm{C}_1} + \left(\frac{\mathrm{C}_1 - 2\mathrm{C}_2}{\mathrm{C}_1 - \mathrm{C}_2} \cdot (1 - \Psi(h_3(\mathrm{B}))) \cdot e^{-2(\mathrm{B}_1(\mathrm{C}_1 - \mathrm{C}_2) + \mathrm{B}_2\mathrm{C}_2)}\right).$$

Inserting $\mathrm{B}_2 = 0$, we indeed find $\exp(-2\mathrm{B}_1\mathrm{C}_1)$, as desired. Also, taking $\mathrm{B}_1 = 0$, we retrieve the formula for the downstream queue, as derived in the previous section.

Remark 12.3.3 We assumed that the input process was a standard Brownian motion, i.e., no drift and $v(t) = t$. It can be verified, however, that the results extend to general Brownian input, which have drift $\mu > 0$ and variance $v(t) = \lambda t$, $\lambda > 0$. Requiring that $\mathrm{C}_1 > \mathrm{C}_2 > \mu > 0$ (thus ensuring stability), it turns out that we have to set $\mathrm{C}_i \leftarrow (\mathrm{C}_i - \mu)/\sqrt{\lambda}$ and $\mathrm{B}_i \leftarrow \mathrm{B}_i/\sqrt{\lambda}$, $i = 1, 2$.

Large buffer asymptotics. In this section we take $\mathrm{B}_1 := \alpha\mathrm{B}$ and $\mathrm{B}_2 := (1 - \alpha)\mathrm{B}$ for some $\alpha \in (0, 1)$, and then we consider the asymptotics for large B. In other words, we investigate

$$q_a(\mathrm{B}) := \mathbb{P}(Q_1 \geq \alpha\mathrm{B}, Q_2 \geq (1 - \alpha)\mathrm{B})$$

for $\mathrm{B} \to \infty$. (We do not treat $\alpha = 0, 1$ here, as these correspond to the marginal distributions of the first and second queue, and these we already know.)

Define

$$\alpha_+ := \frac{c_1}{2c_1 - c_2}; \quad \alpha_- := \frac{c_1 - 2c_2}{2c_1 - 3c_2}.$$

It can be verified that $0 < \alpha_- < \alpha_+ < 1$ if $c_1 > 2c_2$, and $0 < \alpha_+ < 1$ if $c_1 \leq 2c_2$. We use the notation $\zeta(x) := \exp(-x^2/2)/(\sqrt{2\pi}x)$. A useful, and rather straightforward, lemma is the following. It uses the shorthand notation $f(b) \sim g(b)$ as $b \to \infty$ to express that $\lim_{b\to\infty} f(b)/g(b) \to 1$. Clearly, $\Psi(x) \sim \zeta(x)$ as $x \to \infty$, as we have already seen several times in this book.

Lemma 12.3.4 *Let* $B_1 = \alpha B$ *and* $B_2 = (1-\alpha)B$, *with* $\alpha \in (0, 1)$. *As* $B \to \infty$,

$$\Psi(h_1(B)) \sim \zeta(h_1(B));$$

$$\Psi(h_2(B)) \sim \begin{cases} \zeta(h_2(B)) & \text{if } \alpha < \alpha_+; \\ 1/2 & \text{if } \alpha = \alpha_+; \\ 1 & \textit{otherwise}; \end{cases}$$

$$1 - \Psi(h_3(B)) \sim \begin{cases} 1 & \text{if } \alpha < \alpha_- \text{ and } c_1 > 2c_2; \\ 1/2 & \text{if } \alpha = \alpha_- \text{ and } c_1 \geq 2c_2; \\ -\zeta(h_3(B)) & \textit{otherwise}. \end{cases}$$

Now define the following two quantities:

$$\gamma(B) := \frac{1}{\sqrt{2\pi}}\left(\frac{c_2}{c_1 - c_2}\frac{1}{h_1(B)} + \frac{1}{h_2(B)} - \frac{c_1 - 2c_2}{c_1 - c_2}\frac{1}{h_3(B)}\right);$$

$$\delta(B) := \frac{(B_1(c_1 - c_2) + B_2 c_1)^2}{2B_2(c_1 - c_2)}.$$

The next theorem shows the dichotomy that we have seen several times earlier: there is a crucially different behavior for $c_1 > 2c_2$ and $c_1 \leq 2c_2$. Interestingly, the exponential decay rate is continuous in α. It follows directly by applying Lemma 12.3.4.

Theorem 12.3.5 *Suppose that* $c_1 > 2c_2$. *As* $B \to \infty$,

$$q_\alpha(B) \sim \frac{c_1 - 2c_2}{c_1 - c_2}e^{-2(B_1(c_1-c_2)+B_2c_2)}, \quad \text{if } \alpha \in (0, \alpha_-);$$

$$q_\alpha(B) \sim \frac{1}{2}\frac{c_1 - 2c_2}{c_1 - c_2}e^{-2(B_1(c_1-c_2)+B_2c_2)}, \quad \text{if } \alpha = \alpha_-;$$

$$q_\alpha(B) \sim \gamma(B)e^{-\delta(B)}, \quad \text{if } \alpha \in (\alpha_-, \alpha_+);$$

$$q_\alpha(B) \sim \frac{1}{2}e^{-2B_1c_1}, \quad \text{if } \alpha = \alpha_+;$$

$$q_\alpha(B) \sim e^{-2B_1c_1}, \quad \text{if } \alpha \in (\alpha_+, 1).$$

Suppose that $\text{c}_1 \leq 2\text{c}_2$. *As* $\text{B} \to \infty$,

$$q_\alpha(\text{B}) \sim \gamma(\text{B})e^{-\delta(\text{B})}, \quad \text{if } \alpha \in (0, \alpha_+);$$

$$q_\alpha(\text{B}) \sim \frac{1}{2}e^{-2\text{B}_1\text{C}_1}, \quad \text{if } \alpha = \alpha_+;$$

$$q_\alpha(\text{B}) \sim e^{-2\text{B}_1\text{C}_1}, \quad \text{if } \alpha \in (\alpha_+, 1).$$

Interpretation of different regimes by applying sample-path large deviations. Theorem 12.3.5 shows a broad variety of regimes: the asymptotic behavior of $q_\alpha(\text{B})$ critically depends on whether or not $\text{c}_1 > 2\text{c}_2$, and the value of α. Interestingly, these regimes can be explained by applying sample-path large deviations ('Schilder'), as shown below.

To this end, we apply the many-sources scaling: we feed n i.i.d. standard Brownian sources into the tandem system, and also scale the link rates and buffer thresholds by n: nc_1, nc_2, nb_1, and nb_2, respectively. Define $\mathcal{U}$ as

$$\left\{ f \in \Omega : \exists t \geq 0, s \in (0,t) : \forall u \in [0,s] : \begin{array}{l} A[f](-s,0) \geq b_1 + c_1 s \\ A[f](-t,-u) \geq b_2 + c_2 t - c_1 u \end{array} \right\}.$$

Lemma 12.3.6 *With* $B_i(\cdot)$, *for* $i = 1, \ldots, n$, *i.i.d. standard Brownian motions,*

$$\mathbb{P}(Q_{1,n} \geq nb_1, Q_{2,n} \geq nb_2) = p_n[\mathcal{U}] \equiv \mathbb{P}\left(\frac{1}{n}\sum_{i=1}^{n} B_i(\cdot) \in \mathcal{U}\right)$$

Proof. It is clear that $\{Q_{1,n} \geq nb_1\}$ corresponds to the collection of paths such that $A[f](-s,0) \geq b_1 + c_1 s$ for some $s \geq 0$. Also, $\{Q_{2,n} \geq nb_2\}$ reads

$$\begin{aligned} &\{(Q_{1,n} + Q_{2,n}) - Q_{1,n} \geq nb_2\} \\ &= \left\{ \left(\sup_{t \geq 0} \sum_{i=1}^{n} B_i(-t,0) - nc_2 t \right) - \left(\sup_{u \geq 0} \sum_{i=1}^{n} B_i(-u,0) - nc_1 u \right) \geq nb_2 \right\} \\ &= \left\{ \exists t \geq 0 : \forall u \geq 0 : \sum_{i=1}^{n} B_i(-t,-u) \geq nb_2 + nc_2 t - nc_i u \right\}. \end{aligned}$$

The restrictions on the ranges of t, s, u follow from the usual considerations regarding the starting epochs of the corresponding busy periods. □

The previous lemma entails that 'Schilder' can be applied, to get

$$I_c^{(\text{t}),2}(b) := -\lim_{n \to \infty} \frac{1}{n} \log \mathbb{P}(Q_{1,n} \geq nb_1, Q_{2,n} \geq nb_2) = \inf_{f \in \mathcal{U}} \mathbb{I}(f).$$

The infimum of $\mathbb{I}(f)$ over $f \in \mathcal{U}$ can be evaluated as before – it turns out to be a tedious (though relatively straightforward) procedure. We arrive at the following result.

Proposition 12.3.7 *Suppose $c_1 > 2c_2$. Then it holds that*

$$I_c^{(t),2}(b) = \begin{cases} 2(b_1(c_1 - c_2) + b_2c_2) & \text{if } b_1/(b_1 + b_2) \in (0, \alpha_-]; \\ \delta(b) & \text{if } b_1/(b_1 + b_2) \in (\alpha_-, \alpha_+); \\ 2b_1c_1 & \text{if } b_1/(b_1 + b_2) \in [\alpha_+, 1). \end{cases}$$

Suppose $c_1 \le 2c_2$. Then it holds that

$$I_c^{(t),2}(b) = \begin{cases} \delta(b) & \text{if } b_1/(b_1 + b_2) \in (0, \alpha_+); \\ 2b_1c_1 & \text{if } b_1/(b_1 + b_2) \in [\alpha_+, 1). \end{cases}$$

The optimal paths have interesting shapes, as displayed in Figure 12.1. The three shapes correspond to the three cases that occur in the case $c_1 > 2c_2$.

In the first regime, traffic is first generated at a rate $2c_2$, such that queue 2 starts to build up (at rate c_2) but queue 1 does not (as $2c_2 < c_1$). Then the input rate changes to $2c_1 - 2c_2$, such that both queues build up; queue 1 grows at rate

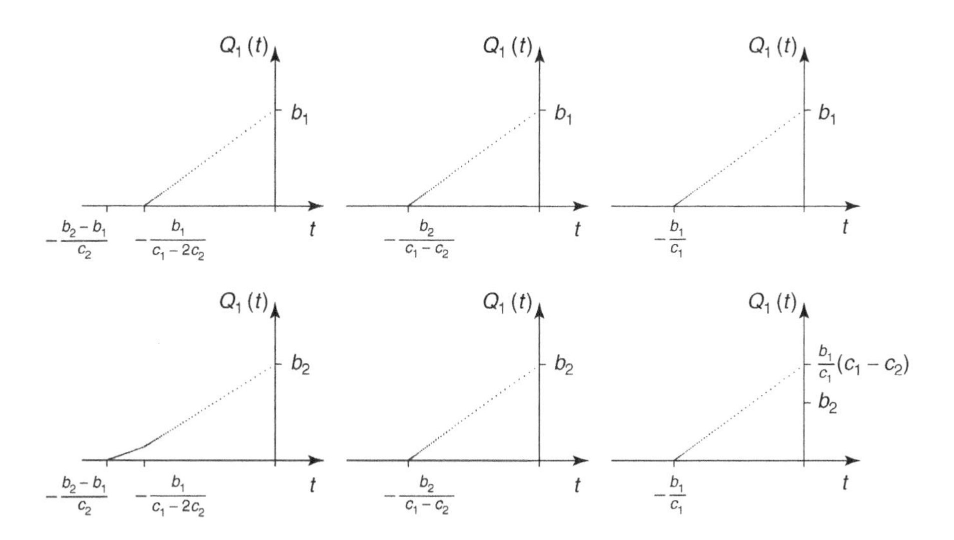

Figure 12.1: The most likely storage path in $\{Q_1 \ge b_1, Q_2 \ge b_2\}$ for $c_1 > 2c_2$; as stated in the first part of Proposition 12.3.7 three different regimes can be distinguished: $b_1/(b_1 + b_2)$ (left) in $(0, \alpha_-]$, (middle) in (α_-, α_+), (right) in $[\alpha_-, 1)$. In case $c_1 \le 2c_2$, the most likely storage path corresponding to $b_1/(b_1 + b_2) \in (0, \alpha_+)$ is as in the second picture, and the path corresponding to $b_1/(b_1 + b_2) \in [\alpha_+, 1)$ is as in the third picture.

$c_1 - 2c_2 > 0$, such that its output rate is c_1, and hence queue 2 grows at rate $c_1 - c_2$. At time 0, the queues reach exactly their 'target values' b_1 and b_2.

In the second regime, traffic is generated at a rate $(b_1/b_2)(c_1 - c_2) + c_1$ all the time. As this is larger than c_1, this means that queue 2 grows at rate $c_1 - c_2$. At time 0, both queues reach their 'target value'.

In the third regime, the input rate is $2c_1$. At time 0, queue 1 reached its 'target value' b_1; queue 2, however, exceeds b_2. Apparently, the most likely path is such that there is some 'surplus' in queue 2.

Bibliographical notes

Most of the material of this chapter was taken from Dębicki, Mandjes, and Van Uitert [66] and Lieshout and Mandjes [175]. As remarked, Brownian motions have independent stationary increments, and hence fall in the class of *Lévy processes*. Therefore we can use various general results from this powerful theory – see for instance [10, 30, 246, 262].

Networks of queues fed by Lévy input are studied in detail by Kella and Whitt [146, 147], but their results (on both the marginal and joint distributions) are in terms of Laplace transforms rather than explicit expressions for the probabilities of interest – for related results, see also Dębicki, Dieker, and Rolski [62].

Part C: Applications

The remaining chapters illustrate how Gaussian queues can be applied in a number of relevant networking examples. It turns out that many of the results derived in Part B are immediately applicable for traffic management and dimensioning purposes.

The first chapter studies the weight-setting problem that arises in generalized processor sharing. It turns out that the, rather explicit, formulae derived in Chapter 10 can be applied directly to find 'optimal weights'.

The next two chapters are about link dimensioning, i.e., the selection of an appropriate value for the link rate such that some prespecified performance criterion is met. Link dimensioning procedures require accurate estimation of the Gaussian input process; in Chapter 15, explanation is provided on how these can be obtained by sampling the buffer content.

The last chapter focuses on bandwidth trading. Often Internet Service Providers (ISPs) are confronted with fluctuating demand from their customers, and therefore it makes sense to buy bandwidth when demand exceeds supply, and to sell bandwidth when supply exceeds demand. This type of 'bandwidth trading' requires, in the first place, an effective mechanism to exchange information about prices and quantities between the ISPs and the 'network owner'. In the second place, the ISPs should be able to estimate how much bandwidth they need to meet the performance requirements of their clients; we show how large-deviations techniques can help here.

Large deviations for Gaussian queues M. Mandjes

Chapter 13

Weight setting in GPS

When selecting appropriate values for the weights ϕ in GPS, as introduced in Chapter 10, it is clear that various objectives could be chosen. In this section, we investigate two such approaches. Following [94], it could be assumed that, for practical reasons, it should be avoided to switch between a large number of different weights – in fact, it could be required that just one set of weights ϕ be used. With any ϕ an *admissible region* $\mathcal{A}(\phi)$ can be associated, i.e., all combinations of sources of both classes such that the required performance is realized. Obviously, the size and shape of $\mathcal{A}(\phi)$ depends critically on the weights ϕ chosen.

The other extreme would be to allow continuous adaptation of ϕ. The resulting admissible region, also called the *realizable region*, say $\mathcal{A}$, contains all the $\mathcal{A}(\phi)$ – in fact $\mathcal{A} = \cup_{\phi}\mathcal{A}(\phi)$. An interesting question is whether there is a value of ϕ for which $\mathcal{A}(\phi)$ is just 'marginally smaller' than $\mathcal{A}$; selecting this weight would somehow maximize the admissible region.

This chapter focuses on the issues raised above. Sections 13.1 and 13.3 relate to the situation that we are asked to find a vector of suitable weights ϕ, i.e., no adaptation of the weights is allowed. Two settings are considered:

- In Section 13.1, we assume that there is a fixed ratio between the numbers of sources in the various classes; specializing to the case of two classes, this is equivalent to considering the situation that there are $n\eta_1$ sources of class 1, and $n\eta_2$ of class 2. Then we analyze how ϕ can be chosen such that n is maximized, under constraints on both overflow probabilities.
- In Section 13.3, we consider the situation that the user population fluctuates just mildly around some 'operating point' $(\overline{n}_1, \overline{n}_2)$. We develop an algorithm to find a ϕ such that some given 'ball' around $(\overline{n}_1, \overline{n}_2)$ is contained in $\mathcal{A}(\phi)$, such that for (n_1, n_2) in this ball the performance requirements are met.

Large deviations for Gaussian queues M. Mandjes

In Section 13.4, we take the opposite approach and allow *infinitely many* weight adaptations. We compare the realizable region $\mathcal{A}$ with the admissible regions $\mathcal{A}(\phi)$ for given ϕ, to quantify the gain of being able to constantly adapt ϕ. In fact, our examples indicate that often it suffices to pick just one weight to cover most of $\mathcal{A}$; interestingly, these weights correspond to priority strategies (i.e., one of the weights equals 0).

In Section 13.2, we present fast and straightforward approximations of the overflow probabilities in the GPS system, which are extensively used in Sections 13.3 and 13.4.

13.1 An optimal partitioning approach to weight setting

Before we turn to the setting of a network node operating under GPS, we first consider the situation of K (not necessarily 2) heterogeneous classes of Gaussian sources sharing common resources B and C. The sources are homogeneous within each class, but there is heterogeneity across classes. Inputs of type i have mean rate μ_i and variance curve $v_i(\cdot)$.

We consider the problem of optimally partitioning (B, C) into $(\mathrm{B}_i, \mathrm{C}_i)$ among the classes: we dedicate resources $(\mathrm{B}_i, \mathrm{C}_i)$ *exclusively* to class i; we require that the overflow probability for traffic of class i is below $e^{-\delta_i}$. In other words, resources not used by class i cannot be claimed by class $j \neq i$. 'Optimal' here refers to maximizing the size of the admissible region by maximizing n for a given *connection mix vector*

$$\eta = (\eta_1, \ldots, \eta_K) := \left(\frac{n_1}{n}, \ldots, \frac{n_K}{n}\right), \quad \text{where } n := \sum_{i=1}^{K} n_i.$$

More precisely, the approach taken in this section is that we maximize the number of admissible sources along a given 'ray':

- We keep the connection mix vector η fixed (without loss of generality, we could assume that the components of η sum to 1).
- We consider the situation that there are $n\eta_i$ sources of type i. We assume that the arrival process of a single source of type i be Gaussian with mean rate μ_i and variance function $v_i(\cdot)$; all sources behave statistically independently.
- Then maximize n under the condition that all classes are guaranteed their performance target by choosing the optimal values of $(\mathrm{B}_i, \mathrm{C}_i)$, with obvious constraints imposed on the bandwidth and buffer consumption: $\sum_{i=1}^{K} \mathrm{C}_i \leq \mathrm{C}$, and $\sum_{i=1}^{K} \mathrm{B}_i \leq \mathrm{B}$.

Clearly, partitioning of resources can support diverse performance requirements by protecting individual classes. Due to the fact that class i can use only resources $(\mathrm{B}_i, \mathrm{C}_i)$, this system loses, compared to GPS, the multiplexing advantage obtained by sharing across classes. We come back to this issue later.

We begin this section by characterizing (and interpreting) the solution to the optimal partitioning problem. Then we indicate how to use this solution for weight setting purposes in GPS.

Characterization of the solution of the optimal partitioning problem. Our first step is to develop an approximation for the overflow probability of class i traffic in the partitioned system. Clearly, the ith queue in this partitioned system is just a FIFO queue, with $n\eta_i$ sources of type i feeding into a queue with resources $(\mathrm{B}_i, \mathrm{C}_i)$. Based on Approximation 5.4.1, we propose the approximation $e^{-\Delta_i}$ for the class-i overflow probability, with

$$\Delta_i \equiv \Delta_i(\mathrm{B}_i, \mathrm{C}_i, n\eta_i) = \inf_{t\geq 0} \frac{(\mathrm{B}_i + (\mathrm{C}_i - n\eta_i\mu_i)t)^2}{2n\eta_i v_i(t)}. \tag{13.1}$$

Now consider the system with K traffic classes with heterogeneous performance requirements $\delta_1, \ldots, \delta_K$. Applying the above approximation, we have to choose $(\mathrm{B}_i, \mathrm{C}_i)$ to satisfy

$$\Delta_i(\mathrm{B}_i, \mathrm{C}_i, n\eta_i) \geq \delta_i. \tag{13.2}$$

Formulation as convex programming problem. To optimize the admissible region (for a fixed connection mix η), we clearly seek to solve the following mathematical programming problem:

Maximize n

subject to

$$\Delta_i(\mathrm{B}_i, \mathrm{C}_i, n\eta_i) \geq \delta_i, \quad \forall\, i = 1, \ldots, K$$

$$\sum_{i=1}^{K} \mathrm{C}_i \leq \mathrm{C}, \quad \sum_{i=1}^{K} \mathrm{B}_i \leq \mathrm{B}. \tag{13.3}$$

We first observe that, at the optimum, all the loss constraints in Equation (13.3) would hold with equality, as otherwise we may reduce B_i and/or C_i for the corresponding class and admit more connections; note that, here, we ignore the fact that n be integral, but this is of minor consequence when $n \gg 1$. We hence suppose that, for fixed n (in particular the optimal value), we may invert Equation (13.2) to obtain the buffer as a function of the other parameters, i.e., $\mathrm{B}_i = \mathrm{B}_i(\mathrm{C}_i, n\eta_i, \delta_i)$. But now recall that, as derived in Section 6.3, this function is *convex* in C_i (under very mild conditions), for fixed n (and evidently also fixed η_i, δ_i).

Now consider the 'intermediate problem' that minimizes the buffer consumption for a given set of sources, by finding the optimal c_i:

$$\text{Minimize} \quad \sum_{i=1}^{K} B_i(c_i, n\eta_i, \delta_i)$$

subject to

$$\sum_{i=1}^{K} c_i \leq c. \tag{13.4}$$

Obviously, to solve the Equation (13.3) we have to find the largest n such that the mathematical program (13.4) has a solution that does not exceed B. Now the crucial observation is that, due to the convexity of the (B_i, c_i)-curves, the optimization (13.4) can be recognized as a standard convex minimization to which the *strong Lagrangian principles* can be applied, see for example [287]. It yields the following *Karush-Kuhn-Tucker* conditions:

$$-\frac{dB_i}{dc_i} = \frac{\partial\Delta_i/\partial c_i}{\partial\Delta_i/\partial B_i} = \lambda$$

for some global (class-independent) nonnegative Lagrange multiplier λ; for general background on these conditions, see for instance [31].

It, hence, follows that the maximum value of n retains feasibility of the following conditions for some λ:

$$\Delta_i(B_i, c_i, n\eta_i) = \delta_i$$

$$-\frac{dB_i}{dc_i} = \frac{\partial\Delta_i/\partial c_i}{\partial\Delta_i/\partial B_i} = \lambda \tag{13.5}$$

$$\sum_{i=1}^{K} c_i \leq c, \quad \sum_{i=1}^{K} B_i \leq B.$$

Note that λ represents the slope of each of the bandwidth-buffer trade-off curves at the optimal operating point. Hence we have shown the interesting property that apparently, under the optimal partitioning $(B_1^\star, c_1^\star), \ldots, (B_K^\star, c_K^\star)$, all the trade-off curves have the same derivative $\lambda^\star$ in the corresponding operating point. This $\lambda^\star$ somehow reflects the optimal trade-off between bandwidth and buffer.

Interpretation. Interestingly, the optimal λ has an insightful interpretation. Denote the optimizing t in Equation (13.1) by $t_i^\star \equiv t_i^\star(B_i, c_i)$; in line with what we have seen earlier, this $t_i^\star$ is to be interpreted as the most likely epoch of overflow of the ith subsystem (or, more precisely, if the ith subsystem reaches overflow at time 0, the corresponding buffer most likely started to build up at time $-t_i^\star$).

Straightforward manipulations show that, under mild regularity conditions

$$\frac{\partial \Delta_i}{\partial \mathrm{C}_i} = \left(\frac{\mathrm{B}_i + (\mathrm{C}_i - n\eta_i\mu_i)t_i^\star}{n\eta_i v_i(t_i^\star)} \right) t_i^\star \ ; \quad \frac{\partial \Delta_i}{\partial \mathrm{B}_i} = \left(\frac{\mathrm{B}_i + (\mathrm{C}_i - n\eta_i\mu_i)t_i^\star)}{n\eta_i v_i(t_i^\star)} \right).$$

Consequently,

$$\frac{\partial \Delta_i / \partial \mathrm{C}_i}{\partial \Delta_i / \partial \mathrm{B}_i} = t_i^\star = \lambda.$$

In other words, the system capacity is maximized, *if all the $t_i^\star$ s of the individual subsystems match*, i.e., when they have the same most likely epoch of overflow. This seems reasonable, since if one subsystem has overflow while other subsystems are still not completely utilized, it is to be expected that one can improve the number of admissible sources by choosing an alternative partitioning $(\mathrm{B}_1, \mathrm{C}_1), \ldots, (\mathrm{B}_K, \mathrm{C}_K)$.

Exercise 13.1.1 This exercise illustrates the optimal partitioning solution (13.3). Consider $n\eta_i$ independent fBm sources (without drift), with Hurst parameter H_i, with $i = 1, \ldots, K$, and assume that even the classes are mutually independent. These sources use a queue with buffering capacity B and service capacity C. Describe the optimal partitioning.

Solution. First consider $n\eta$ fBm sources (without drift), with Hurst parameter H. With $A(\cdot)$ the arrival process of a single source, using $\log \mathbb{E} \exp(\theta A(t)) = \frac{1}{2}\theta^2\sigma^2 t^{2H}$, it easily follows from Equation (13.1) that the buffer-bandwidth trade-off curve is given by

$$\mathrm{B}^{1-H}\mathrm{C}^H = \sqrt{2\delta \cdot n\eta \cdot \sigma^2 \cdot H^H (1-H)^{1-H}}, \tag{13.6}$$

when the Quality of Service (QoS) requirement δ is imposed, cf. the results of Section 6.3. Recall that the buffer-bandwidth trade-off is indeed convex, and is described through a so-called Cobb–Douglas substitution function.

Now consider the situation of K heterogeneous classes. Let class i consist of $n\eta_i$ fBm with parameter H_i. With evident notation, for any i, it follows from Equation (13.6) that

$$\mathrm{B}_i = K_i(n)\mathrm{C}_i^{-\gamma_i}, \text{ where } \gamma_i := \frac{H_i}{1-H_i}, \quad \text{and} \quad K_i(n) = M_i n^{\frac{1}{2(1-H_i)}},$$

for some M_i (independent of n).

From the set of Equations (13.5), we know that there is a class-independent number λ such that

$$\lambda = -\frac{\mathrm{dB}_i}{\mathrm{dC}_i} = K_i(n)\gamma_i \mathrm{C}_i^{-\gamma_i - 1},$$

or

$$c_i(n,\lambda) = \left(\frac{K_i(n)\gamma_i}{\lambda}\right)^{\frac{1}{\gamma_i+1}} = \left(\frac{M_i\gamma_i}{\lambda}\right)^{1-H_i}\sqrt{n}$$

and

$$B_i(n,\lambda) = \left(\frac{K_i(n)\lambda^{\gamma_i}}{\gamma_i^{\gamma_i}}\right)^{\frac{1}{\gamma_i+1}} = \left(\frac{M_i\lambda^{\gamma_i}}{\gamma_i^{\gamma_i}}\right)^{1-H_i}\sqrt{n}.$$

From the buffer and bandwidth constraints $B = \sum_{i=1}^{K} B_i$ and $C = \sum_{i=1}^{K} c_i$ we obtain

$$C \cdot \sum_{i=1}^{K}\left(\frac{M_i\lambda^{\gamma_i}}{\gamma_i^{\gamma_i}}\right)^{1-H_i} = B \cdot \sum_{i=1}^{K}\left(\frac{M_i\gamma_i}{\lambda}\right)^{1-H_i}.$$

From the fact that the left-hand side is increasing in λ, whereas the right-hand side is decreasing in λ, it immediately follows that there is a unique solution (that cannot be characterized in closed form, but can be found through a straightforward bisection procedure). This $\lambda^\star$ determines the optimal allocation $(B_1^\star, c_1^\star), \ldots, (B_K^\star, c_K^\star)$. ◇

Relation to GPS weight setting. The above partitioning model can be used to offer differentiated performance. An alternative way to do so is by means of the mechanism we have studied in Chapter 10. In GPS, each class has its own buffer and (guaranteed) bandwidth, but the bandwidth left unused by one class can be used by another class. In other words, the partitioned system discussed in this section is a 'conservative description' of the corresponding GPS system.

An immediate consequence is that the (B_i, c_i) values of the partitioned system can be used to determine the weights and buffer sizes in the corresponding GPS system: given the optimal split of the partitioned system, i.e., $(B_1^\star, c_1^\star), \ldots, (B_K^\star, c_K^\star)$, the GPS weight ϕ_i for class i can obviously be set conservatively as $\phi_i = c_i/C$. Evidently, the corresponding buffer space must be guaranteed as well. We mention that this can be accomplished using the technique of virtual partitioning as described in [166].

13.2 Approximation of the overflow probabilities

In the previous section, we approximated the GPS system conservatively, by a system in which the resources were explicitly partitioned. As mentioned, this approach does not fully reflect the attractive resource sharing opportunities offered by GPS. We therefore concentrate on Sections 13.3 and 13.4 on weight setting procedures that do incorporate this effect. In these procedures, we rely on an approximation for the overflow probabilities under GPS. It is this approximation, which is based on the theory of Chapter 10 (and, more specifically, the 'trichotomy' of Section 10.6), that is described in the present section.

In the remainder of this chapter, we restrict ourselves to the important case of $K = 2$. This restriction is justified by the observation that the majority of all traffic can broadly be categorized into *streaming* and *elastic* traffic [252]; streaming traffic are predominantly real-time audio and video applications (with strict delay requirements), while elastic traffic corresponds to the transfer of digital documents such as web pages and files (which is less delay-sensitive).

As earlier, inputs of type i have mean rate μ_i and variance curve $v_i(\cdot)$; $\Gamma_i(s,t)$ is defined as the covariance between $A_i(s)$ and $A_i(t)$, with $A_i(\cdot)$ the traffic generated by a source of type i. There are n_i inputs of type i.

In this section we develop an approximation for the overflow probabilities in both queues of the GPS system, where the numbers of sources of type i, n_i are typically large. We denote the stationary buffer content of the ith queue in this GPS model by Q_i, the service rate by C, and the buffer threshold of queue i by B_i. Invoking Remark 9.6.1, the GPS model with $n_1 \neq n_2$ is equivalent to a GPS model with n sources in both classes, mean rates $(n_i/n)\mu_i$ and variance functions $(n_i/n)v_i(\cdot)$.

We scale the buffer threshold and service rate with n such that $nb_i \equiv \mathrm{B}_i$ and $nc \equiv \mathrm{C}$. Now we can apply our earlier results on GPS in the many-sources setting, where we assumed both classes to consist of n sources, with n typically large.

In line with the results of Section 10.6, three regimes should be distinguished. Again, we concentrate on the first queue; the second queue can be treated analogously. Now

$$\Delta_1(n_1, n_2) := -\log \mathbb{P}(Q_1 \geq \mathrm{B}_1) \equiv -\log \mathbb{P}(Q_{1,n} \geq nb_1)$$

can be approximated by $\overline{\Delta}_1(n_1, n_2)$ as follows, based on the theory of Chapter 10. Throughout, we use the notation $\phi_2^o := n_2\mu_2/\mathrm{C}$.

Approximation 13.2.1 *Define*

$$t_{\mathrm{c}} \equiv t_{\mathrm{c}}(n_1, n_2) := \arg\inf_{t>0} \frac{(\mathrm{B}_1 + (\mathrm{C} - n_1\mu_1 - n_2\mu_2)t)^2}{n_1 v_1(t) + n_2 v_2(t)}, \tag{13.7}$$

assuming it is uniquely determined, and

$$\phi_2^c := \frac{n_2\mu_2}{\mathrm{C}} + \sup_{s\in[0,t_{\mathrm{c}}]} \left(\frac{n_2\Gamma_2(t_{\mathrm{c}}, s)}{\mathrm{C}\, s\,(n_1 v_1(t_{\mathrm{c}}) + n_2 v_2(t_{\mathrm{c}}))} \right) (\mathrm{B}_1 + (\mathrm{C} - n_1\mu_1 - n_2\mu_2)t_{\mathrm{c}}).$$

(i) *If $\phi_2 \in [0, \phi_2^o]$, then*

$$\overline{\Delta}_1(n_1, n_2) = \frac{1}{2} \inf_{t>0} \frac{(\mathrm{B}_1 + (\phi_1\mathrm{C} - n_1\mu_1)t)^2}{n_1 v_1(t)}.$$

(ii) If $\phi_2 \in (\phi_2^o, \phi_2^c)$, then

$$\overline{\Delta}_1(n_1, n_2)$$

$$= \frac{1}{2} \inf_{t>0} \sup_{s \in (0,t]} \begin{pmatrix} z_1 \\ z_2 \end{pmatrix}^{\mathrm{T}} \begin{pmatrix} n_1 v_1(t) + n_2 v_2(t) & n_2 \Gamma_2(s,t) \\ n_2 \Gamma_2(s,t) & n_2 v_2(s) \end{pmatrix}^{-1} \begin{pmatrix} z_1 \\ z_2 \end{pmatrix},$$

where

$$\begin{pmatrix} z_1 \\ z_2 \end{pmatrix} \equiv \begin{pmatrix} z_1(t, n_1, n_2) \\ z_2(s, n_1, n_2) \end{pmatrix} := \begin{pmatrix} \mathrm{B}_1 + (\mathrm{C} - n_1\mu_1 - n_2\mu_2)t \\ (\phi_2 \mathrm{C} - n_2\mu_2)s \end{pmatrix}. \tag{13.8}$$

(iii) If $\phi_2 \in [\phi_2^c, 1]$, then

$$\overline{\Delta}_1(n_1, n_2) = \frac{1}{2} \inf_{t>0} \frac{(\mathrm{B}_1 + (\mathrm{C} - n_1\mu_1 - n_2\mu_2)t)^2}{n_1 v_1(t) + n_2 v_2(t)}.$$

Numerical experiments have indicated that usually in regime (ii) of Approximation 13.2.1 the optimizing s is close to the optimizing t. Taking these *a priori* equal to each other, we obtain the following, somewhat more crude, approximation. It coincides with the rough full link approximation proposed in [205].

Approximation 13.2.2 Rough full link approximation. *Let t_{C} be defined as in Equation (13.7), and*

$$\phi_2^c := \frac{n_2\mu_2}{\mathrm{C}} + \left(\frac{n_2 v_2(t_{\mathrm{C}})}{\mathrm{C}\, t_{\mathrm{C}} \,(n_1 v_1(t_{\mathrm{C}}) + n_2 v_2(t_{\mathrm{C}}))} \right) (\mathrm{B}_1 + (\mathrm{C} - n_1\mu_1 - n_2\mu_2)t_{\mathrm{C}}).$$

The regimes (i) and (iii) are as in Approximation 13.2.1, and

(ii) If $\phi_2 \in (\phi_2^o, \phi_2^c)$, then

$$\overline{\Delta}_1(n_1, n_2) = \frac{1}{2} \inf_{t \geq 0} \left(\frac{\mathrm{B}_1 + (\phi_1 \mathrm{C} - n_1\mu_1)t)^2}{n_1 v_1(t)} + \frac{(\phi_2 \mathrm{C} - n_2\mu_2)^2 t^2}{n_2 v_2(t)} \right).$$

Exercise 13.2.3 Suppose that both class 1 and class 2 correspond to Brownian motions. Argue that Approximation 13.2.1 and Approximation 13.2.2 coincide.

Solution. Cases (i) and (iii) of both approximations coincide. The proof of Theorem 11.4.6 has indicated that the optimum in case (ii) of Approximation 13.2.1 is attained for $s = t$. ◇

13.3 Fixed weights

This section focuses on a procedure for finding a weight vector (ϕ_1, ϕ_2) such that both classes receive the required performance, despite (mild) fluctuations in the number of sources present. More precisely, for specified (positive) numbers δ_i, we require that $\Delta_i(n_1, n_2) \geq \delta_i$ $(i = 1, 2)$ for all (n_1, n_2) in a 'ball' $\mathcal{B}(\overline{n}_1, \overline{n}_2)$ around $(\overline{n}_1, \overline{n}_2)$:

$$\mathcal{B}(\overline{n}_1, \overline{n}_2) := \left\{(n_1, n_2) \in \mathbb{N}^2 \mid \gamma_1(n_1 - \overline{n}_1)^2 + \gamma_2(n_2 - \overline{n}_2)^2 \leq 1\right\},$$

for positive γ_1, γ_2. As an aside we mention that it can be easily verified that the procedure described below works, in fact, for any 'target area' $\mathcal{B}$ that is finite and *convex*, rather than just these ellipsoidal sets.

In our procedure, we rely on the expressions for $\overline{\Delta}_i(n_1, n_2)$ as presented in the Section 13.2. To make our algorithm as simple as possible, we use the following expansion of $\overline{\Delta}_i(n_1, n_2)$ around $(n_1, n_2) = (\overline{n}_1, \overline{n}_2)$:

$$\overline{\Delta}_i(n_1, n_2) \approx \overline{\Delta}_i(\overline{n}_1, \overline{n}_2) + (n_1 - \overline{n}_1) \left.\frac{\partial \overline{\Delta}_i(n_1, n_2)}{\partial n_1}\right|_{(n_1,n_2)=(\overline{n}_1,\overline{n}_2)} + (n_2 - \overline{n}_2) \left.\frac{\partial \overline{\Delta}_i(n_1, n_2)}{\partial n_2}\right|_{(n_1,n_2)=(\overline{n}_1,\overline{n}_2)}. \tag{13.9}$$

This approximation requires the evaluation of two partial derivatives, which can be done relatively explicitly, as described in the Appendix.

Relying on Equation (13.9), we have to verify whether for all $(n_1, n_2) \in \mathcal{B}(\overline{n}_1, \overline{n}_2)$ and $i = 1, 2$,

$$\overline{\Delta}_i(\overline{n}_1, \overline{n}_2) + (n_1 - \overline{n}_1, n_2 - \overline{n}_2)^{\mathrm{T}} e_i \geq \delta_i,$$

where

$$e_i \equiv (e_{i1}, e_{i2}) := \left(\left.\frac{\partial \overline{\Delta}_i(n_1, n_2)}{\partial n_1}\right|_{(n_1,n_2)=(\overline{n}_1,\overline{n}_2)}, \left.\frac{\partial \overline{\Delta}_i(n_1, n_2)}{\partial n_2}\right|_{(n_1,n_2)=(\overline{n}_1,\overline{n}_2)} \right).$$

Because of the convex shape of $\mathcal{B}(\overline{n}_1, \overline{n}_2)$, we only have to verify this condition for the two points on the boundary $\partial\mathcal{B}(\overline{n}_1, \overline{n}_2)$ at which the tangents have slopes equal to $-e_{11}/e_{12}$ and $-e_{21}/e_{22}$ respectively. Denoting these points by $(n^\star_{11}, n^\star_{12})$ and $(n^\star_{21}, n^\star_{22})$, it is readily checked that they satisfy

$$(n^\star_{i1}, n^\star_{i2}) := \left(\overline{n}_1 + \sqrt{\left(\gamma_1 + \frac{e_{i2}^2}{e_{i1}^2}\frac{\gamma_1^2}{\gamma_2}\right)^{-1}}, \overline{n}_2 + \sqrt{\left(\gamma_2 + \frac{e_{i1}^2}{e_{i2}^2}\frac{\gamma_2^2}{\gamma_1}\right)^{-1}} \right),$$

$i = 1, 2$. We say that ϕ is feasible if $K_i := \overline{\Delta}_i(\overline{n}_1, \overline{n}_2) + (n^\star_{i1} - \overline{n}_1, n^\star_{i2} - \overline{n}_2)^{\mathrm{T}} e_i \geq \delta_i$ for both $i = 1$ and 2. Notice that K_i is a function of the weights; as $\phi_1 + \phi_2 = 1$, we can write $K_i(\phi_1)$. $K_1(\phi_1)$ will increase in ϕ_1, whereas $K_2(\phi_1)$ will decrease.

This suggests the following solution to the weight setting problem:

1. First find the smallest ϕ_1 such that $K_1(\phi_1) \geq \delta_1$. If this does not exist, then there is no solution.
2. If it does exist, then verify if for this ϕ_1 it holds that $K_2(\phi_1) \geq \delta_2$. If this is true, then the weight setting problem can be solved; if not, then there is no solution (i.e., there is no ϕ such that $\mathcal{B}(\overline{n}_1, \overline{n}_2) \subseteq \mathcal{A}(\phi)$).

Example 13.3.1 We first explain how requirements on the admissible numbers of sources naturally lead to a set of the type $\mathcal{B}(\overline{n}_1, \overline{n}_2)$.

- Our analysis assumes fixed numbers of sources of both types, but in practice this number fluctuates in time: sources arrive, and stay in the system for a random amount of time. Now suppose that sources of both types arrive according to Poisson processes (with rates ν_i, for $i = 1, 2$), and that, if admitted, these would require service for some random duration (with finite means $\mathbb{E}D_i$). If there were no admission control, the distributions of the number of jobs of both types are Poisson with means (and variances!) $\overline{n}_i = \nu_i\, \mathbb{E}D_i$.
- Suppose that the system must be designed such that this mean $(\overline{n}_1, \overline{n}_2) \pm$ twice the standard deviation should be in the admissible region, i.e., should be contained in $\mathcal{A}(\phi)$. This suggests choosing

$$\mathcal{B}(\overline{n}_1, \overline{n}_2) = \left\{ (n_1, n_2) \in \mathbb{N}^2 \;\middle|\; \left(\frac{n_1 - \overline{n}_1}{2\sqrt{\overline{n}_1}}\right)^2 + \left(\frac{n_2 - \overline{n}_2}{2\sqrt{\overline{n}_2}}\right)^2 \leq 1 \right\}.$$

In this example we choose $\overline{n}_1 = 900$ and $\overline{n}_2 = 1600$, which leads to:

$$\begin{aligned}\mathcal{B}(\overline{n}_1, \overline{n}_2) &= \mathcal{B}(900, 1600)\\ &= \left\{ (n_1, n_2) \in \mathbb{N}^2 \mid 16(n_1 - 900)^2 + 9(n_2 - 1600)^2 \leq 57\,600 \right\}.\end{aligned}$$

We suppose that both types of sources correspond to Brownian motions, with $\mu_1 = 0.2$, $\mu_2 = 0.3$, $v_1(t) = 2t$, and $v_2(t) = t$. We rely on explicit results for Brownian motions, as summarized in Appendix 13.2, in particular for the partial derivatives of the $\overline{\Delta}_i(n_1, n_2)$ with respect to the numbers of sources. We choose $\mathrm{C} = 1000$, $\mathrm{B}_1 = 35$, and $\mathrm{B}_2 = 25$.

First suppose that the performance targets are $\delta_1 = 9$ and $\delta_2 = 7$ (roughly corresponding to overflow probabilities $1.2 \cdot 10^{-4}$ and $9.1 \cdot 10^{-4}$). Figure 13.1 shows that no weights ϕ exist to meet this target (to guarantee that the overflow probability in queue 1 is small enough, ϕ_1 should be larger than 0.39, but this implies that $K_2(\phi_1) < 5.7 < \delta_2$). Now suppose that $\delta_1 = 8$ and $\delta_2 = 6$. Then an analogous reasoning gives that ϕ_1 should be chosen in the interval $(0.34, 0.37)$. ◇

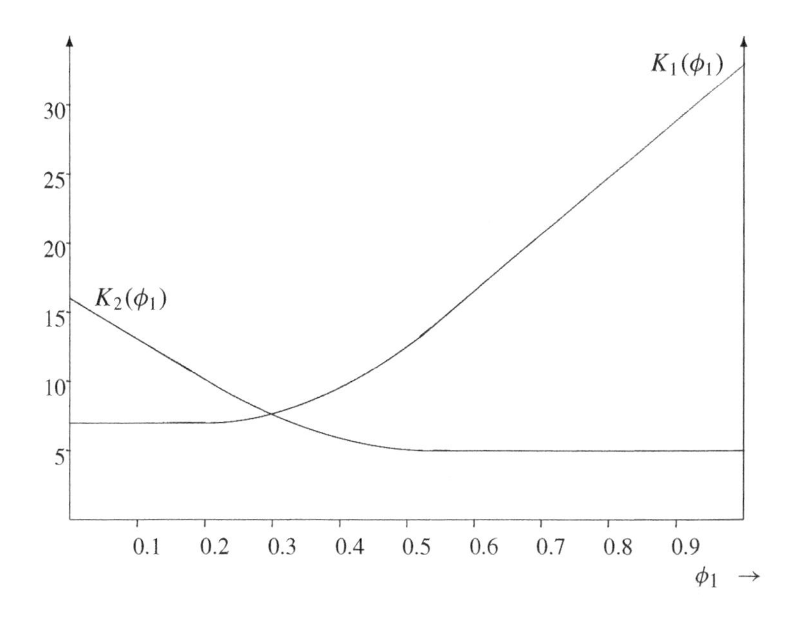

Figure 13.1: The curves $K_i(\phi_1)$ of Example 13.3.1.

13.4 Realizable region

Denote the admissible region for weights $\phi_1 = \phi$ and $\phi_2 = 1 - \phi$ by $\mathcal{A}(\phi)$. In the previous sections, we focused on a procedure for setting the weights, under the assumption that these values could not be adapted 'on the fly'. Clearly, if one could adapt the weight at all times, then the resulting admissible region (to which we refer as the 'realizable region') would grow. On the other hand, the question is how substantial this gain is. The goal of this section is to study this issue. To this end, we consider the realizable region

$$\mathcal{A} = \bigcup_{\phi \in [0,1]} \mathcal{A}(\phi),$$

and verify whether there are weights ϕ for which nearly all of $\mathcal{A}$ is covered by $\mathcal{A}(\phi)$. We do so for a set of scenarios, which are representative in terms of the types of sources, performance targets, buffer sizes, and service capacities.

Explicit expressions for the boundaries of $\mathcal{A}(\phi)$ can be found in case both classes correspond to Brownian motions, see [173, 176]. In our scenarios, however, the inputs are non-Brownian, and hence the boundary of the admissible regions (and thus the realizable region) has to be obtained numerically. Particularly interesting values of ϕ are 0 and 1, as these correspond to the priority queue ($\phi = 0$ gives priority to class 2, $\phi = 1$ to class 1). We therefore compare the realizable region with the admissible region corresponding to the priority cases. Interestingly, the examples given below illustrate that $\mathcal{A}(0)$ or $\mathcal{A}(1)$ (or both) cover virtually the full realizable region.

In the numerical experiments, we have chosen the Gaussian processes that match with a number of relevant and realistic traffic processes known from the literature. We remark that, in addition to the examples presented here, we have considered many other parameter settings. The result that priority strategies cover nearly the entire realizable region appears to remain valid, under quite general circumstances. The numerical computations have been done using Approximation 13.2.2 (rough full link approximation).

Example 13.4.1 Consider two traffic classes sharing a total capacity (c) of 100 Mbps. The first class consists of data traffic, whereas the second class corresponds to voice traffic. Traffic of the first class is modeled asfBm, i.e., $v_1(t) = \alpha t^{2H}$, with $H \in (0, 1)$. The mean traffic rate μ_1 is 0.2 Mbps and its variance function is given by $v_1(t) = 0.0025t^{1.6}$ (such that at timescale $t = 1$ s the standard deviation is 0.05 Mbps). The value of $H = 0.8$ is in line with several measurement studies (one commonly finds a value between, say, 0.7 and 0.85, see Chapter 3).

Traffic of the second class corresponds to the Gaussian counterpart of the Anick-Mitra-Sondhi (AMS) model [9], see [5] and Section 2.5; the variance function is a scaled version of that of iOU traffic. In the AMS model, work arrives from sources in bursts that have peak rate h, and lengths that are exponentially distributed with mean β^{-1}. After each burst, the source is off for a period that is exponentially distributed with mean λ^{-1}. The variance function of a single source is given by (see Section 2.5)

$$v_2(t) = \frac{2\lambda\beta h^2}{(\lambda+\beta)^3}\left(t - \frac{1}{\lambda+\beta}(1 - \exp(-(\lambda+\beta)t))\right). \tag{13.10}$$

We choose $h = 0.032$, $\lambda = 1/0.65$, and $\beta = 1/0.352$ in Equation (13.10), in line with the parameters for coded voice given in [270]. Hence, the mean traffic rate of a source of class 2 (μ_2) is 0.011 Mbps. Note that traffic of class 1 is lrd (i.e., the autocorrelations are nonsummable), whereas traffic of class 2 is srd.

We allow an overflow probability of 10^{-6} for the first class and 10^{-3} for the second class (corresponding to $\delta_1 \approx 13.8$ and $\delta_2 \approx 6.9$). We choose B_1 such that $\mathrm{B}_1/\mathrm{c} = 0.05$ (i.e., 50 ms) and B_2 such that $\mathrm{B}_2/\mathrm{c} = 0.01$ (i.e., 10 ms). Hence we allow a (relatively) large delay but small loss for the data traffic, and a small delay but (relatively) large loss for the voice traffic.

Figure 13.2(a) depicts the admissible region for the priority cases ($\mathcal{A}(0)$ and $\mathcal{A}(1)$), the realizable region $\mathcal{A}$ and the 'stable region' ($\mathcal{T}$ (i.e., all combinations such that $n_1\mu_1 + n_2\mu_2 < \mathrm{c}$). Obviously, $\mathcal{A} \subseteq \mathcal{T}$, but they almost coincide. Furthermore, the boundaries of $\mathcal{A}(0)$, $\mathcal{A}(1)$, and $\mathcal{A}$ almost match (the boundaries of $\mathcal{A}(0)$ and $\mathcal{A}(1)$ are hardly visible). That is, most of $\mathcal{A}$ can be obtained by giving priority to class 1 or 2. In fact, any weight of $\phi \in [0, 1]$ yields an admissible region $\mathcal{A}(\phi)$ that closely resembles $\mathcal{A}$.

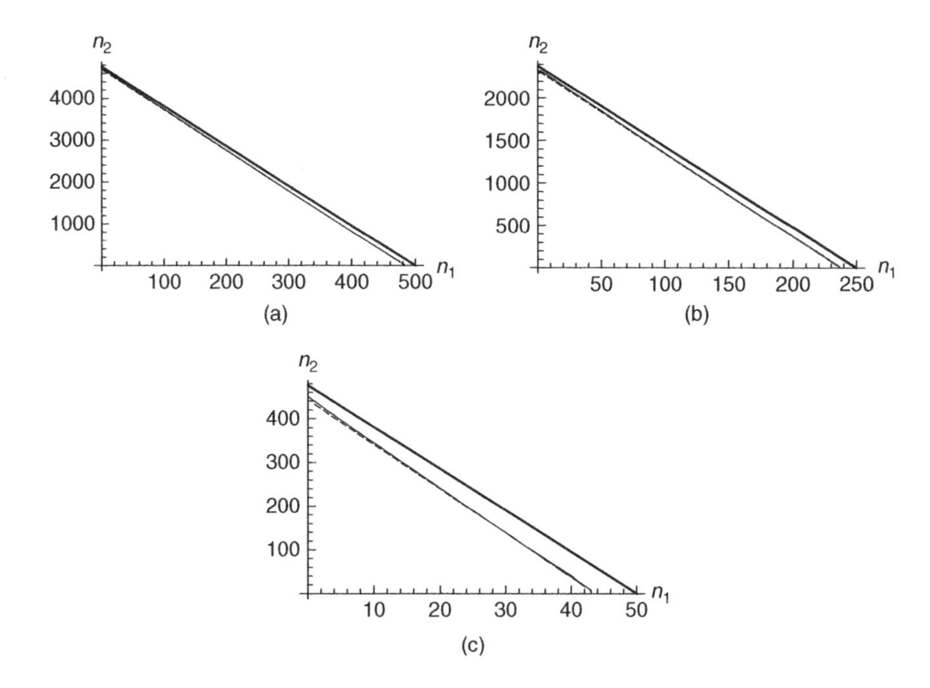

Figure 13.2: Realizable regions of Example 13.4.1. The boundary of $\mathcal{A}(1)$ is given by the small dashed line. The long dashed line represents the boundary of $\mathcal{A}(0)$. The solid line above the boundary of $\mathcal{A}(0)$ and $\mathcal{A}(1)$ is the boundary of $\mathcal{A}$. The thick line on top is the boundary of $\mathcal{T}$. (a): $\mathrm{C} = 100$ Mbps; (b): $\mathrm{C} = 50$ Mbps. (c): $\mathrm{C} = 10$ Mbps.

We have also experimented with other values for C as depicted in Figure 13.2(b) and 13.2(c). As the value of C becomes smaller (with still $\mathrm{B}_1/\mathrm{C} = 0.05$, $\mathrm{B}_2/\mathrm{C} = 0.01$, and all other parameters left unchanged), the difference between the boundary of $\mathcal{A}$ and $\mathcal{T}$ becomes clearer. Note that $\mathcal{A}$ still closely resembles $\mathcal{A}(0)$ and $\mathcal{A}(1)$. This indicates that GPS scheduling is only marginally more effective than a strict priority discipline ($\phi = 0$ or $\phi = 1$).

As the values of δ_1 and δ_2 increase, the performance requirements become more stringent and therefore the difference between the regions $\mathcal{A}$ and $\mathcal{T}$ becomes more substantial. For large values of C this is hardly visible, and therefore we show this for the case that $\mathrm{C} = 10$ Mbps (with the parameter values corresponding to Figure 13.2(c).

Figures 13.2(c) and 13.4(a,b) show the expected impact of the 'performance requirements' δ_i, $i = 1, 2$. Although the difference between $\mathcal{A}$ and $\mathcal{T}$ becomes larger as δ_1 and δ_2 increase, $\mathcal{A}$ continues to be closely approximated by $\mathcal{A}(0)$ or $\mathcal{A}(1)$. Compare Figure 13.2(c) with Figure 13.4(a), and observe that if the δ_is are doubled, then $\mathcal{A}$ decreases by less than 15%. ◇

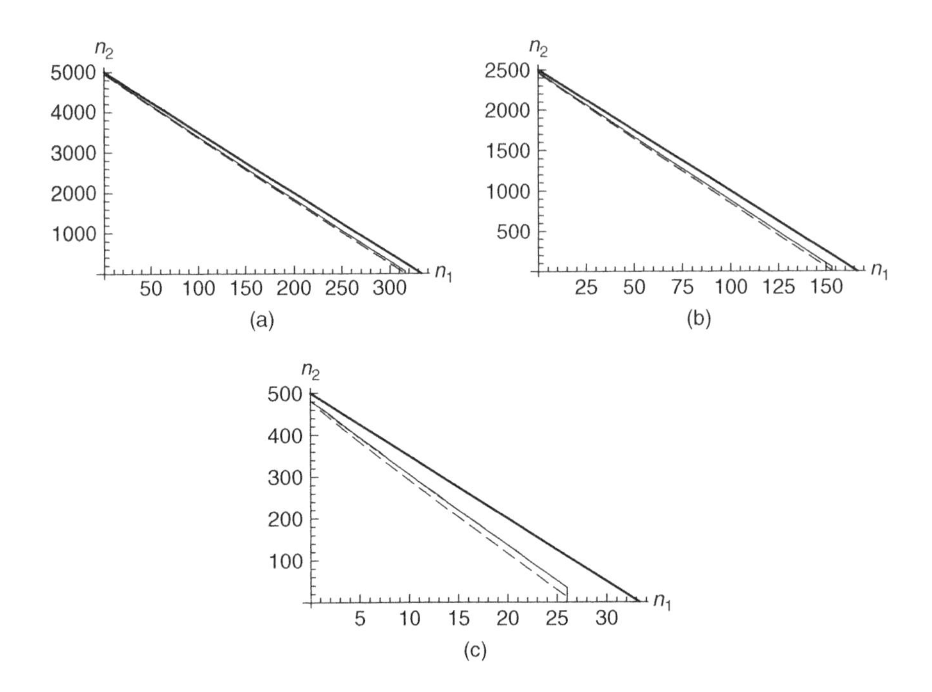

Figure 13.3: Realizable regions of Example 13.4.2. The boundary of $\mathcal{A}(1)$ is given by the small dashed line. The long dashed line represents the boundary of $\mathcal{A}(0)$. The solid line above the boundary of $\mathcal{A}(0)$ and $\mathcal{A}(1)$ is the boundary of $\mathcal{A}$. The thick line on top is the boundary of $\mathcal{T}$. (a): $\mathrm{C} = 1000$ Mbps; (b): $\mathrm{C} = 500$ Mbps. (c): $\mathrm{C} = 100$ Mbps.

Example 13.4.2 In this example, we let the two traffic classes share a total capacity of 1 Gbps. The traffic of the first class is data traffic with a higher access rate, and traffic of the second class with a lower access rate. The mean traffic rate of a source of the first (second) class is 3 Mbps (0.2 Mbps). The variance functions are given by $0.5625t^{1.6}$ and $0.0025t^{1.6}$, respectively, such that at timescale $t = 1$ s the standard deviations are 0.75 Mbps and 0.05 Mbps, respectively. We allow an overflow probability of 10^{-8} (10^{-3}) for the first (second) class ($\delta_1 \approx 18.4$ and $\delta_2 \approx 6.9$). The buffer thresholds are such that $\mathrm{B}_1/\mathrm{C} = 0.04$ and $\mathrm{B}_2/\mathrm{C} = 0.01$.

Figure 13.3(a) shows the resulting realizable region. Once again, most of $\mathcal{A}$ is covered by the admissible region of a priority strategy. Furthermore, also the influence of C is as mentioned earlier, as can be seen in Figure 13.3(b,c). For large values of C, the boundaries of $\mathcal{A}(0)$, $\mathcal{A}(1)$, $\mathcal{A}$ and $\mathcal{T}$ almost coincide. As C decreases, the difference between the boundaries of $\mathcal{A}$ and $\mathcal{T}$ becomes significant. Also note that the difference between the boundaries of $\mathcal{A}(0)$ and $\mathcal{A}(1)$ becomes visible for small values of C. In all the experiments as depicted in Figure 13.3, $\mathcal{A}$ nearly coincides with $\mathcal{A}(1)$.

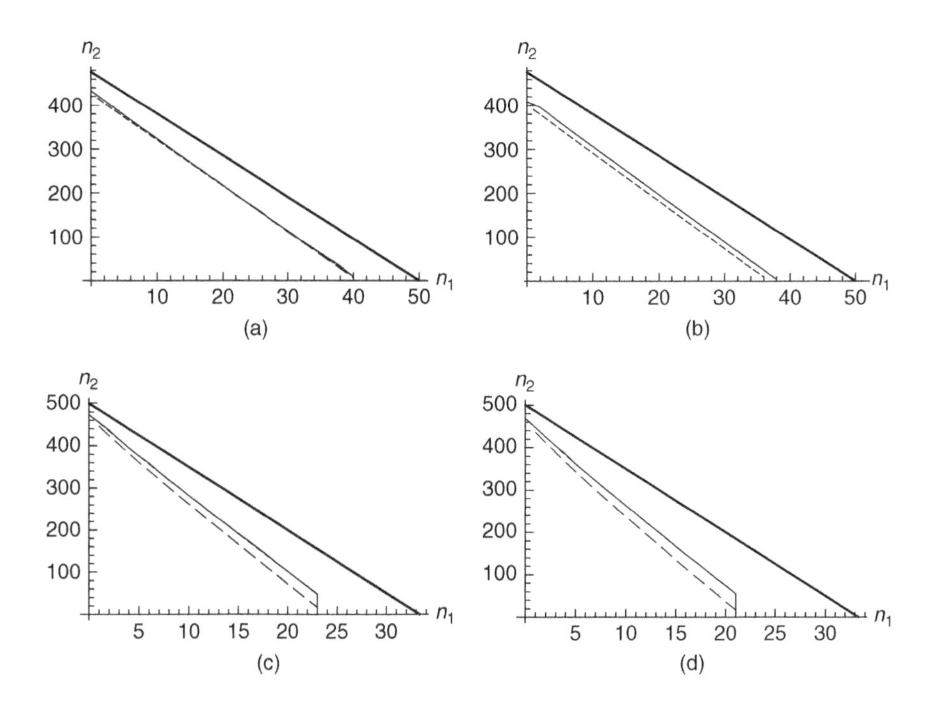

Figure 13.4: Fig. (a,b) [Top]: Realizable regions of Example 13.4.1; (c,d) : Realizable regions of Example 13.4.2. The boundary of $\mathcal{A}(1)$ is given by the small dashed line. The long dashed line represents the boundary of $\mathcal{A}(0)$. The solid line above the boundary of $\mathcal{A}(0)$ and $\mathcal{A}(1)$ is the boundary of $\mathcal{A}$. The thick line on top is the boundary of $\mathcal{T}$. (a): $\delta_1 = 27.6$ and $\delta_2 = 13.8$; (b): $\delta_1 = 41.4$ and $\delta_2 = 27.6$; (c): $\delta_1 = 36.8$ and $\delta_2 = 13.8$; (d): $\delta_1 = 55.2$ and $\delta_2 = 20.7$.

For $c = 100$ Mbps (setting of Figure 13.3(c)), Figure 13.4(c,d) depicts the sensitivity with respect to the δ_is. Again, $\mathcal{A}$ becomes considerably smaller when the performance requirements become more stringent (i.e., increasing δ_i, $i = 1,2$). Furthermore, the boundary of $\mathcal{A}$ still seems to closely match the boundary of $\mathcal{A}(1)$. ◇

In the above examples, we saw the interesting effect that $\mathcal{A}(0)$ or $\mathcal{A}(1)$ covers $\mathcal{A}$ nearly completely. The next question is whether this conclusion holds under more general circumstances. To verify this, [173] consider a broad range of other, highly heterogeneous, scenarios. To quantify the efficiency of the priority policies, the following performance metric was introduced. Denote

$$\varrho(\phi) := \frac{\#\{(n_1, n_2) \in \mathcal{A}(\phi)\}}{\#\{(n_1, n_2) \in \mathcal{A}\}};$$

the denominator in this expression reflects the size of the admissible region if the weights could be adapted at all times. Then define $\varrho := \max\{\varrho(0), \varrho(1)\}$. The closer ϱ is to 1, the better the admissible region of one of the strict priority policies approximates the realizable region. In nearly all experiments in [173], it was found that ϱ was above 99%. On the other hand, however, no simple guideline was found to determine which class should be given priority.

The above numerical results indicate that, in terms of efficiency (i.e., maximizing the admissible region) hardly anything is lost by using priority scheduling rather than GPS (with weight adaptation); this conclusion is very much in line with the simulation results obtained in [88]. It suggests that simple priority scheduling strategies may suffice for practical purposes.

It is worth pointing out one important caveat. Priority scheduling carries the potential risk of complete starvation, which may cause flow control protocols like Transmission Control Protocol (TCP) to suffer a severe degradation in throughput performance. One of the main merits of GPS is that it prevents such starvation effects when other classes misbehave. In the above arguments, however, we implicitly assume the classes to satisfy the declared statistical properties, and, in particular, to conform to the specified traffic intensity. In other words, the merit of GPS may not be so much in handling normal statistical variations as it is in offering protection in case the actual statistical features deviate from the traffic specification.

Bibliographical notes

The optimal partitioning approach of Section 13.1 was developed in Kumaran and Mandjes [162]; note that the analysis of [162] is not restricted to Gaussian sources. Sections 13.2 and 13.3 are based on material in Mandjes and Van Uitert [200], whereas Section 13.4 is taken from Lieshout, Mandjes and Borst [173, 176]. We remark that [173] contains substantially more numerical experiments (with representative parameters), providing convincing evidence for the claim that nearly the full realizable region can be achieved by simple priority policies. The rough full link approximation, i.e., Approximation 13.2.2, was proposed by Mannersalo and Norros [205].

In Example 13.4.1 (the Gaussian counterpart of) the AMS model was considered. This type of models attracted substantial attention in the 1980s, see [159, 217] and the survey paper [160]; the steady-state buffer content distribution was expressed in terms of a system of linear differential equations. An interesting contribution to this class of models is [256], which presents a solution in terms of a Wiener–Hopf factorization.

In Chapter 10, we already provided a number of references on the computation of the performance of GPS. The reverse problem of selecting suitable weight factors so as to satisfy given performance requirements has received surprisingly less attention in the literature. Dukkipati *et al.* [85] and Panagakis *et al.* [233] derive algorithms for allocating optimal weights to leaky-bucket constrained traffic sources with deterministic performance requirements (zero loss and finite delay bound). A closely related problem concerns the characterization of the admissible region, see for instance [298] for a large-deviations based analysis for short-range inputs. Elwalid and Mitra [94] consider a two-class GPS system with leaky-bucket regulated traffic sources, and argue that the boundary of the admissible region is approximately linear. In addition, they show that most of the realizable region is achieved by selecting only one or two weight values. Further results along these lines may be found in Kumaran *et al.* [165].

Appendix

In this appendix, we determine expressions for the partial derivatives of $\overline{\Delta}_1(n_1, n_2)$ to the numbers of sources, as required in the weight setting algorithm of Section 13.3. The Cases (i), (ii), (iii) below correspond to the regimes in Approximation 13.2.1.

(i) Based on Theorem 4.1.8, in the regime $\phi_2 \in [0, \phi_2^o]$,

$$\overline{\Delta}_1(n_1, n_2) = \inf_{t>0} \sup_{\theta \in \mathbb{R}} \left(\theta(\mathrm{B}_1 + (\phi_1 \mathrm{C} - n_1\mu_1)t) - \frac{1}{2}\theta^2 n_1 v_1(t) \right).$$

The inner supremum is attained for

$$\theta^\star \equiv \theta^\star(t) = \frac{\mathrm{B}_1 + (\phi_1 \mathrm{C} - n_1\mu_1)t}{n_1 v_1(t)}.$$

Denoting the optimizing t by $t^\star$, we derive

$$\frac{\partial \overline{\Delta}_1(n_1, n_2)}{\partial n_1} = -\theta^\star \mu_1 t^\star - \frac{1}{2}(\theta^\star)^2 v_1(t^\star), \qquad \frac{\partial \overline{\Delta}_1(n_1, n_2)}{\partial n_2} = 0.$$

(ii) Similarly, in the regime $\phi_2 \in [\phi_2^o, \phi_2^c]$, $\overline{\Delta}_1(n_1, n_2)$ can be rewritten as

$$\inf_{t>0} \sup_{s\in(0,t]} \sup_{\theta\in\mathbb{R}^2} \left(\theta^{\mathrm{T}} \begin{pmatrix} z_1 \\ z_2 \end{pmatrix} - \frac{1}{2}\theta^{\mathrm{T}} \begin{pmatrix} n_1 v_1(t) + n_2 v_2(t) & n_2\Gamma_2(s,t) \\ n_2\Gamma_2(s,t) & n_2 v_2(s) \end{pmatrix} \theta \right),$$

where it is recalled that z_1 depends on t and z_2 on s, see Equation (13.8). The optimizing θ is given by

$$\theta^\star = \begin{pmatrix} n_1 v_1(t) + n_2 v_2(t) & n_2\Gamma_2(s,t) \\ n_2\Gamma_2(s,t) & n_2 v_2(s) \end{pmatrix}^{-1} \begin{pmatrix} z_1(t, n_1, n_2) \\ z_2(s, n_1, n_2) \end{pmatrix}.$$

Straightforward computations give that, with the optimizing s, t denoted by $s^\star, t^\star$,

$$\frac{\partial \overline{\Delta}_1(n_1, n_2)}{\partial n_1} = -\theta_1^\star \mu_1 t^\star - \frac{1}{2}(\theta_1^\star)^2 v_1(t^\star),$$

$$\frac{\partial \overline{\Delta}_1(n_1, n_2)}{\partial n_2} = -\theta_1^\star \mu_2 t^\star - \theta_2^\star \mu_2 s^\star - \frac{1}{2}\theta^{\star\mathrm{T}} \begin{pmatrix} v_2(t^\star) & \Gamma_2(s^\star, t^\star) \\ \Gamma_2(s^\star, t^\star) & v_2(s^\star) \end{pmatrix} \theta^\star.$$

(iii) In the third regime $\phi_2 \in [\phi_2^c, 1]$,

$$\overline{\Delta}_1(n_1, n_2) = \inf_{t>0} \sup_{\theta \in \mathbb{R}} \left(\theta z_1(t, n_1, n_2) - \frac{1}{2}\theta^2 (n_1 v_1(t) + n_2 v_2(t)) \right).$$

The inner supremum is attained for

$$\theta^\star = \frac{z_1(t, n_1, n_2)}{n_1 v_1(t) + n_2 v_2(t)}.$$

Denoting the optimizing t by $t^\star$, we derive

$$\frac{\partial \overline{\Delta}_1(n_1, n_2)}{\partial n_1} = -\theta^\star \mu_1 t^\star - \frac{1}{2}(\theta^\star)^2 v_1(t^\star),$$

$$\frac{\partial \overline{\Delta}_1(n_1, n_2)}{\partial n_2} = -\theta^\star \mu_2 t^\star - \frac{1}{2}(\theta^\star)^2 v_2(t^\star).$$

Now we consider the special case that both types of sources correspond to Brownian motions. We assume $v_1(t) = \lambda_1 t$, $v_2(t) = \lambda_2 t$. We again consider the three regimes separately. We have explicit formulae for the 'critical' values of ϕ_2:

$$\phi_2^c = 1 - \frac{n_1\lambda_1 - n_2\lambda_2}{n_1\lambda_1 + n_2\lambda_2}\left(1 - \frac{n_1\mu_1 + n_2\mu_2}{\mathrm{C}}\right) - \frac{n_1\mu_1}{\mathrm{C}}; \qquad \phi_2^o = \frac{n_2\mu_2}{\mathrm{C}}.$$

(i) In this case

$$t^\star = \frac{\mathrm{B}_1}{\phi_1 \mathrm{C} - n_1\mu_1}; \qquad \overline{\Delta}_1(n) = 2\,\frac{\phi_1 \mathrm{C} - n_1\mu_1}{n_1\lambda_1}\,\mathrm{B}_1.$$

This yields:

$$\frac{\partial \overline{\Delta}_1}{\partial n_1} = -2\mathrm{B}_1 \frac{\phi_1 \mathrm{C}}{n_1^2 \lambda_1}; \qquad \frac{\partial \overline{\Delta}_1}{\partial n_2} = 0.$$

(ii) In this case

$$t^\star = \mathrm{B}_1 \Bigg/ \sqrt{(\phi_1\mathrm{C} - n_1\mu_1)^2 + (\phi_2\mathrm{C} - n_2\mu_2)^2 \frac{n_1\lambda_1}{n_2\lambda_2}}\,;$$

$$\overline{\Delta}_1(n_1, n_2) = \frac{1}{2}\left(\frac{(\mathrm{B}_1 + (\phi_1\mathrm{C} - n_1\mu_1)t^\star)^2}{n_1\lambda_1 t^\star} + \frac{(\phi_2\mathrm{C} - n_2\mu_2)^2}{n_2\lambda_2}\, t^\star\right).$$

Also $s^\star = t^\star$. This yields:

$$\frac{\partial\overline{\Delta}_1}{\partial n_1} = -(\mathrm{B}_1 + (\phi_1\mathrm{C} - n_1\mu_1)t^\star)\,\frac{\mu_1}{n_1\lambda_1} - \frac{1}{2}\frac{(\mathrm{B}_1 + (\phi_1\mathrm{C} - n_1\mu_1)t^\star)^2}{n_1^2\lambda_1 t^\star};$$

$$\frac{\partial\overline{\Delta}_1}{\partial n_2} = -(\phi_2\mathrm{C} - n_2\mu_2)t^\star\,\frac{\mu_2}{n_2\lambda_2} - \frac{1}{2}\frac{(\phi_2\mathrm{C} - n_2\mu_2)^2}{n_2^2\lambda_2}\, t^\star.$$

(iii) In this case

$$t^\star = \frac{\mathrm{B}_1}{\mathrm{C} - n_1\mu_1 - n_2\mu_2}; \qquad \overline{\Delta}_1(n_1, n_2) = 2\,\frac{\mathrm{C} - n_1\mu_1 - n_2\mu_2}{n_1\lambda_1 + n_2\lambda_2}\,\mathrm{B}_1.$$

This yields:

$$\frac{\partial\overline{\Delta}_1}{\partial n_1} = -\frac{2\mathrm{B}_1\mu_1}{n_1\lambda_1 + n_2\lambda_2} - 2\mathrm{B}_1\lambda_1\,\frac{\mathrm{C} - n_1\mu_1 - n_2\mu_2}{(n_1\lambda_1 + n_2\lambda_2)^2};$$

$$\frac{\partial\overline{\Delta}_1}{\partial n_2} = -\frac{2\mathrm{B}_1\mu_2}{n_1\lambda_1 + n_2\lambda_2} - 2\mathrm{B}_1\lambda_2\,\frac{\mathrm{C} - n_1\mu_1 - n_2\mu_2}{(n_1\lambda_1 + n_2\lambda_2)^2}.$$

Chapter 14

A link dimensioning formula and empirical support

As argued in the introduction, bandwidth dimensioning (or, provisioning) procedures require a thorough understanding of the relation between the characteristics of the offered traffic, the link speed, and the resulting performance. The availability of such a relation enables the selection of a link capacity that guarantees that the aggregate rate of the offered traffic exceeds the link capacity less than some predefined (small) fraction of time. We usually refer to such a performance target as 'Quality of Service' (QoS).

In such a 'QoS by provisioning' approach (i.e., allocating enough bandwidth to meet the QoS requirements of all applications present), all traffic streams are treated in the same way. An alternative is to use traffic differentiation mechanisms, such as those of the *IntServ* approach [294] developed within the Internet Engineering Task Force (IETF). IntServ enables stringent QoS guarantees, by relying on per-flow admission control, but suffers from the inherent scalability problems. Therefore, it may be applied at the edge of the network (where the number of flows is relatively low), but not likely in the core.

The *DiffServ* [155, 284] architecture can be seen as a hybrid approach between pure provisioning and IntServ: agreements are made for aggregates of flows rather than microflows, thus solving the scalability problems. However, then the lack of admission control demands adequate bandwidth provisioning, in order to actually realize the QoS requirements. Thus, both in the pure provisioning approach and in DiffServ, i.e., the most promising QoS-enabling mechanisms, a prominent role is played by bandwidth provisioning procedures.

Bandwidth provisioning has several other advantages over traffic differentiation mechanisms; see also, e.g., [102]. In the first place, the complexity of the network routers can be kept relatively low, as no advanced scheduling and prioritization

Large deviations for Gaussian queues M. Mandjes

capabilities are needed. Secondly, traffic differentiation mechanisms require that the parameters involved are 'tuned well', in order to meet the QoS needs of the different classes. This usually requires the selection of various parameters (for instance, weights in GPS queuing algorithms; see Chapter 13).

Bandwidth provisioning has a number of obvious drawbacks as well: the lack of any QoS differentiation mechanism dictates that all flows should be given the most stringent QoS requirement, thus reducing the efficiency of the network (in terms of maximum achievable utilization). However, it is expected that this effect would be mitigated if there is a high degree of aggregation, even in the presence of heterogeneous QoS requirements across users, as argued in, e.g., the introduction of [102] and in [151].

Clearly, the challenge for a network operator is to provision bandwidth such that an appropriate trade-off between efficiency and QoS is achieved; without sufficient bandwidth provisioning, the performance of the network will drop below tolerable levels, whereas by provisioning too much the performance hardly improves and is potentially already better than needed to meet the users' QoS requirements, thus leading to inefficient use of resources.

The bandwidth provisioning procedures currently used in practice are usually very crude. A common procedure is to (i) use Multi Router Traffic Grapher (MRTG) [230] to get coarse measurement data (e.g., 5-min intervals), (ii) determine the average traffic rate during these 5-min intervals, and (iii) estimate the required capacity by some quantile of the 5-min measurement data; a commonly used value is the 95% quantile. This procedure is sometimes 'refined' by focusing on certain parts of the day (for instance office hours, in the case of business customers), or by adding safety and growth margins. The main shortcoming of this approach is that it is not clear how the coarse measurement data relates to the traffic behavior at timescales relevant for QoS. More precisely, a crucial question is whether the coarse measurements give any useful information on the capacity needed: QoS degradation experienced by the users may be caused by fluctuations of the offered traffic on a much smaller timescale, e.g., seconds (file transfers, web browsing) or even less (interactive, real-time applications).

Goal of this chapter. The above remarks explain the need for accurate and reliable provisioning procedures that require a *minimal measurement effort*. Therefore, our goal is to develop an 'interpolation' formula that predicts the bandwidth requirement on relatively short timescales (say the order of 1 s), by using large timescale measurements (e.g., in the order of 5 min). In our approach, we express QoS in terms of the probability (to be interpreted as fraction of time) that, on a predefined timescale T, the traffic supply exceeds the available bandwidth. The bandwidth c should be chosen such that this probability does not exceed some given bound ϵ.

We emphasize that the timescale T and performance target ϵ are case specific: they are parameters of our model, and can be chosen on the basis of the specific needs of the most demanding application involved. We remark that in this setting buffers are not explicitly taken into account; evidently, there is a relation between the timescale T in which the traffic rate exceeds the link rate and the buffer size needed to absorb the excess traffic.

Our approach relies on minimal modeling assumptions. Notably, we assume that the underlying traffic model is Gaussian; empirical evidence for this assumption can be found in Section 3.1 and, e.g., [102, 156, 211]. For the special case that the access rates of individual users are constrained (with a peak rate r), we can use (the Gaussian counterpart of) M/G/∞ type of input processes (see Section 2.5), leading to an elegant, explicit formula for the required bandwidth. As before, the M/G/∞ input model corresponds to a flow arrival process that is Poisson with rate λ, and flow durations that are independent and identically distributed as some random variable D (with $\delta := \mathbb{E}D$). We find that, by measuring a load $\mu = \lambda\delta r$ (in Mbps), the required bandwidth (to meet the QoS criterion) has the form

$$\mathrm{C}(T, \epsilon) = \mu + \alpha\sqrt{\mu}. \tag{14.1}$$

It is clear that the μ can be estimated by coarse traffic measurements (e.g., 5- or 15-min measurements). The α depends on the characteristics of the individual flows, and its estimation requires detailed (i.e., on timescale T) measurements. In many situations, however, there are reasons to believe that the α is fairly constant in time; the estimate needs to be updated only when one expects that the flow characteristics have changed (for instance due to the introduction of new applications).

We expect that the provisioning approach advocated in this chapter would extend to several other types of networking environments. In situations with large numbers of (more or less) i.i.d. users, the Gaussian assumption will apply due to central limit type of arguments, and hence the procedure followed goes through. In this chapter, we have validated the approach in an IP setting, but, evidently, these tests could also have been performed in other types of (packet-switched) networks.

Apart from its simplicity, the bandwidth dimensioning formula (14.1) has a number of attractive features. In the first place it is *transparent*, in that the impact of changing the 'QoS parameters' (i.e., T and ϵ) on α is explicitly given. Secondly, the provisioning rule is to some extent *insensitive*: α does not depend on λ, but depends just on characteristics of the individual flows, i.e., the flow duration D and the peak rate r. This property enables a simple estimate of the additionally required bandwidth to be obtained if in a future scenario traffic growth is mainly due to a change in λ (e.g., due to growth of the number of subscribers). Furthermore, the analytical expression for α provides valuable insight into the impact of changes in D and r.

Our bandwidth provisioning rule has been empirically investigated through the analysis of extensive traffic measurements in various network environments with different aggregation levels, user populations, etc.

Before we further explain our dimensioning approach, we first reflect on a number of related papers. The approach of [102] is related to ours, in that it uses bandwidth provisioning based on traffic measurements to deliver QoS. An important difference, however, is that in their case the performance metric is packet delay (whereas we consider the probability of exceeding the link rate in intervals of prespecified length). Also, in [102] measurements are used to fit the Gaussian model, and subsequently this model is used to estimate the bandwidth needed; this is an essential difference compared to our work, where our objective is to minimize the required measurement input/effort, such that bandwidth provisioning is done on the basis of only coarse measurements. Another closely related paper is [100], where several bandwidth provisioning rules are empirically validated.

The remainder of this chapter is organized as follows. In Section 14.1, we describe the objectives of this study and the proposed modeling approach in detail; next, we provide the analysis leading to our bandwidth provisioning rule. Numerical results of our modeling and analysis are presented and discussed in Section 14.2. In Section 14.3, the bandwidth provisioning rule is assessed through extensive measurements performed in several operational network environments. Section 14.4 provides implementation aspects. In the bibliographical notes, we present some background on dimensioning.

14.1 Objectives, modeling, and analysis

The typical network environment we focus on in this chapter is an IP network with a considerable number of users generating mostly TCP traffic (from, e.g., web browsing, downloading music and video, etc.). Then, the main objective of bandwidth provisioning is to take care that the links are more or less 'transparent' to the users, in that the users should not (or almost never) perceive any degradation of their QoS due to a lack of bandwidth.

Clearly, this objective will be achieved when the link rate is chosen such that only during a small fraction of time ϵ the aggregate rate of the offered traffic (measured on a sufficiently small timescale T) exceeds the link rate. The values to be chosen for the QoS parameters T and ϵ typically depend on the specific needs of the application(s) involved. Informally, the more interactive the application, the smaller the T and ϵ that should be chosen. In more formal terms, our objective can be stated as follows: the fraction ('probability') of sample intervals of length T in which the aggregate offered traffic exceeds the available link capacity C should be below ϵ, for prespecified values of T and ϵ. In other words, recalling that $A(t)$ denotes the amount of traffic offered in an arbitrary time interval of length t,

$$\mathbb{P}(A(T) \geq \mathrm{C}T) \leq \epsilon. \tag{14.2}$$

For provisioning purposes, the crucial question is, for given T and ϵ, find the minimally required bandwidth $\mathrm{C}(T, \epsilon)$ to meet the target.

In the remainder of this section, we derive explicit, tractable expressions for our target probability $\mathbb{P}(A(T) \geq \mathrm{C}T)$; see Equation (14.2). We first do this for a given traffic input process $(A(t), t \geq 0)$, where the only explicit assumption imposed is that it has stationary increments. It is noted that this assumption will likely hold on timescales that are not too long (up to, say, hours); on longer timescales there is no stationarity due to diurnal patterns and growth (or decline) of the number of subscriptions (timescale of weeks, months, ...). Once we have an expression for (an upper bound to) $\mathbb{P}(A(T) \geq cT)$, we can find the minimal c required to make sure that this probability is kept below ϵ. We thus find the required bandwidth $c(T, \epsilon)$. It is expected that this function would decrease in both T and ϵ (as increasing T or ϵ makes the service requirement less stringent).

Then, we make the traffic arrival process more specific: motivated by Chapter 3, we assume Gaussian input. It turns out that then that the formula for $c(T, \epsilon)$ simplifies considerably. Finally, we focus on the model in which the input corresponds to the Gaussian counterpart of the M/G/∞ input model (see Section 2.5), and find formula (14.1).

General arrival process. We can apply the Chernoff bound (see Theorem 4.1.2):

$$\mathbb{P}(A(T) \geq cT) \leq \min_{\theta \geq 0} \left(\mathbb{E}e^{\theta A(T) - \theta c T} \right). \tag{14.3}$$

This bound is unfortunately rather implicit, as it involves both the computation of the moment-generating function $\mathbb{E}\exp(\theta A(T))$ and an optimization over $\theta \geq 0$. Note that $c(T, \epsilon)$ could be chosen as the smallest number c such that the right-hand side of Equation (14.3) is smaller than ϵ:

$$c(T, \epsilon) := \min \left\{ c : \min_{\theta \geq 0} \left(\mathbb{E}e^{\theta A(T) - \theta c T} \right) \leq \epsilon \right\}.$$

Rearranging the terms, we find that equivalently we are looking for the smallest c such that there is a $\theta \geq 0$ such that

$$c \geq \frac{\log \mathbb{E}e^{\theta A(T)} - \log \epsilon}{\theta T}.$$

This c is obviously equal to the infimum of the right-hand side over $\theta \geq 0$:

$$c(T, \epsilon) = \min_{\theta \geq 0} \frac{\log \mathbb{E}e^{\theta A(T)} - \log \epsilon}{\theta T}. \tag{14.4}$$

Gaussian traffic. Assuming that $A(T)$ contains the contributions of many individual users (who act more or less independently), it is justified (based on the central limit theorem) to assume that $A(T)$ is Gaussian, provided T is not too small; see Chapter 3. In other words, $A(T)$ is now distributed $\mathcal{N}(\mu T, v(T))$, for some load μ (in Mbps) and variance $v(T)$ (in Mb^2). For this Gaussian case, we now show that we can determine the right-hand side of Equation (14.4) explicitly.

The first step is to recall that $\log \mathbb{E} e^{\theta A(T)} = \theta \mu T + \frac{1}{2}\theta^2 v(T)$. The calculation of the minimum in Equation (14.4) is now straightforward:

$$
\begin{aligned}
c(T, \epsilon) &= \mu + \min_{\theta \geq 0} \left(\frac{\theta v(T)}{2T} - \frac{\log \epsilon}{\theta T} \right) \\
&= \mu + \frac{1}{T}\sqrt{(-2 \log \epsilon) \cdot v(T)},
\end{aligned} \tag{14.5}
$$

and the minimum is attained at $\theta = \sqrt{(-2\log \epsilon)/v(T)}$. We remark that, evidently, $c(T, \epsilon)$ can also be found by first recalling the Chernoff bound for Gaussian traffic

$$
\mathbb{P}(A(T) \geq cT) \leq \exp\left(-\frac{1}{2} \frac{(c-\mu)^2 T^2}{v(T)} \right); \tag{14.6}
$$

then it is easily checked that Equation (14.5) is the smallest c such that the value in Equation (14.6) is below ϵ. As for any input process with stationary increments, $v(\cdot)$ cannot increase faster than quadratically (in fact, a quadratic function $v(\cdot)$ corresponds to perfect positive correlation), $\sqrt{v(T)}/T$ is decreasing in T, and hence also the function $c(T, \epsilon)$–the longer the T, the easier it is to meet the QoS requirement. Also, the higher the ϵ, the easier it is to meet the requirement, which is reflected by the fact that the function decreases in ϵ.

Exercise 14.1.1 Let jobs arrive according to a Poisson process, and let the sizes of these jobs be i.i.d. We call such an input process a *compound Poisson process.* (i) Characterize the required service capacity for a compound Poisson input process. Find an explicit expression for the case of exponentially distributed jobs. (ii) Compare this with the required service capacity of the Gaussian counterpart.

Solution. (i) With D denoting, as usual, a random variable that represents the job size, we have that $\log \mathbb{E}\exp(\theta A(T)) = \lambda T(\mathbb{E}e^{\theta D} - 1)$. We therefore obtain

$$
c = \inf_{\theta \geq 0} \left(\frac{\lambda}{\theta} \left(\mathbb{E}e^{\theta D} - 1 \right) - \frac{\log \epsilon}{\theta T} \right).
$$

Now, consider the case where D is exponential with mean $\delta := 1/\nu$. Then, it is straightforward to derive that the optimizing θ is

$$
\theta^\star = \frac{\nu}{1 + \sqrt{(\lambda T)/f}}; \quad f := -\log \epsilon.
$$

After some algebra we obtain

$$
c = \frac{\lambda}{\nu} + \frac{2}{\nu}\sqrt{\frac{\lambda f}{T}} + \frac{f}{\nu T} = \frac{1}{\nu}\left(\sqrt{\lambda} + \sqrt{\frac{f}{T}} \right)^2. \tag{14.7}
$$

Observe that this bandwidth is indeed larger than the offered load λ/ν.

(ii) It can be checked that $v(T) = \lambda T \cdot \mathbb{V}\mathrm{ar} D$, so that

$$c = \lambda\,\mathbb{E}D + \sqrt{\frac{(-2\log\epsilon)\lambda\mathbb{V}\mathrm{ar}D}{T}}.$$

For exponential jobs this reduces to

$$c = \frac{\lambda}{\nu} + \frac{1}{\nu}\sqrt{\frac{2\lambda f}{T}}.$$

It is easily verified that this value is smaller than that in Equation (14.7). ◇

Remark 14.1.2 Effective bandwidth. There is some reminiscence between formula (14.5) and the effective bandwidth concept proposed earlier in the literature (see, e.g., [92, 95, 136, 149]) but there are major differences as well.

One of the key attractive properties of effective bandwidths is their 'additivity'. To explain in more detail, consider two sources, to each of which a bandwidth is assigned that is parameterized by the QoS-criterion, say $c_1(\epsilon)$ and $c_2(\epsilon)$. Then, effective-bandwidth theory says that the bandwidth needed by the superposition of the two sources can be estimated by the sum of the individual effective bandwidths. In self-evident notation,

$$c_{1+2}(\epsilon) = c_1(\epsilon) + c_2(\epsilon).$$

As said, there are differences between these effective bandwidths and our formula for $c(T,\epsilon)$. Most importantly, it can be argued that interpreting Equation (14.5) as an effective bandwidth would lead to a bandwidth allocation that is too pessimistic: noting that

$$\sqrt{v_1(T) + v_2(T)} \leq \sqrt{v_1(T)} + \sqrt{v_2(T)},$$

we observe that the amount of bandwidth to be provisioned for the aggregate input could be substantially less than the sum of the individually required bandwidths.

In Remark 2 of [27], the relation with the equivalent capacity formulae of [122] is discussed. ◇

The formula for $c(T,\epsilon)$ indicates that, given that we are able to estimate the load μ and the variance $v(T)$ on the 'advertised' timescale T, we have found a straightforward provisioning rule. In the following, we focus on the special case of (the Gaussian counterpart of) M/G/∞ input; in this (still quite general) case, the expressions simplify even further.

Gaussian counterpart of M/G/∞ input. From Equation (2.4) we see immediately that, in case of M/G/∞ input, it holds that $v(T) = \lambda g(T, r, D)$ for some function g; importantly, this g depends on the input process only through r and the characteristics of the distribution of the duration D, but is independent of the arrival rate λ. Equivalently, we can write

$$v(T) = \lambda\delta r\, h(T, r, D) = \mu\, h(T, r, D),$$

for some h that does not depend on λ. As a result, bandwidth provisioning formula (14.5) simplifies to

$$c(T, \epsilon) = \mu + \alpha\sqrt{\mu}, \quad \text{where} \quad \alpha \equiv \alpha(T, r, D) = \frac{1}{T}\sqrt{(-2\log\epsilon)\cdot h(T, r, D)}. \tag{14.8}$$

For instance, for exponentially distributed D, we have that

$$\alpha = \left(\frac{T}{\delta}\right)^{-1}\sqrt{(-2\log\epsilon)\cdot 2r(e^{-T/\delta} - 1 + T/\delta)}.$$

Observe that in this case α depends on T only through the ratio T/δ. For other distributions, α can be computed in a similar way.

Remark 14.1.3 If T is small (i.e., small compared to δ), then α becomes insensitive in the flow duration D. This can be seen as follows. From Equation (2.4), it can be derived that $v(T)/T^2 \to \mu r$ if $T \downarrow 0$. Then, Equation (14.5) yields

$$c(T, \epsilon) \approx \mu + \sqrt{(-2\log\epsilon)\cdot\mu r},$$

exclusively depending on μ, for small T. This result can be derived differently, by noting that for $T \downarrow 0$, the performance criterion reduces to requiring that the number of active users does not exceed c/r. It is well known that the number of active users in this M/G/∞ system has a Poisson distribution with mean $\lambda\delta$; this explains the insensitivity.

Remark 14.1.4 The case of exponential flow lengths can be easily extended to, e.g., hyperexponentially distributed flows; we say that a random variable X is hyperexponentially distributed [275, p. 446] if with probability $p \in (0, 1)$ it is distributed exponentially with mean δ_1, and else exponentially with mean δ_2. Then the hyperexponential case is just the situation with two flow types feeding independently into the link (each type has its own exponential flow length distribution); note that the variance of the total traffic is equal to the sum of the variances of the traffic generated by each of the different exponential flow types.

Remark 14.1.5 The above approach assumes that traffic arrives as 'fluid': it is generated at a constant rate r. It is perhaps more realistic to assume that, during the flow's 'life time', traffic arrives as a Poisson stream of packets (of size s);

the rate of the Poisson process is γ, where γs is equal to r. Denoting the above fluid-based variance function by $v_{\rm f}(T \mid r)$ and the packet-based variance function by $v_{\rm p}(T \mid \gamma, s)$, it can be verified that

$$v_{\rm p}(T \mid \gamma, s) = v_{\rm f}(T \mid r) + \mu s T. \tag{14.9}$$

irrespective of the flow duration distribution. Importantly, the provisioning formula $c(T, \epsilon) = \mu + \alpha\sqrt{\mu}$ remains valid (for an α that does not depend on λ).

14.2 Numerical study

We now illustrate a number of interesting features of our provisioning formula (14.5). We do so by using the traffic parameters and QoS parameters displayed in Table 14.1, unless specified otherwise.

Experiment 1: Fluid model vs. packet-level model. Figure 14.1 shows the required capacity obtained by the packet-level and fluid model as a function of T, for various mean flow durations δ. It is seen that for large values of the timescale T, both models obtain the same required capacity. This can be understood by looking at the extra term $\mu s T$ of Equation (14.9), which influence on $c(T, \epsilon)$ becomes negligible for increasing T, cf. Equation (14.5).

For $T \downarrow 0$, the required capacity obtained by the packet-level model behaves like

$$c(T, \epsilon) \sim \mu + \sqrt{\mu s}\frac{1}{\sqrt{T}}\sqrt{-2\log\epsilon}$$

and hence $c(T, \epsilon) \uparrow \infty$ as $T \downarrow 0$, whereas the required capacity of the fluid model converges to $\mu + \sqrt{(-2\log\epsilon)\cdot\mu r}$, as was already argued in Remark 14.1.3. The fast increase in the required capacity in the packet-level model for a decreasing

Traffic Model			QoS Parameters		
δ	1	s	T	1	s
Distribution of D	Exponential		ϵ	0.01	-
ρ	10	Mbps			
Model	Fluid				

Fluid Model			Packet-level Model		
r	1	Mbps	γ	83.3	Packet/s
			s	1500	Bytes

Table 14.1: Default parameter settings for the numerical results.

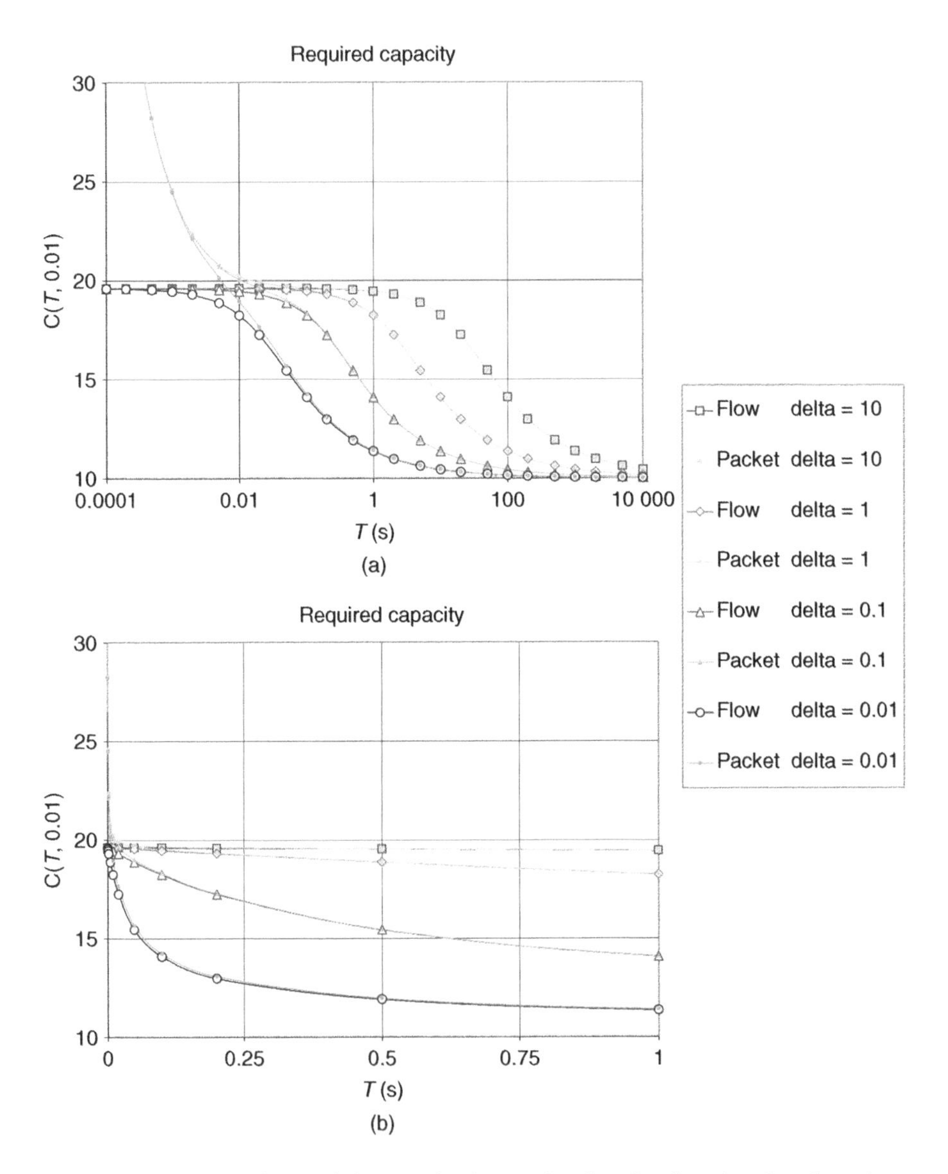

Figure 14.1: Comparison of the required capacity for the flow-level and packet-level models as a function of the timescale T. (a) Logarithmic axis. (b) Linear axis.

timescale T was also observed in e.g., [102]. Note that, in fact, the required capacity is not influenced by the absolute value of T, but rather by the ratio of T/δ. The graph in Figure 14.1(b) shows the same results as that in (a), but now on a linear axis and only for $T \in [0, 1]$. In the remainder of this section, we restrict ourselves to the fluid-based model, as we will focus on situations with values of $T/\delta > 0.1$, for which the required capacity is almost identical in both models.

Experiment 2: Impact of the flow duration distribution. Next, we investigate the impact of the flow duration distribution on the required capacity. Figure 14.2 contains four graphs with results for hyperexponentially distributed flow durations D. Each graph shows, for a particular value of the mean flow size δ, the required capacity as a function of the offered load μ, for different Coefficients of Variation (CoV) of D. These graphs show that the required capacity is almost insensitive to the CoV for the long- ($\delta = 10$ s) and short-flow durations ($\delta = 0.01$ s). For the other cases ($\delta \in \{0.1, 1\}$ s), the required capacity is somewhat more sensitive to the CoV. The graphs show that for hyperexponentially distributed flow durations less capacity is required if the CoV increases.

It should also be noticed that the required capacity for $T = 0$, also shown in Figure 14.2, corresponds to the often-used M/G/∞ bandwidth provisioning approach; cf. the discussion in Remark 14.1.3 and the discussion on Experiment 1. The numerical results show that particularly for short flow durations significantly less capacity is required than suggested by the classical M/G/∞ approach; for longer flows, this effect is less pronounced.

Experiment 3: Impact of QoS parameter ϵ. Figure 14.3 shows the required capacity as a function of the QoS requirement ϵ, which specifies the fraction of intervals in which the offered traffic may exceed the link capacity. A larger value of ϵ corresponds to relaxation of the QoS requirement, and hence less capacity is needed. Obviously, for $\epsilon \uparrow 1$, the required capacity converges to the long-term average load $\mu = 10$. For $\epsilon \downarrow 0$, the required capacity increases rapidly to infinity (according to $\sqrt{-2\log\epsilon}$).

Experiment 4: Impact of the CoV of the flow duration distribution. To investigate the impact of the flow duration characteristics, we have computed the required capacity for exponential, hyperexponential, and Pareto distributed flow durations with different CoV values; see Figure 14.4(a). It shows that the required capacity is almost insensitive to the flow-duration distribution. Note that for hyperexponentially distributed flow durations the required capacity slightly decreases for increasing CoV, while for Pareto distributed flow durations the required capacity slightly increases for increasing CoV.

Experiment 5: Impact of the access rate. Finally, Figure 14.4(b) studies the effect of the access rate r on the required capacity. Three values of the access rate r and the mean flow duration δ are chosen such that the mean flow size $\delta \cdot r$ remains constant. As expected, the required capacity increases considerably when r becomes larger (i.e., the traffic burstiness grows). The results in this graph for hyperexponential and Pareto flow sizes confirm the conclusions of Experiments 2 and 4 that the required capacity is almost insensitive to the flow-duration distribution.

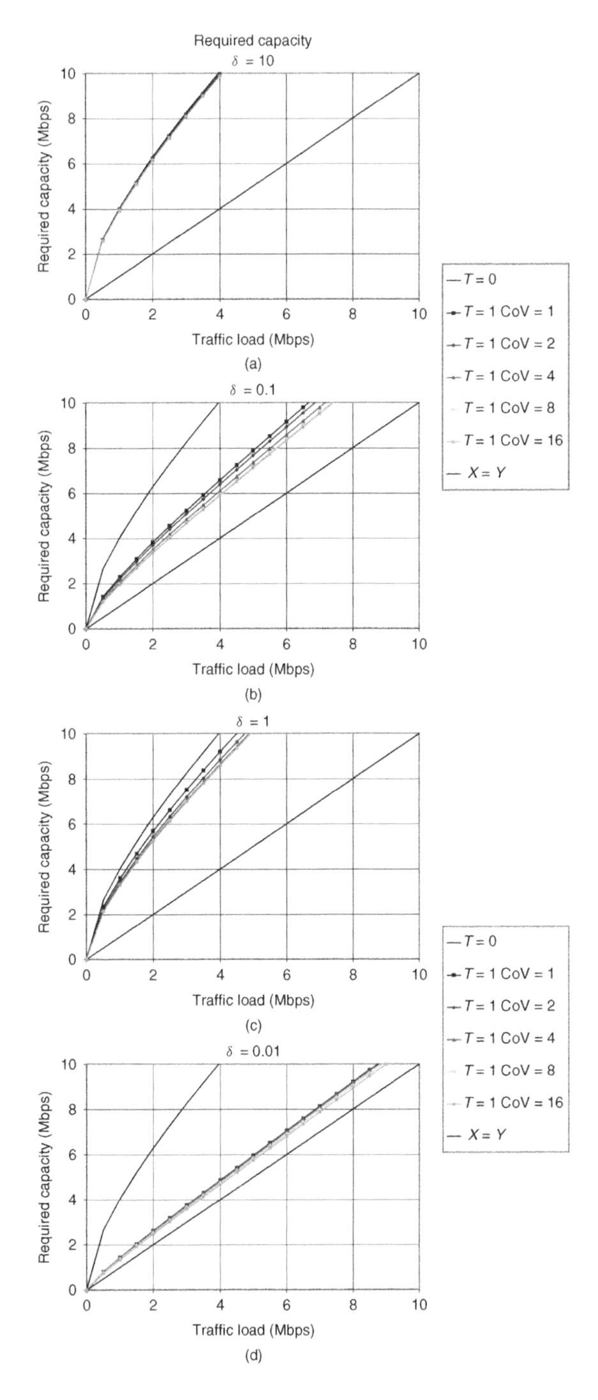

Figure 14.2: Required capacity for hyper-exponential flow durations with different means and CoVs.

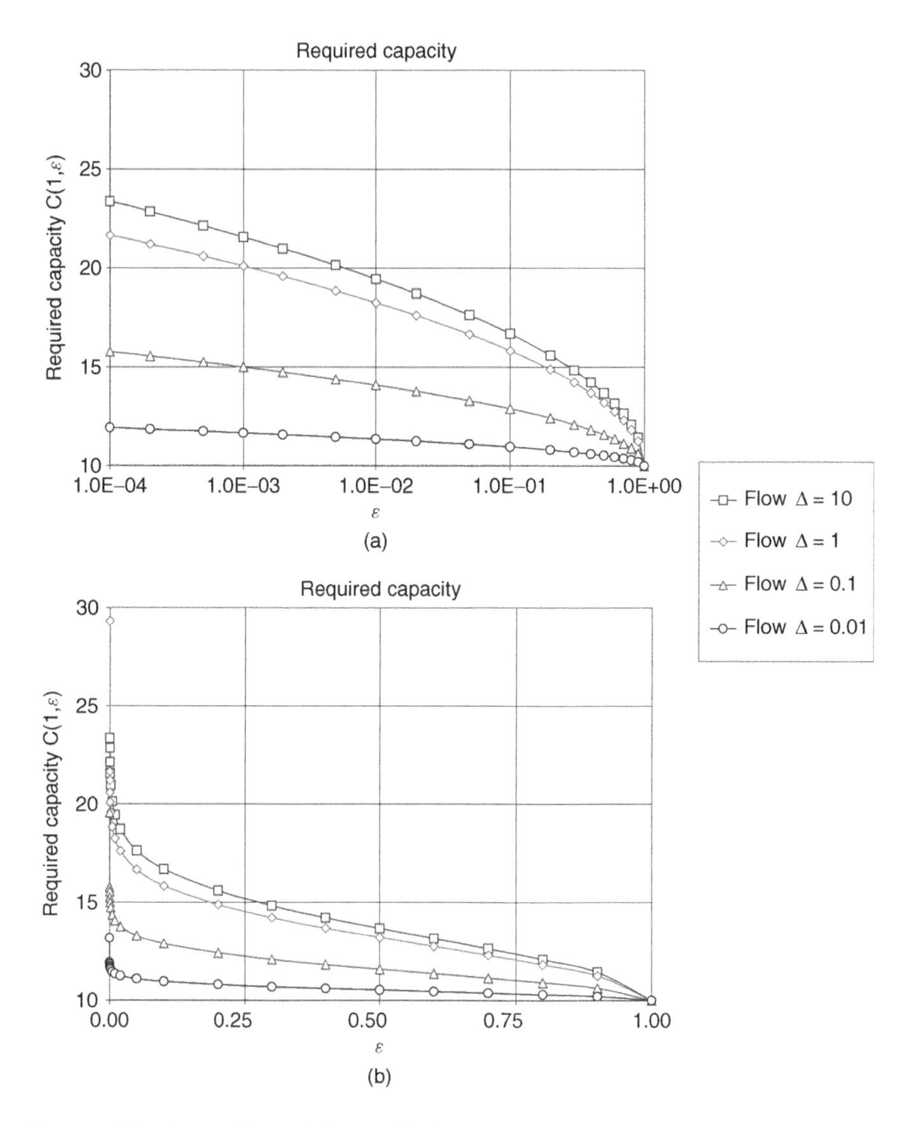

Figure 14.3: Comparison of the required capacity for the flow-level and packet-level models as a function of the QoS parameter ϵ.

14.3 Empirical study

In this section, we analyze measurement results obtained in operational network environments in order to validate the modeling approach and bandwidth provisioning rule presented in Section 14.1. In particular, we will investigate the relation

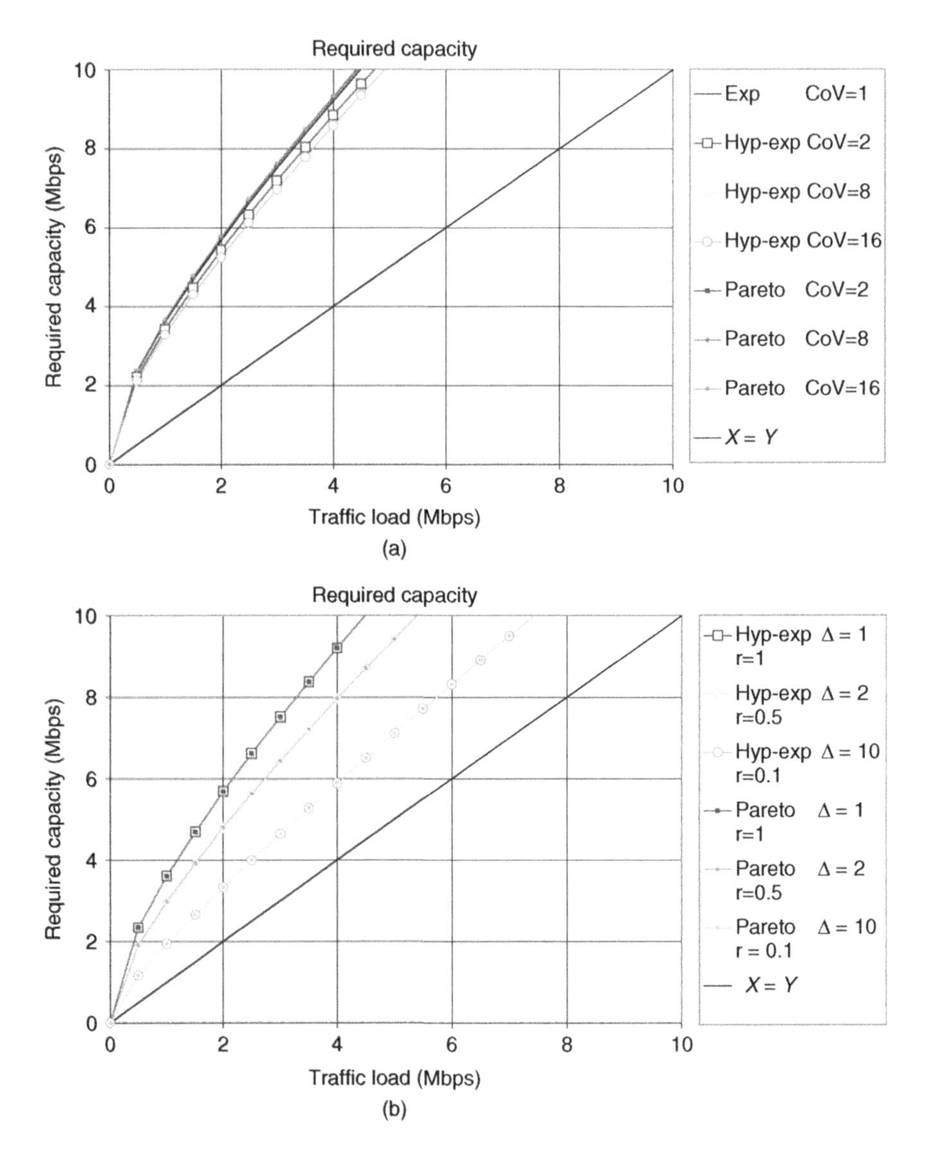

Figure 14.4: (a) Required capacity for different flow duration distribution and CoVs. (b) Required capacity for different access rates.

between measured traffic load values $\hat{\mu}$ during 5-min periods (long enough to assume stationarity) and the traffic fluctuations at a 1-s timescale within these periods. Clearly, if our M/G/∞ traffic modeling assumptions of Section 14.1 apply and if differences in the load μ are caused by changes in the flow arrival rate λ (i.e., the flow size characteristics remain unchanged during the measurement period), then, as a function of μ, for given (T, ϵ), the required bandwidth c_μ should satisfy

$$c_\mu = \mu + \alpha\sqrt{\mu}$$

for some fixed value of α.

To assess the validity of this relation for a specific networking environment, we have carried out measurements in a national IP network providing Internet access to residential Asymmetric Digital Subscriber Line (ADSL) users. Here, the main assumptions made in Section 14.1 in order to justify use of the M/G/∞ input model seem to be satisfied, i.e., the flow peak rates are limited due to the ADSL access rates (which are relatively small compared to the network link rates), and the traffic flows behave more or less independently of each other (the IP network links are generously provisioned and, hence, there is hardly any interaction among the flows). Details on the measurement setup can be found in [27].

We choose the sample size $T = 1$ s, motivated by the fact that this can be assumed to be the timescale that is most relevant for the QoS perception of end users of typical applications like web browsing. Elementary transactions, such as retrieving single web pages, are normally completed in intervals roughly in the order of 1 s. If the network performance is seriously degraded during one or several seconds, then this will affect the quality as perceived by the users.

Time was split into 5-min chunks over which the load μ was determined. In addition, for each 5-min period, the 99% quantile of the 1-s measurements was determined. This quantile was assumed to indicate the minimum capacity c that is needed to fulfill the QoS requirement $\mathbb{P}(A(T) \geq cT) \leq \epsilon$, with $\epsilon = 1\%$ and $T = 1$ s.

The graph in Figure 14.5(a) results from the measurements on 11 links at various locations. For orientation purposes, we have added a straight line representing the unity relation ('$y = x$'). It is remarkable how the 99% quantiles almost form a solid curve. We fitted a function $\mu + \alpha\sqrt{\mu}$, such that roughly 95% of the 99% quantiles are lower or equal to this function. The reason for fitting an upper bound, instead of finding the function that gives the minimum least square deviation, is that eventually we intend to use this function for capacity planning: then it is better to overestimate the required bandwidth than to underestimate it. The graph shows an extremely nice fit for the function $c = \mu + 1.0\sqrt{\mu}$ (with c and μ expressed in Mbps).

At the time of the measurements, the busiest links did not carry more traffic than 20 Mbps during the busiest hours, so we could not verify that the found upper bound also holds for higher traffic volumes. To overcome this problem at least partly, we synthesized artificial traffic measurements by taking the superposition of the traffic measured on several (unrelated) links. The graph in Figure 14.5(b) shows the results of this experiment. As expected on theoretical grounds, the fitted function $c = \mu + 1.0\sqrt{\mu}$ remains valid.

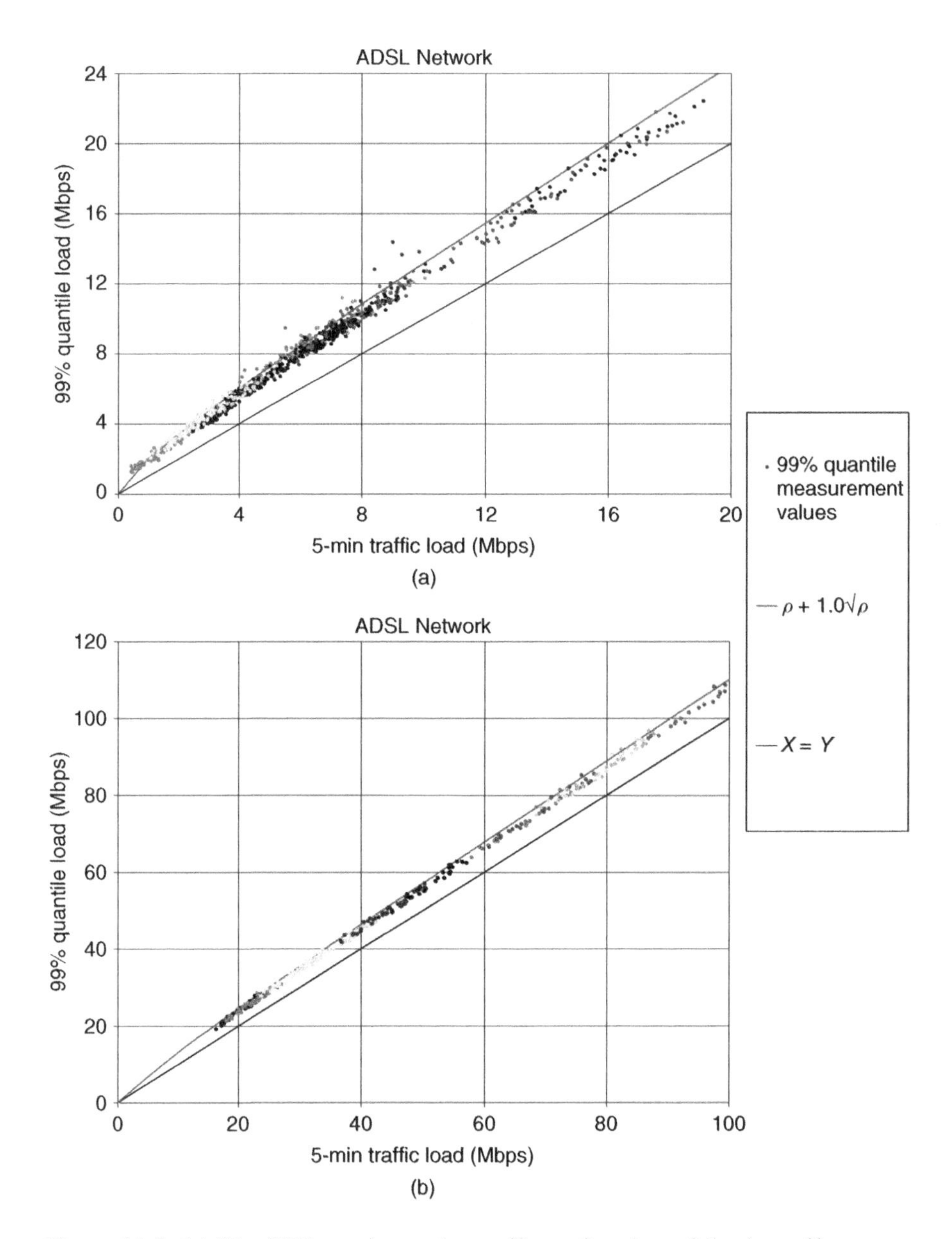

Figure 14.5: (a) The 99% maximum 1-s traffic as function of 5-min traffic mean. (b) Synthesized traffic measurements for higher traffic volumes.

14.4 Implementation aspects

Our formula (14.8) for the required bandwidth $c(T, \epsilon)$ can be used to develop bandwidth provisioning procedures. Obviously, the first step in this procedure is to verify whether the main M/G/∞ modeling assumptions are satisfied in the network environment under consideration, such that formula (14.8) can indeed be applied. The next step is then to estimate μ and α.

Clearly, μ can be estimated through coarse traffic measurements, as it is just the average load; α, however, contains (detailed) traffic characteristics on timescale T (viz., the variance $v(T)$). In the following text, we propose to estimate α directly. An alternative to this approach, however, would be to fit the flow-size distribution, such that α can be computed by inserting this into the explicit formulae of Section 14.1. Recall that Experiment 2 of Section 14.2 showed that the CoV of the flow-size has just a modest impact on α. Also, fitting the full flow-size distribution has the evident drawback that it requires per-flow measurements. Therefore, we prefer direct estimation of α.

In Section 14.1, it was noticed that α depends on the flow peak rate r and on the parameters of the flow duration D, but, importantly, α does not depend on the flow arrival rate λ; α can be considered as a characteristic of the individual flows. This 'dichotomy' between μ and α gives rise to efficient provisioning procedures. Consider the following two typical situations:

- Situations in which there is a set of links that differ (predominantly) in the number of connected users; across the links, the individual users have essentially the same type of behavior (in terms of the distribution D and the access rate r). Then, α can be estimated by performing detailed measurements at (a part of) the existing links. When a new link is connected, one could obtain an estimate $\hat{\mu}$ of the load by performing coarse measurements (e.g., every 5 min, by using the MRTG tool [230]). Then, the provisioning rule $\hat{\mu} + \hat{\alpha}\sqrt{\hat{\mu}}$ can be used. An example is the ADSL scenario described in Section 14.3, in which one could use $\hat{\alpha} \approx 1.0$ to dimension a new link.
- Growth scenarios in which it is expected that the increase in traffic is (mainly) due to a growing number of subscribers (i.e., the λ), while the user behavior remains unchanged. Here, it suffices to perform infrequent detailed measurements at timescale T, yielding an estimate $\hat{\alpha}$ of α. If a future load $\hat{\mu}$ is envisaged, the required bandwidth can be estimated by the provisioning rule $\hat{\mu} + \hat{\alpha}\sqrt{\hat{\mu}}$.

The estimate of α has to be updated after a certain period (perhaps in the order of months). This should correspond to the time at which it is expected that the 'nature' of the use of resources changes (due to, e.g., new applications, etc.). The explicit formulas for α derived in Section 14.1 are also useful when examining the impact of changes in the user behavior or the QoS parameters. For instance, the impact of an upgrade of the access speed r can be evaluated. Also, one could assess the effect of imposing a stronger or weaker performance criterion ϵ: when replacing ϵ_1 by ϵ_2, α needs to be multiplied by $\sqrt{\log \epsilon_2 / \log \epsilon_1}$.

The measurement period of 5 min mentioned above for estimating the load μ is motivated by the fact that this is the timescale on which measurements in an operational network can be (and are) performed on a routine basis. A higher frequency would be desirable, but this would put a high load on the processing capacity of routers, the transport capacity of management links, etc., particularly if there are many routers and ports involved. On the other hand, measurements performed at lower frequencies (for instance one hour to several hours) are too coarse, as traffic is not likely to be stationary over such long periods. Therefore, 5 min will usually be a suitable trade-off, as this is feasible to measure, and at the same time a reasonable period during which the traffic can still be assumed stationary.

Bibliographical notes

A vast amount of research has been done over the last century in the area of dimensioning of (tele-)communication networks. In order to put this chapter into perspective, here we give a brief overview of the major developments in this area.

Circuit-switched networks. In the traditional telephony setting, the amount of offered traffic (or the *load*, say μ) is expressed in terms of the (dimensionless) unit *Erlang*. Because of the *circuit-switched* nature of those telephony networks, each call is assumed to occupy a fixed amount of bandwidth capacity. In order to provide an adequate performance level, network operators strive to keep the probability of call blocking (i.e., a user not being able to set up a connection to the remote end) below some threshold δ (typically in the order of 1%). Hence, here the crucial question in the context of dimensioning is, what is the lowest (trunk) capacity c such that the probability that calls are blocked, for given traffic load μ, is below δ? This basic question has long been answered, and in fact its analysis goes back to 1909 [96]. Later, many variants of this model also were solved (models with trunk reservation, impatient customers, reattempts, etc.).

Multiservice networks. Later, starting from the early 1980s, the notion of multiservice networks emerged. Multiservice networks deliver more services than just voice calls like the traditional telephony networks, e.g., high-speed data transmission, or video conferencing. An example of a multiservice network is Broadband Integrated Service Digital Network (B-ISDN) (see, e.g., [231]). As various services occupy different amounts of bandwidth capacity (constant within a single call, however), the traditional assumption that each source occupies a fixed amount of bandwidth no longer holds.

Therefore, the so-called *multirate models* were developed, in which the total amount of offered traffic was decomposed into traffic loads μ_i ($i = 1, \ldots, N$) for each service and associated bandwidth requirements c_i. Clearly, the above dimensioning question can be altered to incorporate the idea of various services – the new target is to find the lowest bandwidth c that ensures that the probability that a new call of type i is blocked is below δ_i, for all $i = 1, \ldots, N$, for given (small) numbers $\delta_1, \ldots, \delta_N$. Kaufman and Roberts developed efficient techniques for calculating blocking probabilities in such a multirate setting; see, e.g., [145, 148, 251, 253] for more background on multirate models and associated network dimensioning issues.

Variable bit rate traffic–effective bandwidth. As it turned out in the late 1980s and early 1990s, the above assumption that sources generate traffic at a constant rate was no longer valid–instead, traffic streams are in general of a variable bit rate (VBR) type: the bit rate fluctuates over time [231]. The notion of 'effective bandwidth' (or 'equivalent bandwidth') was introduced to overcome the problem with nonconstant rates: it assigns to each traffic stream a number between the mean and peak rate, reflecting the minimum bandwidth capacity that such a traffic stream requires to meet a certain performance criterion. In VBR models, this performance criterion is usually specified at the 'packet level', e.g., in terms of a packet loss probability. Following that example, the effective bandwidth, say $c_i(\epsilon)$, incorporates some (maximum) packet loss probability ϵ–the lower the ϵ, clearly, the higher the $c_i(\epsilon)$, as more capacity is occupied when the probability of packet loss decreases (i.e., the performance criterion becomes more stringent). This effective bandwidth $c_i(\epsilon)$ can now be dealt with in the same way as the bandwidth requirement c_i in the multirate models described above.

Therefore, when we assume that traffic of class i occupies $c_i(\epsilon)$ bandwidth–guaranteeing a maximum loss probability ϵ at the packet-level–it is possible to compute the blocking probabilities of all classes at the call-level (and, hence, to dimension this system such that these are below predefined values $\delta_1, \ldots, \delta_N$). Thus, one could say that two types of performance are offered simultaneously: at the packet-level (loss fractions below ϵ) and at the call-level (blocking probabilities below δ_i). See, e.g., [92, 122, 136, 149, 253] for more information on VBR models and associated network link dimensioning issues; other key articles on effective bandwidth are [53, 279, 110, 149, 154, 285].

Dimensioning for IP-based networks. The use of the VBR models described earlier (and the effective-bandwidth concept) is not by definition limited to, say, use in ATM networks. VBR models, however, still use the notion of calls and are not designed to deal with very heterogeneous traffic mixes in which individual flows may contain just a few packets. Internet traffic is generally known to be of highly heterogeneous nature (thus, it may be hard to distinguish separate 'classes' of traffic with their own bandwidth requirements), and 'operates' with packets instead of circuits. In addition, measurements have indicated that Internet traffic tends to be extremely unpredictable (see Section 3.1 and [172, 239])–significantly more bursty than traditional VBR models (such as the exponential on-off model) allow.

Hence, one cannot straightforwardly apply VBR models to accurately capture the behavior of Internet traffic.

The fundamentally different nature of the Internet (packet-oriented), compared to the traditional telephony system, as well as ATM (both circuit-oriented), inspired researchers to come up with traffic models, performance criteria (e.g., packet delay or packet loss bounds, instead of call blocking probability), and network dimensioning frameworks that focus primarily on Internet traffic and achieving packet-level performance targets.

Related literature. Studies that are closely related to the research presented in this chapter, are by Fraleigh [101] and Papagiannaki [234]. In those theses, dimensioning for highly aggregated network links ('backbone links') is studied, with a focus on delay-sensitive applications (such as Voice-over-IP); see also [102]. Therefore, they aim at dimensioning such that the delay incurred on the network (due to queuing) does not exceed a certain threshold. The results of Fraleigh and Papagiannaki are validated using traffic measurements on operational 'backbone' networks.

We also mention the dimensioning approach by Bonald, Olivier, and Roberts [36]–they concentrate on data traffic at the flow-level (in fact, at TCP-connections), aiming at dimensioning such that the throughput rates of TCP-connections remain above a certain threshold. Another dimensioning study in which traffic is modeled at the flow-level is by Barakat, Thiran, Iannaccone, Diot, and Owezarski [21], who aim at keeping congestion below a certain threshold.

A few words on the relation with the work presented in this chapter. Here, we have primarily aimed at achieving *link transparency*, i.e., choose c such that $\mathbb{P}(A(T) \geq \mathrm{c}T) \leq \epsilon$, which is clearly a different objective than the delay requirement of [101, 234]. For T sufficiently small, our criterion also bounds the delay in the network–in our research, we assume that buffers are used to absorb traffic bursts on timescales smaller than T. Thus, the impact of queuing in buffers (which contributes to delay) is also limited to small timescales. Another difference between the approach in [101, 234] and ours is that measurements have indicated that our research is applicable to network links with both (relatively) low and high aggregation in terms of users, instead of only 'backbone' links; see also the detailed study on the applicability of the Gaussian traffic model [211].

Chapter 15

Link dimensioning: indirect variance estimation

In the previous chapter we presented a method for bandwidth provisioning, i.e., for finding the minimum bandwidth C such that a certain given performance criterion was met.

A major advantage of this method is that it is simple (the key formula has the appealing form $\mathrm{C} = \mu + \alpha\sqrt{\mu}$, for load μ and coefficient α), transparent (the impact of the QoS criterion on α is explicitly given), and robust (there is some sort of insensitivity: α depends on the input traffic only through the characteristics of the individual flows, and not through the arrival rate of flows). There are, however, also a number of significant drawbacks:

1. In fact, for general Gaussian traffic the formula was, see Equation (14.5),

   $$\mathrm{C}(T, \epsilon) = \mu + \frac{1}{T}\sqrt{(-2\log\epsilon)\cdot v(T)}, \tag{15.1}$$

   and only in case of (the Gaussian counterpart of) the M/G/∞ input model the formula reduces to $\mu + \alpha\sqrt{\mu}$.

   For situations in which the traffic is indeed Gaussian, but not of the M/G/∞ type, however, the procedure sketched in Section 14.4 to estimate the required bandwidth by estimating α is obviously of no value; in that case it would be better to have an estimate of $v(T)$, and to plug this into provisioning formula (15.1).

2. The procedure for estimating α had a somewhat 'ad hoc character': the right value for α was found by fitting the curve $\mu + \alpha\sqrt{\mu}$, and of course this can be done in many ways.

Large deviations for Gaussian queues M. Mandjes

3. The performance criterion used, i.e., $\mathbb{P}(A(T) \geq \mathrm{c}T) \leq \epsilon$, could serve as a first benchmark, but is perhaps somewhat simplistic, as it does not involve any queuing.

   An alternative could be to consider the criterion: minimize c such that we meet $\mathbb{P}(Q \geq \mathrm{B}) \leq \epsilon$, for some predefined small number ϵ. Here B could be the (physical) buffer size of the queue, but an alternative could be to choose B/C as the maximum tolerable delay.

Drawbacks 1. and 2. indicate that it would be useful to have a method for efficiently estimating $v(T)$. If timescale T is sufficiently large, this can be done by sampling the arrival process over intervals of length T, and compute the sample variance, as in formula (3.1). When T is really small, however, which will be the case for some demanding applications, it is hard to do precise measurements on this timescale [213], and as a result it is hard to estimate $v(T)$.

Drawback (3) indicates that there is a need for finding the equivalent of Formula (15.1), but now for the new performance criterion. It turns out that this can be done similarly to the computations in Section 6.3. We rely on Approximation 5.4.1: find the smallest $\mathrm{c} := \mathrm{c}(\epsilon)$ such that

$$\mathbb{P}(Q \geq \mathrm{B}) \approx \exp\left(- \inf_{t \geq 0} \frac{(\mathrm{B} + (\mathrm{c} - \mu)t)^2}{2v(t)} \right) \leq \epsilon.$$

In other words, for all $t \geq 0$

$$\frac{(\mathrm{B} + (\mathrm{c} - \mu)t)^2}{2v(t)} \geq -\log \epsilon, \quad \text{or} \quad \mathrm{c} \geq \mu + \frac{\sqrt{(-2\log\epsilon)\cdot v(t)} - \mathrm{B}}{t},$$

so that we get the provisioning formula

$$\mathrm{c}(\epsilon) = \mu + \sup_{t \geq 0} \frac{\sqrt{(-2\log\epsilon)\cdot v(t)} - \mathrm{B}}{t}; \tag{15.2}$$

notice that its form is somewhat similar to Formula (15.1). The interesting aspect of formula (15.2) is that the computation of $\mathrm{c}(\epsilon)$ requires knowledge of the *entire* variance function $v(\cdot)$.

The above observations explain why it is of crucial importance to find accurate and fast techniques to estimate the variance function $v(\cdot)$. Here it is noted that for the 'crude' criterion $\mathbb{P}(A(T) \geq \mathrm{c}T) \leq \epsilon$ it suffices to estimate $v(\cdot)$ just for the timescale T, but the comparatively more advanced criterion $\mathbb{P}(Q \geq \mathrm{B}) \leq \epsilon$ requires knowledge of the entire variance function – see Equation (15.2).

This chapter describes a method to estimate $v(\cdot)$. This is done by coarse-grained sampling of the buffer occupancy, estimating the buffer content distribution (BCD) function, and inverting this into the variance function. Interestingly, in the context of

Gaussian queues, under mild conditions, the BCD uniquely determines the variance function.

Importantly, the procedure described above eliminates the need for traffic-measurements on small timescales. In this sense, we remark that our proposed procedure is rather counterintuitive: without doing measurements on timescale T, we are still able to accurately estimate $v(T)$. In fact, one of the attractive features of our 'inversion approach' is that it yields the *entire* variance curve $v(\cdot)$ (of course, up to some finite time horizon), rather than just $v(T)$ for some prespecified T.

Section 15.1 sketches the theoretical foundations behind the inversion approach. These are used in Section 15.2, where the primary focus is on implementation. A detailed error analysis of the inversion approach, using synthetic traffic, is performed in Section 15.3. Extensive tests with real traffic are presented in Section 15.4; two effects are decoupled: (i) the quality of the provisioning formula (14.5) given perfect knowledge of the variance function $v(\cdot)$; (ii) the impact of errors in the estimation of $v(\cdot)$ on the estimation of the required bandwidth.

15.1 Theoretical foundations

The conventional way to estimate $v(T)$ (for some interval length T) is what we refer to as the 'direct approach'. This method is based on traffic measurements for disjoint intervals of length T, and just computes their sample variance as in formula (3.1). It is recalled that the convergence of this estimator could be prohibitively slow when traffic is lrd [25], but the approach has two other significant drawbacks:

1. When measuring traffic using windows of size T, it is clearly possible to estimate $v(T)$, $v(2T)$, $v(3T)$, etc. However, these measurements obviously do not give any information on $v(\cdot)$ on timescales *smaller* than T. Hence, in the direct approach, to estimate $v(T)$ measurements should be done at granularity T or less. This evidently leads to a substantial measurement effort.

2. The provisioning formula (15.2) requires knowledge of the *entire* variance function $v(\cdot)$, whereas the direct approach described above just yields an estimate of $v(T)$ on a prespecified timescale T. Therefore, a method that estimates the entire curve $v(\cdot)$ is preferred.

This section presents a powerful alternative to the direct approach; we refer to it as the *inversion approach*, as it 'inverts' the BCD to the variance curve. This inversion approach overcomes the problems identified earlier. We rely on the framework of large deviations for Gaussian queues.

Define

$$\mathcal{I}_{\mathrm{C}}(\mathrm{B}) := \inf_{t\geq 0} \frac{(\mathrm{B} + (\mathrm{C} - \mu)t)^2}{2v(t)}, \tag{15.3}$$

such that Approximation 5.4.1 reduces to $-\log \mathbb{P}(Q \geq \text{B}) \approx \mathcal{I}_{\text{C}}(\text{B})$. This implies that, for all B and t, it holds that $\mathcal{I}_{\text{C}}(\text{B}) \leq (\text{B} + (\text{C} - \mu)t)^2/(2v(t))$. Noticing that both $v(\cdot)$ and $\mathcal{I}_{\text{C}}(\cdot)$ are nonnegative, this results in

$$v(t) \leq \inf_{\text{B}>0} \frac{(\text{B} + (\text{C} - \mu)t)^2}{2\mathcal{I}_{\text{C}}(\text{B})},$$

for all $t > 0$, cf. Inequality (6.14). Interestingly, we now show that often the inequality in the previous display is actually an equality.

Define the 'most likely epoch of overflow' for a given buffer value $\text{B} > 0$: $t^\star_{\text{B}}$ is the optimizer in the right-hand side of Equation (15.3); note that $t^\star_{\text{B}}$ is not necessarily unique. Define the set $\mathcal{T}$ as follows:

$$\mathcal{T} := \{t > 0 : \exists \text{B} > 0 : t = t^\star_{\text{B}}\}.$$

The following theorem presents conditions under which the above upper bound is tight.

Theorem 15.1.1 Inversion. *Assume that* $v(\cdot) \in C^1([0, \infty))$.
(i) For any $t > 0$,

$$v(t) \leq \inf_{\text{B}>0} \frac{(\text{B} + (\text{C} - \mu)t)^2}{2\mathcal{I}_{\text{C}}(\text{B})};$$

(ii) There is equality for all $t \in \mathcal{T}$ *;*
(iii) If $2v(t)/v'(t) - t$ *grows from 0 to* ∞ *when* t *grows from 0 to* ∞*, then* $\mathcal{T} = (0, \infty)$.

Proof. Claim (i) was proven above.

Now consider a $t \in \mathcal{T}$. Then, by definition, there exists $\text{B} = \text{B}_t > 0$ such that $\mathcal{I}_{\text{C}}(\text{B}) = (\text{B} + (\text{C} - \mu)t)^2/(2v(t))$. We thus obtain claim (ii).

Now consider claim (iii). We have to prove that for all $t > 0$ there exists a $\text{B} > 0$ such that $t = t^\star_{\text{B}}$. Evidently, $t^\star_{\text{B}}$ solves $2v(t)(\text{C} - \mu) = (\text{B} + (\text{C} - \mu)t)v'(t)$, or, equivalently,

$$\text{B} \equiv \text{B}^\star_t := \left(2\frac{v(t)}{v'(t)} - t\right)(\text{C} - \mu).$$

Hence, it is sufficient if $\text{B}^\star_t$ grows from 0 to ∞ when t grows from 0 to ∞. □

Note that even if condition (iii) does not apply, as was found in some recent traces, see [299], $v(t)$ will be bounded from above by the infimum over B, due to (i). Remarkably, the above theorem says, in broad terms, that for Gaussian sources the BCD uniquely determines the variance function. This property is exploited in the following heuristic, which follows from Approximation 5.4.1 in conjunction with Theorem 15.1.1.

Approximation 15.1.2 *The following estimate of the function $v(t)$ (for $t > 0$) can be made using the* BCD:

$$v(t) \approx \inf_{\mathrm{B}>0} \frac{(\mathrm{B} + (\mathrm{C} - \mu)t)^2}{-2\log \mathbb{P}(Q \geq \mathrm{B})}. \tag{15.4}$$

Hence, if we can estimate $\mathbb{P}(Q > \mathrm{B})$, then 'inversion formula' (15.4) can be used to retrieve the variance; notice that the infimum can be computed for any t, and consequently we get an approximation for the entire variance curve $v(\cdot)$ (of course, up to some finite horizon). These ideas are exploited in the procedure described in the next section, but we first illustrate the inversion formula in a number of exercises.

Exercise 15.1.3 (i) Show that Approximation. 15.1.2 is exact for Brownian motion.
(ii) Show that Approximation 15.1.2 is exact for Brownian bridge, for $t \leq \frac{1}{2}$.
(iii) Show that for fBm indeed $\mathcal{T} = (0, \infty)$.

Solution. (i) Recall that $-\log \mathbb{P}(Q \geq \mathrm{B}) = 2\mathrm{BC}$. Now it easily verified that indeed

$$\inf_{\mathrm{B}\geq 0} \frac{(\mathrm{B} + \mathrm{C}t)^2}{4\mathrm{BC}} = t;$$

the optimal B equals Ct.
(ii) Recall that $-\log \mathbb{P}(Q \geq \mathrm{B}) = 2\mathrm{B}(\mathrm{B} + \mathrm{C})$. Hence we have to compute

$$\inf_{\mathrm{B}\geq 0} \frac{(\mathrm{B} + \mathrm{C}t)^2}{4\mathrm{B}(\mathrm{B} + \mathrm{C})}.$$

If $t \in [0, \frac{1}{2}]$, then $\mathrm{B}_t^\star = \mathrm{C}t/(1 - 2t)$, and inserting this into the objective function indeed yields $v(t) = t(1 - t)$. Now note that $\mathcal{T} = [0, \frac{1}{2}]$. Apparently it is not possible to retrieve $v(t)$ for $t \notin [0, \frac{1}{2}]$.
(iii) It holds that

$$2\frac{v(t)}{v'(t)} - t = \left(\frac{1}{H} - 1\right)t,$$

which indeed, for any $H \in (0, 1)$, grows from 0 to ∞. ◇

Exercise 15.1.4 Consider the variance curve $v(t) = \max\{\alpha_1 t^{2H_1}, \alpha_2 t^{2H_2}\}$, with $H_2 > H_1$, as in Equation (3.3). Determine $\mathcal{T}$.

Solution. For any $\text{B}, \text{C} > 0$,

$$\mathcal{I}_\text{C}(\text{B}) = \inf_{t\geq 0} \frac{(\text{B}+\text{C}t)^2}{2\max\{\alpha_1 t^{2H_1}, \alpha_2 t^{2H_2}\}}$$
$$= \min\left\{\inf_{t\geq 0}\frac{(\text{B}+\text{C}t)^2}{2\alpha_1 t^{2H_1}}, \inf_{t\geq 0}\frac{(\text{B}+\text{C}t)^2}{2\alpha_2 t^{2H_2}}\right\}$$
$$= \min\left\{\frac{1}{2\alpha_1}\left(\frac{\text{B}}{1-H_1}\right)^{2-2H_1}\left(\frac{\text{C}}{H_1}\right)^{2H_1}, \frac{1}{2\alpha_2}\left(\frac{\text{B}}{1-H_2}\right)^{2-2H_2}\left(\frac{\text{C}}{H_2}\right)^{2H_2}\right\}.$$

This function is continuous in B; the value of B for which both expressions between brackets match, is

$$\text{B}_\text{crit} := \text{C}\cdot\left(\sqrt{\frac{\alpha_1}{\alpha_2}}\cdot\frac{(1-H_1)^{1-H_1}H_1^{H_1}}{(1-H_2)^{1-H_2}H_2^{H_2}}\right)^{1/(H_2-H_1)}.$$

For B smaller than B_crit, $t_\text{B}^\star$ looks like $(\text{B}/\text{C})\cdot h_1$, and for B larger than B_crit it looks like $(\text{B}/\text{C})\cdot h_2$, where $h_i := H_i/(1-H_i)$. Hence,

$$\mathcal{T} = (0,\infty)\setminus\left(\frac{\text{B}_\text{crit}}{\text{C}}\cdot h_1, \frac{\text{B}_\text{crit}}{\text{C}}\cdot h_2\right).$$

In other words, the inversion cannot retrieve $v(t)$ from $\mathcal{I}_\text{C}(\text{B})$ for all $t > 0$. ◇

Remark 15.1.5 Observe that, in fact, $\mathbb{P}(Q > \text{B})$ is also a function of the service speed C (as the random variable Q depends on C). Interestingly, as $v(t)$ obviously does not depend on C, Equation (15.4) entails that, if the approximation is correct, the minimum in the right-hand side should also not depend on C.

The unique correspondence between input and buffer content does not only apply for Gaussian processes, but can be found in many other situations as well. A well-known example is the M/G/1 queue. Assume Poisson arrivals stream (rate λ) of jobs, with service requirement distributed as random variable B (where its Laplace transform is denoted by $b(s) := \mathbb{E}e^{-sB}$), and service speed C. Let the load ϱ be defined as $\lambda\,\mathbb{E}B$. Denote by $q(\cdot;\text{C})$ the Laplace transform of the buffer content: $q(s;\text{C}) := \mathbb{E}e^{-sQ}$; the parameter 'C' is added to emphasize the dependence on C. The Pollaczek-Khintchine formula says

$$q(s;\text{C}) = \left(1-\frac{\varrho}{\text{C}}\right)\cdot\frac{s}{s-(\lambda/\text{C})\cdot(1-b(s))}.$$

This can be inverted to

$$b(s) = \frac{\text{C}s}{\lambda q(s;\text{C})}\cdot\left(\left(1-\frac{\varrho}{\text{C}}\right)+\left(\frac{\lambda}{\text{C}s}-1\right)q(s;\text{C})\right).$$

We see similar phenomena as above: (i) the BCD uniquely defines the distribution of the input process (the distribution of B); (ii) the left-hand side of the previous display clearly does not depend on C, so apparently the C in the right-hand side also cancels.

Also for these M/G/1 queues inversion algorithms (i.e., mechanisms to estimate the input process from samples of the buffer content) have been developed, see for instance [127, 264, 265]. ◇

15.2 Implementation issues

In this section, we show how the theoretical results of the previous section can be used to estimate $v(\cdot)$. First, we propose an algorithm for estimating the (complementary) buffer- content distribution (in the sequel abbreviated to BCD), such that, by applying Approximation. 15.1.2, the variance curve $v(\cdot)$ can be estimated. Second, we specialize in our demonstration to the case of synthetic input, i.e., traffic generated according to some stochastic process; we choose fBm input, but we emphasize that the procedure could be followed for any other process. Finally, we compare, for fBm, our estimation for $v(\cdot)$ with the actual variance function, yielding the first impression of the accuracy of our approach (a more detailed numerical evaluation follows in the subsequent sections).

The inversion procedure consists of two steps: (i) determining the BCD, and (ii) 'inverting' the BCD to the variance curve $v(\cdot)$ by applying Approximation. 15.1.2. We propose the following algorithm:

Algorithm 15.2.1 Inversion approach.
1. Collect 'snapshots' of the buffer contents: $q_1, \ldots, q_N$; here q_i denotes the buffer content as measured at time $\tau_0 + i\tau$, for some $\tau > 0$. Estimate the BCD by the empirical distribution function of the q_i, i.e., estimate $\mathbb{P}(Q > \text{B})$ by

$$\phi(B) := \frac{\#\{i : q_i > B\}}{N}.$$

2. Estimate $v(t)$, for any $t \geq 0$, by

$$\hat{v(t)} = \inf_{\text{B}>0} \frac{(B + (\text{C} - \mu)t)^2}{-2 \log \phi(B)}.$$

In the above algorithm, snapshots of the buffer content are taken at a constant frequency. To get an accurate estimate of the BCD, both τ and N should be chosen sufficiently large. We come back to this issue in Section 15.3. Notice that we chose a fixed polling frequency (i.e., τ^{-1}) in our algorithm, but this is not strictly necessary; obviously, the BCD-estimation procedure still works when the polling epochs are not equally spaced. In the remainder of this section, we demonstrate the inversion approach of Algorithm 15.2.1 through a simulation with synthetic (fBm) input. The simulation of the queue fed by fBm yields an estimate for the BCD; this estimated BCD is inverted to obtain the estimated variance curve, which is compared with the actual variance curve.

Simulation procedure. Concentrating on slotted time, we generate traffic according to some stochastic process. In this example we focus on the case of fBm input, but it is stressed that the procedure could be followed for any other stochastic process. The traffic stream is fed into a simulated queue with link rate c. The buffering dynamics are simulated as follows:

1. Using an fBm simulator [76], fBm is generated with a specific Hurst parameter $H \in (0, 1)$, yielding a list of 'offered traffic per time slot'.
2. For every slot, the amount of offered traffic is added to the queue, and an amount equal to $c\tau$ is drained from the queue (while assuring the queue's content is nonnegative).
3. Every τ slots, the queue's content is observed, yielding N snapshots that are then used to estimate $\mathbb{P}(Q \geq b)$ (cf. Algorithm 15.2.1).

In this demonstration of the inversion procedure, we generate an fBm traffic trace with Hurst parameter $H = 0.7$ and length 2^{24} slots. The link capacity c is set to 0.8, and we take snapshots of the buffer content every $\tau = 128$ slots.

Estimating the variance curve. We now discuss the output of the inversion procedure for our simulated example with fBm traffic. First we estimate the BCD; a plot is given in Figure 15.1(a). For presentation purposes, we plot the (natural) logarithm of the BCD, i.e., $\log \mathbb{P}(Q \geq b)$. The BCD in Figure 15.1(a) is 'less smooth' for larger values of b. This is due to the fact that large buffer levels are rarely exceeded, leading to less accurate estimates. Second, we estimate the variance $v(t)$ for t equal to the powers of 2 ranging from 2^0 to 2^7, using the BCD, i.e., by using Algorithm 15.2.1. The resulting variance curve is shown in Figure 15.1(b) ('inversion approach'). The minimization (over b) was done by straightforward numerical techniques. To get an impression of the accuracy of the inversion approach, we have also plotted in Figure 15.1(b) the variance curve as can be estimated directly from the synthetic traffic trace (i.e., the 'direct approach' introduced in Section 15.1), as well as the real variance function for fBm traffic, i.e., $v(t) = t^{2H}$. Figure 15.1(b) shows that the three variance curves are remarkably close to each other. This is a first indication that the inversion approach is an accurate way to estimate the burstiness.

15.3 Error analysis of the inversion procedure

In the previous section, the inversion approach was demonstrated. It was shown to perform well for fBm with $H = 0.7$, under a specific choice of N and τ. Evidently,

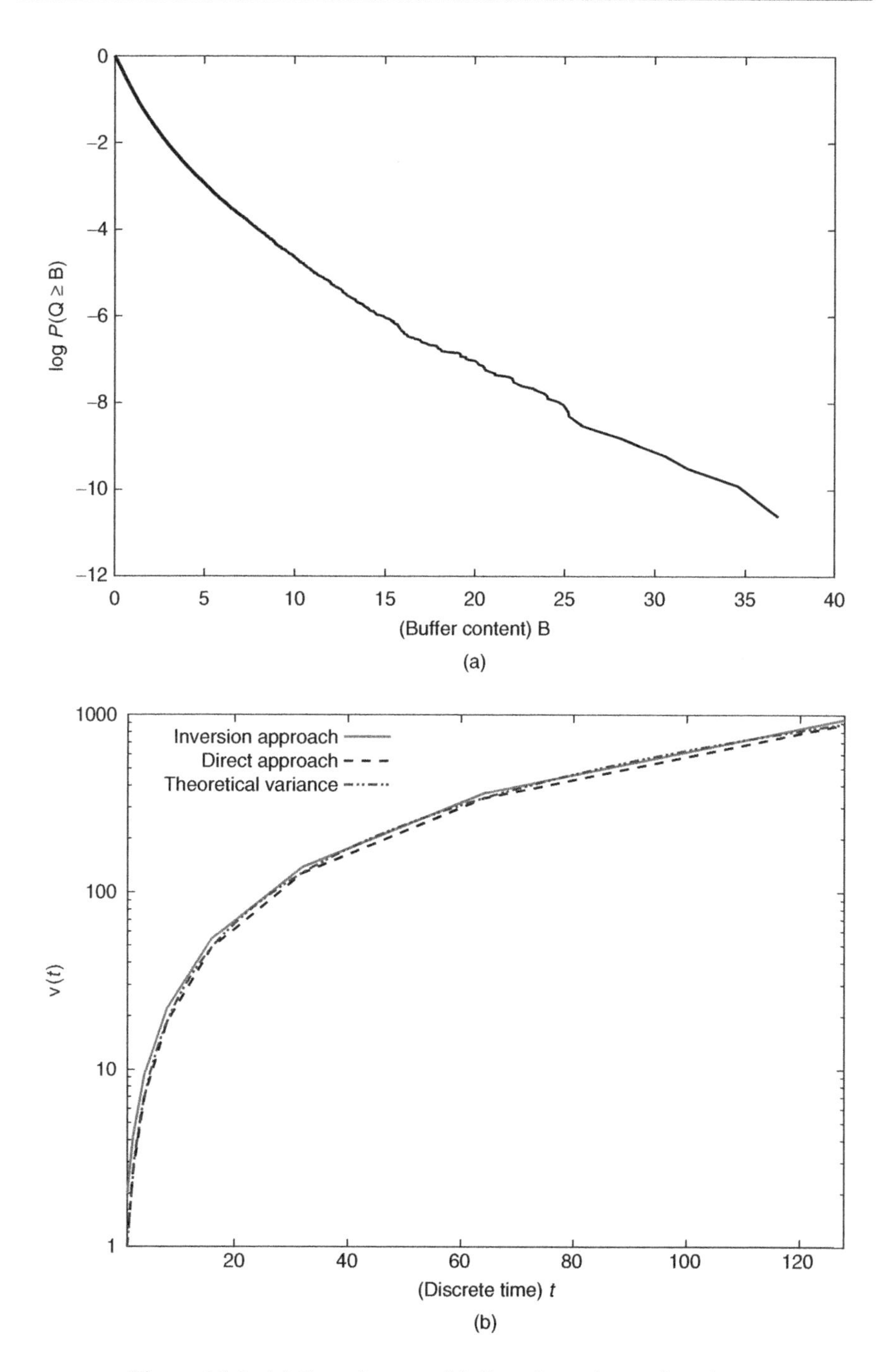

Figure 15.1: (a) Sample BCD; (b) Sample variance functions.

the key question is whether the procedure still works under other circumstances. To this end, we first identify the three possible sources of errors:

- The inversion approach is based on the *Approximation* (5.4.1).
- $\mathbb{P}(Q \geq \text{B})$ is *estimated*; there could still be an estimation error involved. In particular, we wonder what the impact of the choice of N and τ is.
- The procedure *assumes* perfectly Gaussian traffic, although real network traffic may not be (accurately described by) Gaussian.

We will now quantitatively investigate the impact of each of these errors on our 'indirect approach'. These investigations are performed through simulation as outlined in Section 15.2.

Approximation of the BCD. The inversion is based on an approximation of $\mathbb{P}(Q \geq \text{B})$, see Approximation 5.4.1. Consequently, evidently, errors in Approximation 5.4.1 might induce errors in the inversion. We first determine the infimum in the right-hand side of Equation (15.4), which we consider as a function of B. In line with the previous section, we choose fBm input: $\mu = 0$ and $v(t) = t^{2H}$. As we have seen earlier, we can rewrite Approximation (5.4.1), viz.:

$$\log \mathbb{P}(Q \geq \text{B}) \approx -\frac{1}{2}\left(\frac{\text{B}}{1-H}\right)^{2-2H}\left(\frac{\text{C}}{H}\right)^{2H}.$$

We verify how accurate the approximation is, for two values of H: the pure Brownian case $H = 0.5$, and a case with lrd $H = 0.7$ (in line with earlier measurement studies of network traffic). Several runs of fBm traffic are generated (with different random seeds), 2^{24} slots of traffic per run. We then simulate the buffer dynamics. For $H = 0.5$ we choose link rate $\text{C} = 0.2$, for $H = 0.7$ we choose $\text{C} = 0.8$; these choices C are such that the queue is nonempty sufficiently often (in order to obtain a reliable estimate of the BCD). Figure 15.2 shows for the various runs the approximation of the BCD, as well as their theoretical counterpart. It can be seen that, in particular, for small B the empirically determined BCD almost perfectly fits the theoretical approximation.

Estimation of the BCD. As we estimate the BCD on the basis of snapshots of the buffer content, there will be some error involved. The impact of this error is the subject of this section. It could be expected that the larger N (more observations) and τ (less correlation between the observations), the better the estimate.

First, we investigate the impact of N. The simulator is run as in previous cases (with $H = 0.7$), with the difference that we only use the first x % of the snapshot samples to determine $\mathbb{P}(Q \geq \text{B})$. Figure 15.3(a) shows the estimation of the BCD, for various x ranging from 0.1 to 100. The figure shows that, in particular for relatively small B, a relatively small number of observations suffices to get an accurate estimate of the BCD.

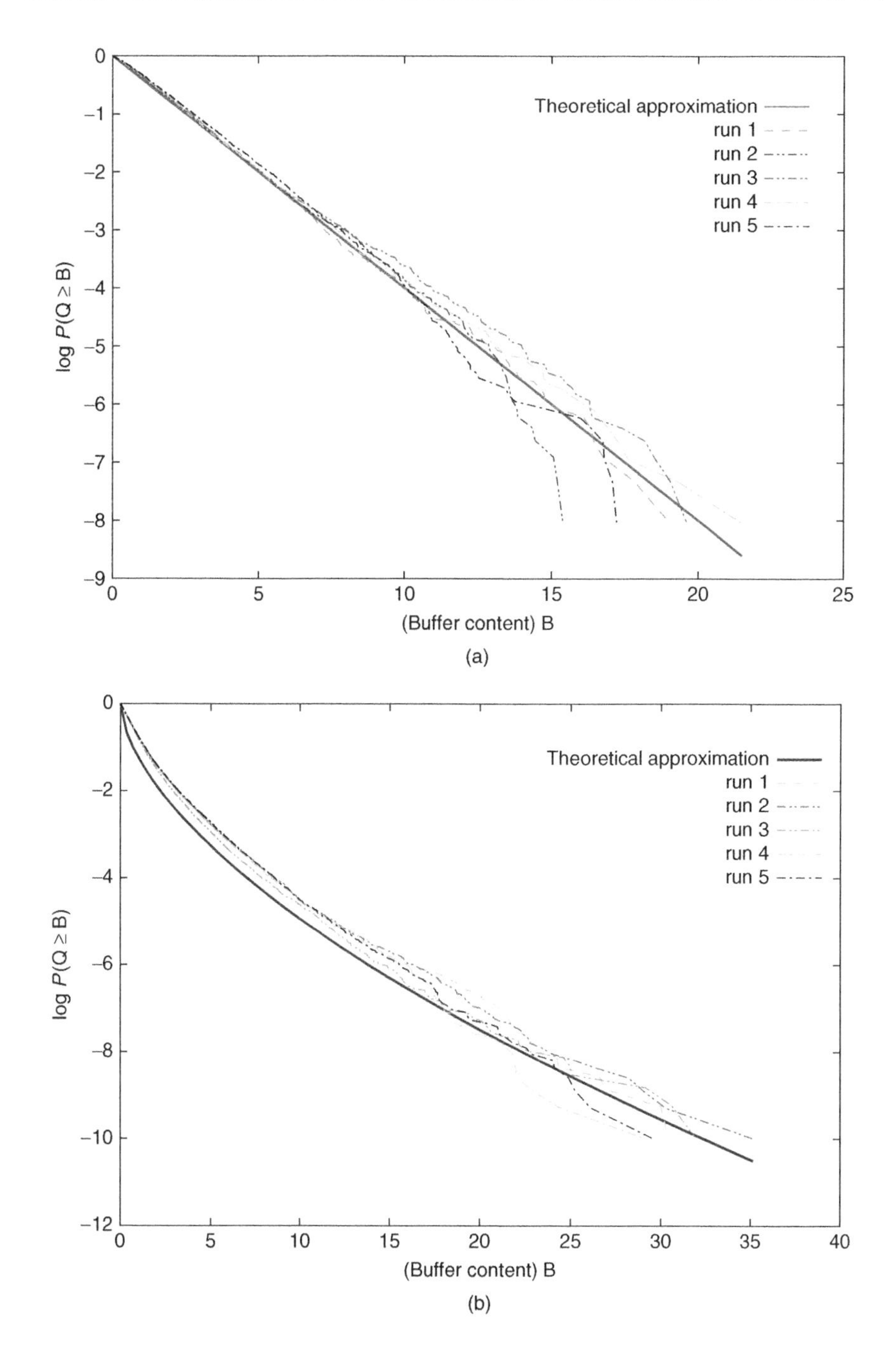

Figure 15.2: $\mathbb{P}(Q \geq \text{B})$ and theoretical approximation. (a) $H = 0.5$; (b) $H = 0.7$.

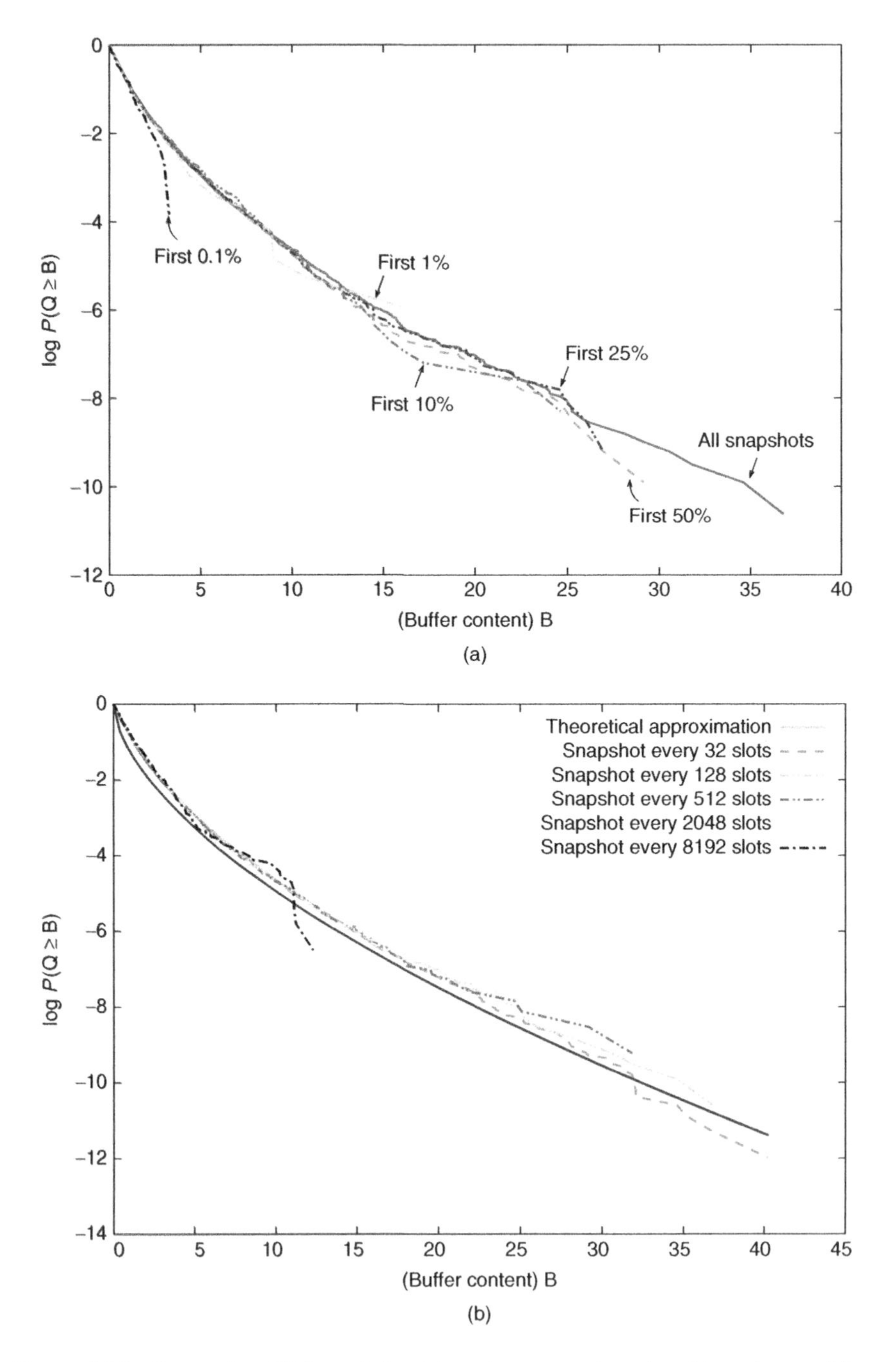

Figure 15.3: Comparing $\mathbb{P}(Q \geq \text{B})$ for (a) various trace lengths; (b) various polling intervals.

Second, we investigate the impact of the interval length between two consecutive snapshots τ. Figure 15.3(b) shows the determined BCD for τ ranging from observing every 32 to every 8192 slots. It can be seen that, particularly for small B the fit is quite good, even when the buffer content is polled only relatively rarely.

The impact of the Gaussianity assumption. Approximation (15.4) explicitly assumes that the traffic process involved is Gaussian. Various measurement studies find that real network traffic on the Internet is (accurately described by) Gaussian, see Section 3.1; it was observed, however, that particularly on small timescales, traffic may not always be Gaussian [156]. Therefore we now investigate how sensitive our 'inversion approach' is with respect to Gaussianity of the input traffic.

We study the impact of non-Gaussianity by mixing, for every slot, a fraction α of the generated fBm traffic with a fraction $1 - \alpha$ traffic from an alternative (non-Gaussian) stream, before the mixture is fed to the (virtual) queue. Note that the variance of the mixture is

$$v(t) = \alpha^2 v_{[\mathrm{fBm}]}(t) + (1 - \alpha)^2 v_{[\mathrm{alt}]}(t).$$

We vary α from 1 to 0, to assess the impact of the non-Gaussianity.

The alternative input model that we choose here is an M/G/∞ input model (i.e., *not* its Gaussian counterpart). In the M/G/∞ input model, jobs arrive according to a Poisson(λ) process. The job durations are i.i.d., and during their duration each job generates traffic at a constant rate r. In line with measurements studies, we choose Pareto(β) jobs. As the objective is to assess the impact of varying the parameter α, we have chosen to select the parameters of the M/G/∞ model such that the fBm and M/G/∞ traffic streams are 'compatible', in that their mean and variance $v(\cdot)$ are similar, which is achieved as follows.

The means of both traffic stream are made compatible by adding a drift to the fBm inputs equal to the mean of the M/G/∞ traffic stream, i.e., $\lambda r/(\beta - 1)$. The Gaussianity of the fBm input is clearly not affected by the addition of such a drift.

To make the variances of both traffic streams compatible, we make use of a derivation in earlier work of the exact variance function $v(t)_{[\mathrm{alt}]}$, see Equation (2.6). It is not possible to achieve the desired 'compatibility' of the variance on all timescales. As long-range dependence is mainly a property of long timescales, we choose to focus on these. For larger timescales, the variance of the M/G/∞ traffic stream roughly looks like

$$v_{[\mathrm{alt}]}(t) \approx 2r^2\lambda \frac{t^{3-\beta}}{(3 - \beta)(2 - \beta)(\beta - 1)},$$

assuming $\beta \in (1, 2)$. Clearly, we can now estimate the remaining parameters and compute the variance of the traffic mixture.

The next step is to run the simulation, for different values of α,. We then determine the (theoretical) variance curve of the traffic mixture, and compare it to the variance curve found through our 'indirect approach'. In Figure 15.4(a)

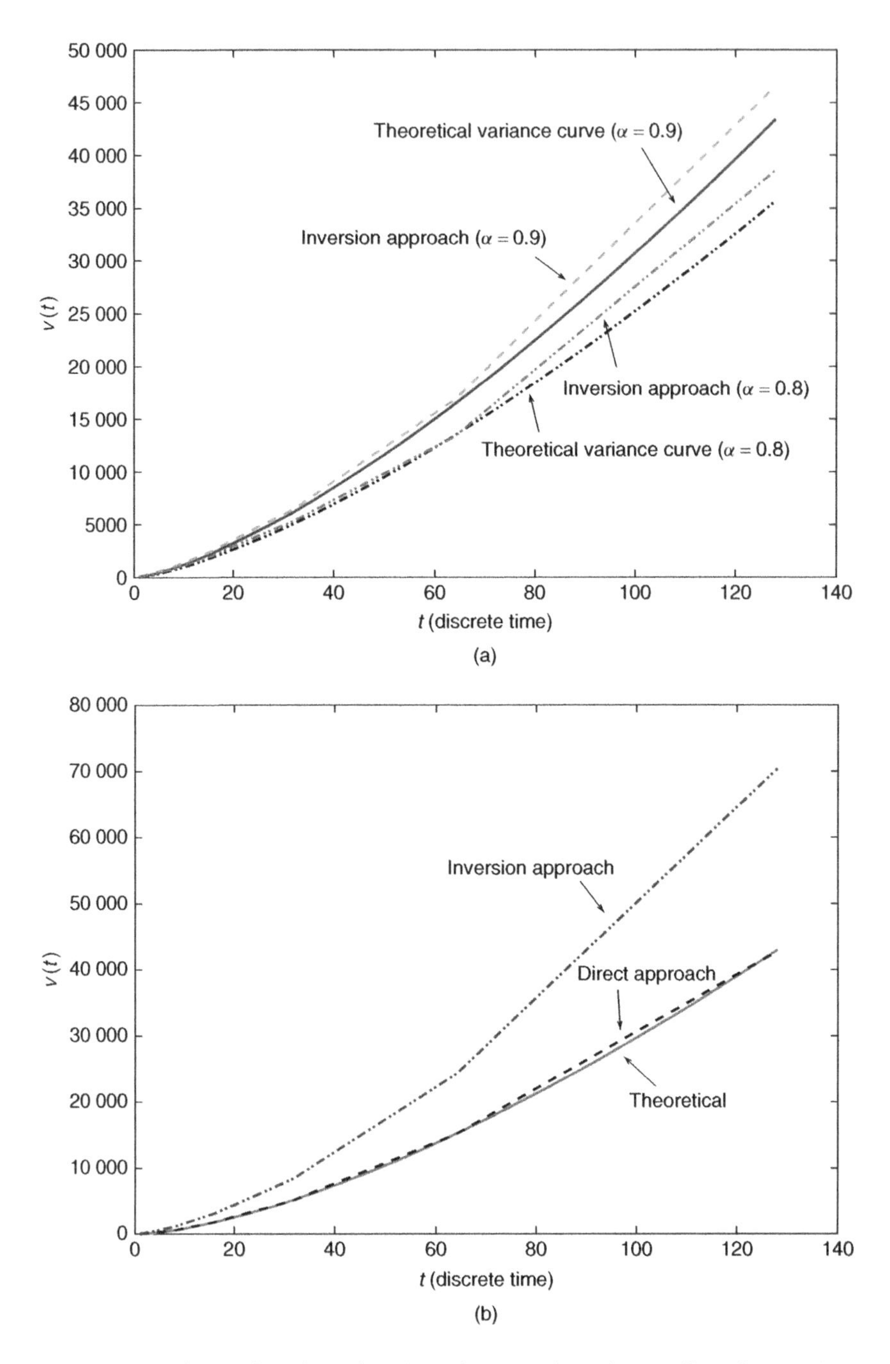

Figure 15.4: Variance functions for Gaussian/non-Gaussian traffic mixes, (a) $\alpha = \{0.8, 0.9\}$; (b) $\alpha = 0$.

we focus on the 'nearly-Gaussian' cases $\alpha = 0.8$ and $\alpha = 0.9$, which are plotted together with their theoretical counterparts. The figure shows that the presence of non-Gaussian traffic has some, but no crucial impact on our inversion procedure.

We also consider the (extreme) case of $\alpha = 0$, i.e., no Gaussian traffic at all, to see if our inversion procedure still works. In Figure 15.4(b) the various variance curves are shown: the theoretical curve, the curve based on the 'direct approach', as well as the curve based on the inversion approach. Although not a perfect fit, the curves look similar and still relatively close to each other (but, of course, the fit is worse than for $\alpha = 0.8$ and 0.9). Note that the non-Gaussian traffic may 'have some Gaussian characteristics' if there is a large degree of aggregation, by virtue of central-limit type of arguments, which may explain that the fit is still reasonable.

We conclude that our simulation experiments show the 'robustness' of the inversion procedure. Despite the approximations involved, with a relatively low measurement effort, the variance curve is estimated accurately, even for traffic that is not perfectly Gaussian. Given the evident advantages of the inversion approach over the 'direct approach' (minimal measurement effort required, retrieval of the entire variance curve $v(\cdot)$, etc.), the former method is to be preferred. In the next section, we verify whether this conclusion also holds for real (i.e., not artificially generated) network traffic.

15.4 Validation

As we have seen in the previous section, the inversion approach shows rather good performance: even if the underlying traffic deviates from Gaussian, still a relatively good estimate of the variance function is obtained. It remains unclear however whether (i) plugging in this variance function into the provisioning formula, i.e., either (15.1) or (15.2), also leads to good estimates of the required bandwidth; (ii) such conclusions remain valid for large sets of representative real traces (rather than artificially generated data).

The proposed provisioning approach consists of two steps:

1. Estimate the variance function $v(\cdot)$ by estimating the BCD and performing the inversion.

2. Insert this estimated variance into the provisioning formula, i.e., (15.1) or (15.2).

The idea behind this validation section is to systematically assess the accuracy of our provisioning approach. We do so by decoupling two effects.

- *Validation of the required bandwidth formula.* In the first place, we test the quality of the provisioning formula. The question is then: suppose we are given perfect information about the variance function, and we plug this into our provisioning formula, how good is the resulting estimate for the required bandwidth? Of course we do not have perfect information about $v(\cdot)$; instead we use the variance as estimated through the direct approach.

- *Impact of estimation errors in $v(\cdot)$ on required bandwidth.* In the second place, we test how the errors in the estimate of $v(\cdot)$, caused by the inversion procedure, have impact on the estimate of the required bandwidth.

The 'decoupling' allows us to gain precise insight into – and thus a good validation of – both steps of our link provisioning procedure.

Measurement locations. We have performed our measurements at various locations, with very distinct types of users. The five locations we have considered are (U) a university residential network (15 traces, 1800 hosts), (R) a research institute (185 traces, 250 hosts), (C) a college network (302 traces, 1500 hosts), (A) an ADSL access network (50 traces, 2000 hosts), and (S) a server hosting provider (201 traces, 100 hosts). Traces correspond to 15 min intervals. At these locations traffic is generated at average rates of 170, 6, 35, 120, and 12 Mbps, respectively.

Validation of the required bandwidth formula. As mentioned above, we first assume that we have perfect information on the variance, and test how good our theoretical provisioning formulae are. We focus here on the performance criterion $\mathbb{P}(A(T) \geq \mathrm{c}T) \leq \epsilon$, i.e., provisioning formula (15.1).

We choose to determine the average traffic rate μ per trace (recall that each trace contains 15 min of traffic), and set T to 1 s, 500 ms, and 100 ms (and thus determine the variance at those timescales by applying the direct method). We set ϵ to 1%. As emphasized earlier, the settings for T and ϵ used here are meant as examples – network operators can choose the setting that suits their (business) needs best. Then we insert the resulting estimates, say $\hat{\mu}$ and $\hat{v}_{\mathrm{dir}}(T)$, into Equation (15.1). A trace consists of n observations $a_1, \ldots, a_n$, where nT corresponds to 15 min.

In order to validate the accuracy of the resulting bandwidth estimate, we introduce the notion of 'realized exceedance', denoted by $\hat{\epsilon}$. We define the realized exceedance as the fraction of intervals of length T in which the amount of offered traffic a_i exceeds $\mathrm{c}T$. In other words,

$$\hat{\epsilon} := \frac{\#\{i \in \{1, \ldots, n\} \mid a_i > \mathrm{c}T\}}{n}.$$

We have constructed our provisioning rule (15.1) such that 'exceedance' (as in $A(T) > \mathrm{c}T$) may be expected in a fraction ϵ of all intervals. There are, however, (at least) two reasons why $\hat{\epsilon}$ and ϵ may not be equal in practice. Firstly, the bandwidth provisioning formula (15.1) assumed 'perfectly Gaussian' traffic, but as we have seen, this is not always the case. Evidently, deviations from 'perfectly Gaussian' traffic (in other words, violation of the modeling assumption) may have an impact on the estimated c. Secondly, in the derivation of Equation (15.1) we used Approximation 5.4.1, which contained a conservative element (Chernoff bound) as well as an 'aggressive' element (principle of largest term), and it is not clear up front which effect dominates.

To assess the accuracy of the provisioning formula (15.1) for real traffic, it is clearly interesting to compare ϵ with $\hat{\epsilon}$. Thus, we study the difference between ϵ

and $\hat{\epsilon}$ to get an insight into the deviation between the outcome of the model-based formula and the real traffic. The validation study, as described above, is performed using the hundreds of traces that we collected at measurement locations {U, R, C, A, S}.

The third and fourth column of Table 15.1 present the average differences between the targeted ϵ and the realized exceedance $\hat{\epsilon}$ at each location, as well as the standard deviations. This is done for three different timescales T. Clearly, the differences between ϵ and $\hat{\epsilon}$ are small, and hence the required bandwidth to meet a prespecified performance target of ϵ is estimated rather accurately.

For bandwidth provisioning 'rules of thumb'-purposes, it is interesting to get an idea of the required 'dimensioning factor', i.e., the (estimated) required bandwidth capacity compared to the average load (in other words, $d := \mathrm{C}/\mu$). The dimensioning factors, averaged over all traces at each location, as well as their standard errors, are given in the fifth and sixth column of Table 15.1 for $T = 1$ s, 500 and 100 ms. It shows, for instance, that at location U, some 33% extra bandwidth capacity would be needed in addition to the average traffic load μ, to cater for 99% ($\epsilon = 0.01$) of all traffic peaks on a timescale of $T = 1$ second. At location R, relatively more extra bandwidth is required to meet the same performance criterion: about 191%. Such differences between those locations can be explained by looking at the network environment: at location R, a single user can significantly influence the aggregated

Location	T	avg $\lvert\epsilon - \hat{\epsilon}\rvert$	stderr $\lvert\epsilon - \hat{\epsilon}\rvert$	d	σ_d
Location U	1 s	0.0095	0.0067	1.33	0.10
	500 ms	0.0089	0.0067	1.35	0.09
	100 ms	0.0077	0.0047	1.42	0.09
Location R	1 s	0.0062	0.0060	2.91	1.51
	500 ms	0.0063	0.0064	3.12	1.57
	100 ms	0.0050	0.0053	3.82	1.84
Location C	1 s	0.0069	0.0047	1.71	0.44
	500 ms	0.0066	0.0043	1.83	0.49
	100 ms	0.0055	0.0041	2.13	0.67
Location A	1 s	0.0083	0.0027	1.13	0.03
	500 ms	0.0083	0.0024	1.14	0.03
	100 ms	0.0079	0.0020	1.19	0.03
Location S	1 s	0.0052	0.0050	1.98	0.78
	500 ms	0.0049	0.0055	2.10	0.87
	100 ms	0.0040	0.0059	2.44	1.01

Table 15.1: Required bandwidth: estimation errors and dimensioning factor ($\epsilon = 0.01$).

traffic, because of the relatively low aggregation level (tens of concurrent users) and the high access link speeds (100 Mbps, with a 1 Gbps backbone); at location U, the user aggregation level is much higher, and hence, the traffic aggregate is 'more smooth'. As could be expected, the dimensioning factor increases when the performance criterion (in terms of either ϵ or T) becomes more stringent.

To give a few examples of the influence of the performance parameters T and ϵ on the required bandwidth capacity, we plot curves for the required bandwidth capacity at $T = 10, 50, 100$ and 500 ms, and ϵ ranging from 10^{-5} to 0.1, in Figure 15.5. In these curves, μ and $v(T)$ are (directly) estimated from an example

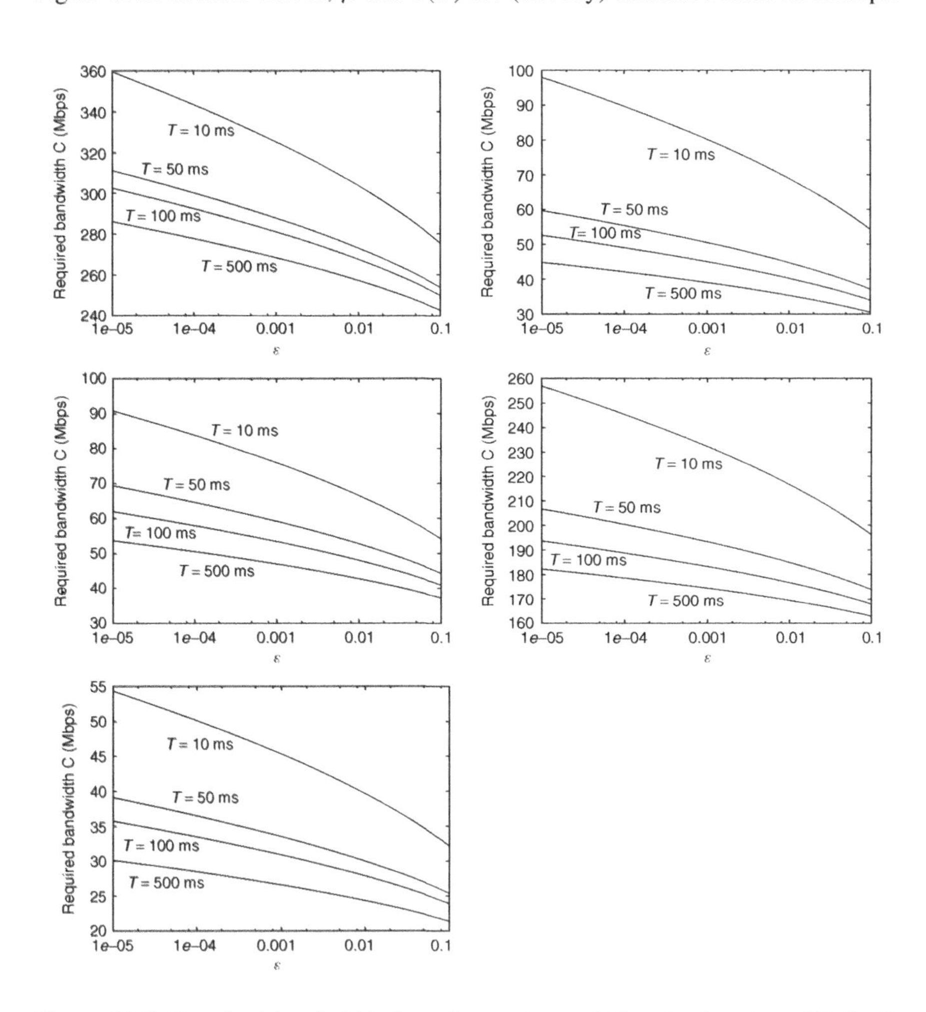

Figure 15.5: Required bandwidth for other settings of T, ϵ for locations {U, R, C, A, S}.

traffic trace collected at each of the locations {U, R, C, A, S}. It can be seen from Figure 15.5 that the required bandwidth c decreases in both T and ϵ.

So far, we verified whether the required bandwidth is accurately estimated for these case studies with different settings of T and ϵ. It should be noted however, that one cannot do similar tests for all possible combinations (T, ϵ): for $\epsilon = 10^{-5}$ and $T = 500$ ms for instance, there are only 1800 samples in our traffic trace (which has a length of 15 min), while our criterion says that 1 out of 10^5 samples is allowed to correspond to an 'exceedance'.

Another remark that should be made here is that, for locations with only limited aggregation in terms of users (say some tens concurrent users), combined with a small timescale of $T = 10$ ms, the traffic is no longer Gaussian (i.e., the linear correlation coefficient γ_n is substantially smaller than 1, see Section 3.1). As a consequence, the accuracy of our required bandwidth estimation decreases.

Next, we assess the impact of estimation errors in $v(\cdot)$ on required bandwidth.

Impact of estimation errors in $v(\cdot)$ on required bandwidth. In Section 15.3 we have seen for artificially generated traffic that the inversion worked well, and the first part of the present section assessed the quality of the bandwidth provisioning formula (15.1). We now combine these two elements: we perform the inversion, and we insert the resulting variance into formula (15.1), and see how well this predicts the required bandwidth.

For every trace we estimate the variance $v(T)$ through Algorithm 15.2.1. We do so by using a simulator, in which the trace is fed into a queue that is served at a constant rate. Notice that the service speed of this queue, say c_q, is still a degree of freedom.

A few words on how c_q should be chosen. Clearly, when c_q is too small, say $c_q < \mu$, the system is not stable in that it will never be able to serve all the traffic offered. Hence, we should choose $c_q \geq \mu$. On the other hand, if c_q is chosen to be much larger than μ, the queue's occupancy will, obviously, be 0 at most observation times. As this would lead to an unreliable estimation of the BCD (if any), setting c_q too large is to be avoided. We have performed numerous experiments to find a viable c_q, and it turns out that $c_q \approx 1.3 \cdot \mu$ often works well. We also saw that the specific choice of c_q is often not so critical.

Then we compare the resulting estimate, say $\hat{v}_{\text{inv}}(T)$, with $\hat{v}_{\text{dir}}(T)$. We introduce ν as an indicator of the accuracy of the estimation:

$$\nu := \frac{\hat{v}_{\text{inv}}(T)}{\hat{v}_{\text{dir}}(T)}.$$

For each trace we compute c_{dir} by inserting $\hat{v}_{\text{dir}}(T)$ into formula (15.1), and c_{inv} analogously. We also determine the 'empirical' minimally required bandwidth c_{emp}, which denotes the minimum bandwidth that suffices to meet the performance criterion:

$$c_{\text{emp}} := \min \left\{ c \ : \ \frac{\#\{i \in \{1, \ldots n\} \mid a_i > cT\}}{n} \leq \epsilon \right\}.$$

Also, we introduce Δ as an indicator of the 'goodness' of the estimation of the required bandwidth through the inversion approach. We compare this c_{inv} with both c_{dir} and c_{emp}. Hence,

$$\Delta_{var} := \frac{c_{inv}}{c_{dir}}, \quad \text{and} \quad \Delta_{cap} := \frac{c_{inv}}{c_{emp}}.$$

Thus, if Δ_{var} is close to 1, our methodology to (indirectly) estimate the burstiness of network traffic leads to similar required bandwidth capacity levels as in the direct approach. If Δ_{cap} is close to 1, our 'indirect' burstiness estimation approach ultimately yields a similar required bandwidth capacity level as the minimum bandwidth that, according to the traces of real traffic, suffices to meet the performance criterion.

Table 15.2 lists the validation results of several traces. The timescale on which the variances $v(T)$ are determined and estimated, is $T = 100$ ms. It can clearly be seen from Table 15.2 that the variances are rather accurately estimated. Table 15.3 continues the validation results, by comparing the estimated required capacity (with $\epsilon = 0.01$, $T = 100$ ms) computed via both the direct and indirect approaches. Also the empirically found minimally required bandwidth is tabulated.

As can be immediately seen from the values of Δ_{var}, the required bandwidth capacity as estimated through the indirect approach to estimate the variance, is remarkably close to the direct approach: on average, the differences are less than 1%. Also, comparison with the empirical minimally required bandwidth, through Δ_{cap}, shows that our coarse-grained measurement (i.e., indirect) procedure leads to estimations for the required bandwidth that are, remarkably, on average less than 4% off.

Comparing the respective values for ν and Δ_{var} in Tables 15.2 and 15.3, it becomes clear that an estimation error in $v(\cdot)$, indeed, has only limited impact on the error in c.

Trace	μ	$v_{dir}(T)$	$v_{inv}(T)$	ν
Location U #1	207.494	3.773	4.026	1.067
#2	238.773	7.313	7.690	1.052
Location R #1	18.927	0.492	0.483	0.981
#2	3.253	0.058	0.062	1.062
Location C #1	23.894	0.633	0.644	1.018
#2	162.404	10.650	12.379	1.162
Location A #1	147.180	0.939	1.064	1.133
#2	147.984	0.745	0.747	1.003
Location S #1	14.254	0.200	0.201	1.004
#2	2.890	0.023	0.023	1.022

Table 15.2: Validation results for the burstiness estimation methodology (μ is in Mbps, v is in Mb^2).

Trace	c_{emp}	c_{dir}	$c_{indirect}$	Δ_{var}	Δ_{cap}
Location U #1	258.398	266.440	268.385	1.007	1.039
#2	302.663	320.842	322.934	1.007	1.067
Location R #1	37.653	40.221	40.020	0.995	1.063
#2	10.452	10.568	10.793	1.021	1.033
Location C #1	44.784	48.033	48.250	1.005	1.077
#2	265.087	261.444	269.182	1.030	1.015
Location A #1	171.191	176.588	178.480	1.011	1.043
#2	168.005	174.178	174.218	1.000	1.037
Location S #1	27.894	27.843	27.873	1.001	0.999
#2	7.674	7.482	7.532	1.007	0.981

Table 15.3: Validation results for the burstiness estimation methodology (continued) (c is in Mbps).

(a)

Location	avg Δ_{var}	stderr Δ_{var}	avg Δ_{cap}	stderr Δ_{cap}
U	1.00	0.01	1.01	0.07
R	0.96	0.10	0.90	0.19
C	1.00	0.03	1.04	0.11
A	1.00	0.01	1.04	0.02
S	1.00	0.03	0.99	0.10

(b)

Location	avg Δ_{var}	stderr Δ_{var}	avg Δ_{cap}	stderr Δ_{cap}
U	1.00	0.01	1.03	0.06
R	1.00	0.02	1.00	0.10
C	1.00	0.02	1.05	0.08
A	1.00	0.01	1.04	0.01
S	1.00	0.01	1.01	0.05

Table 15.4: Validation results for the burstiness estimation methodology (overall results) – upper table: all traces; lower table: traces with $\gamma > 0.9$.

Finally, to assess the overall accuracy of our provisioning approach, we have computed the Δ values as described above for *all* our traces at every location – see the upper part of Table 15.4. We have tabulated the average values of Δ_{var} and Δ_{cap}, as well as their standard error terms. Clearly, the required bandwidth estimations are remarkably accurate. In the lower part of Table 15.4, we have tabulated the same metrics, but only used the traces that are 'fairly Gaussian', in that their linear correlation coefficient is above 0.9. This improves the results even further – leading to the conclusion that errors in the required bandwidth estimation using the indirect approach to estimate the variance, are primarily caused by non-Gaussianity of the offered traffic.

Remark 15.4.1 We now study the impact of changing performance parameters (T and ϵ) on the required bandwidth. As illustrated in Figure 15.5, it is possible to express the estimated required bandwidth capacity as function of ϵ and T. Having such a function at our disposal, and one or two actual estimates of the required bandwidth, it is possible to 'extrapolate' such estimates to other settings of ϵ and T. This allows for investigation of the impact of, say, a more stringent performance target on the required capacity. We first assess the impact of a change in ϵ and then of a change in T.

Suppose that, for a given T, a proper required bandwidth estimate is known, $\mathrm{c}(T, \epsilon_1)$, for some ϵ_1 and estimated μ. From formula (15.1) it follows that $\mathrm{c}(T, \epsilon_1) = \mu + \delta_1 \cdot \xi$, for some number $\xi > 0$ and $\delta_1 := \sqrt{-2 \log \epsilon_1}$. Evidently, we can estimate ξ by $(\mathrm{c}(T, \epsilon_1) - \mu)/\delta$. Then, to find the required bandwidth estimate for some *other* performance target ϵ_2, it is a matter of inserting these μ and ξ into

$$\mathrm{c}(T, \epsilon_2) = \mu + \xi\sqrt{-2 \log \epsilon_2}.$$

We give an example of this using the top-left graph (location U) in Figure 15.5. At the $T = 100$ ms timescale, taking $\epsilon_1 = 0.01$, $\mu = 207$ Mbps, it follows that $\mathrm{c}(T, \epsilon_1) \approx 266$ Mbps. Hence, $\xi \approx 19.4$. Suppose we are interested in the impact on the required bandwidth capacity if we reduce ϵ with a factor 1000, i.e., $\epsilon_2 = 10^{-5}$. Estimating the new required bandwidth capacity through the formula above yields that $\mathrm{c}(T, \epsilon_2) \approx 300$ Mbps, which indeed corresponds to the required bandwidth as indicated by the curve in Figure 15.5. Hence, in casual terms, the additional bandwidth required to cater for 1000 times as many 'traffic peaks' is, in this scenario, just some 34 Mbps.

Secondly, we look at the impact of a change in T on the required bandwidth. Compared to the above analysis for ϵ, we now have the extra complexity of the variance $v(T)$ in formula (15.1), which evidently changes with various T. We therefore impose the additional assumption that traffic can be modeled as fBm: $v(T) \approx \sigma \cdot T^{2H}$, where H is the Hurst parameter, and σ is some positive scaling constant. Using this variance function, formula (15.1) can be rewritten as $\mathrm{c}(T, \epsilon) = \mu + \delta \cdot \bar{\xi}(T)$, with $\bar{\xi}(T) = \sqrt{\sigma} \cdot T^{H-1}$.

Now suppose that for two different time intervals, namely $T_1 = T$ and $T_2 = \beta T$ (for some $\beta > 0$; ϵ is held fixed), the required bandwidth is known. This enables us to compute $\bar{\xi}(T)$ and $\bar{\xi}(\beta T)$, as above. But then

$$\frac{\bar{\xi}(\beta T)}{\bar{\xi}(T)} = \frac{\sqrt{\sigma} \cdot (\beta T)^{H-1}}{\sqrt{\sigma} \cdot T^{H-1}} = \beta^{H-1},$$

or, in other words,

$$g := \frac{1}{\log \beta} \cdot \log \left(\frac{\bar{\xi}(\beta T)}{\bar{\xi}(T)} \right)$$

is constant in β (and has value $H-1$). Again we consider, as an example, location U, with $\epsilon = 10^{-3}$. For $T = 100$ ms we obtain from $c(T, \epsilon) \approx 279$ that $\overline{\xi}(T) = 19.37$. Now take $\beta = 0.5$; from $c(\beta T, \epsilon) \approx 290$ we obtain $\overline{\xi}(\beta T) = 22.3$. It follows that $g = -0.20$. Suppose we now wish to dimension for $T_3 = \beta' T$ with $\beta' = 0.1$ (i.e., $T = 10$ ms), we obtain $\overline{\xi}(\beta' T) = \overline{\xi}(T)(\beta')^g \approx 30.7$, so that $c(\beta' T, \epsilon) = \mu + \sqrt{-2\log\epsilon} \cdot \overline{\xi}(\beta' T) \approx 321$. It can be verified that this is in agreement with the required bandwidth as indicated by the curve in Figure 15.5. $\diamond$

Bibliographical notes

This chapter is based on Mandjes and Van de Meent [196, 198]. Here [196] focuses on the inversion and its inherent inaccuracies, and [198] on using the inversion in dimensioning procedures. Remark 15.4.1 is taken from [212].

Chapter 16

A framework for bandwidth trading

This chapter describes a framework that facilitates bandwidth trading between a network and a number of 'large users' (to be thought of as Internet service providers (ISPs)). We will explain why measurements should play a pivotal role in the bandwidth allocation procedure. Then, large-deviations techniques are used to interpret these measurements. The large-deviations results we rely on are different from those we have seen earlier in this book; particularly Sanov's theorem and its inverse will appear to be extremely useful.

16.1 Bandwidth trading

For reasons of scalability, many on-line systems are organized in a decentralized way. It is clearly of much interest to devise procedures that help improve their performance but are, at the same time, relatively simple to implement. Importantly, in a decentralized setting, control procedures should be based on 'local information' only. Thus, a rather restrictive constraint is imposed on the set of feasible control strategies.

In the context of modern communications networks, several examples of on-line systems with decentralized control can be found. Consider, for instance, a (broadband) network that sells resources (bandwidth in the setting of this chapter) to ISPs (in this chapter'users'). These ISPs, in turn, sell Internet connectivity to clients in the corporate and residential market (in this chapter 'end users'). The ISPs need the network's bandwidth in order to offer the end users an acceptable Quality of Service (QoS), as agreed upon in a Service Level Agreement (SLA). The amount of bandwidth each ISP chooses to buy in fact maximizes its *utility*,

Large deviations for Gaussian queues M. Mandjes

where utility expresses the extent to which the end users can be guaranteed their SLA, compensated by the price to be paid by the ISP to the network. The network has a conflicting objective, however: when selling resources to the ISPs, it usually maximizes *profit*. Since the amount of bandwidth on each link of the network is finite, the ISPs are essentially competing for this scarce resource.

The question that naturally arises is that of the coordination of the competition between the users. A complicating issue is that in practice the users cannot communicate directly with each other. Instead, they rely on individual communication with the network, for each of the links that they traverse. More precisely, the network sets the prices (which they charge the users per supplied unit of bandwidth), and each user reacts to these by revealing its preferred amount of bandwidth, without knowing the reaction of the other users; then the network can adapt the prices, etc. It is clear that it would be desirable that these (iterative) price and bandwidth adaptations are such that convergence is guaranteed to the utility-maximizing allocation and profit-maximizing price. In this chapter, we show how such procedures could be developed.

We consider a network that is made up of a set of nodes that are connected by links, on which finite capacity resides. Routes, defined as origin–destination pairs of nodes, are sets of consecutive links through which traffic is transported; we can identify users (as introduced earlier) with routes. In this setting, each link can be used by multiple competing routes.

Evidently, the allocation reached reflects both the routes' (users) and the network's objectives. One could term these objectives 'aligned', if it is in the network's interest to set prices that ensure that the equilibrium allocation also maximizes the users' aggregate utility; this is often referred to as 'social welfare optimization'. An example of such an aligned setting is studied by Kelly [150, Section 2]; its convergence to an equilibrium solution (by applying an iterative optimization-based flow control scheme) was analyzed in detail by Low and Lapsley [179], and Kelly *et al.* [153], whereas Johari and Tan [142], Kelly [151, 152] and Vinnicombe [281] addressed related stability issues. In fact, the setup of [150] enforces that the users and the network coordinate their actions so that the social optimum is reached. One can alternatively consider the situation in which the network imposes a utility function on each route that achieves a certain fairness criterion [219].

It is clear that it is not realistic to assume that the network and the users always have aligned objectives. As explained above, this chapter focuses on such a 'misaligned situation', in which the network acts as a monopolist and wishes only to maximize its profit (subject to the link capacity constraints), while each individual user optimizes its utility function (or, more precisely, *compensated* utility, defined as utility minus costs; utility is a function of the bandwidth allocated). We study a model in which the network acts as a price-setter: it chooses for each link ℓ a price p_ℓ that it charges for each unit of bandwidth. The users act as price-takers, and adjust their desired bandwidths according to the prevailing price. In this scenario, both parties' objectives have an impact on the final allocation of capacity, and corresponding prices. It is therefore natural to expect that the resulting prices and allocations will deviate from the ones under social welfare maximization.

The differences between the outcomes of social welfare optimization and monopoly markets have been well researched in the economic literature; see, in particular, the seminal work by Pigou [241]. Informally, one could say that monopoly markets are usually beneficial for the supplying party (the network, in our situation), but create a 'welfare loss' for the participants (the users, in our situation). Recently, the term *the price of anarchy* was introduced to quantify this welfare loss [257]; it reflects the inefficiency of the equilibrium solution. Huang, Ozdağlar, and Acemoğlu [135] showed, in a specific setting, that for specific types of utility functions, the monopoly equilibrium coincides with the social welfare equilibrium, with regard to bandwidth allocations. Ozdağlar [232] considered the price of anarchy in the context of routing and flow control in a network consisting of multiple parallel links, where a cost per unit bandwidth and a congestion-based cost were imposed on each user. For a utility curve satisfying stated conditions, she derived a bound on the efficiency of the solution by comparing the equilibria reached in the monopolist and social welfare optimization contexts, respectively.

Instead of considering the price of anarchy, in this chapter we focus our attention on devising an iteration scheme by which the network and users achieve their respective equilibrium solutions. We impose the condition that the solution be found in a distributed manner; importantly, each user does not have any information about the other users' preferences. We derive an optimization-based scheme for price updates (network) and bandwidth updates (users), which is provably convergent. In the first part of this chapter, it is assumed that the users' utilities are functions of the bandwidth allocated.

As argued above, the users (ISPs) guarantee their clients (end users) a certain performance, for instance, in terms of a packet loss probability. For this reason, it may be more natural to consider a setup in which the user's utility curve is a function not of the bandwidth x itself, but rather of the loss probability $\delta \equiv \delta(x)$ that is realized when the user is allocated x. Therefore, to be able to use such a framework, the users must have a procedure to estimate the loss probability $\delta(x)$ as a function of the allocation x. It is clear, however, that in fact $\delta(x)$ depends not just on x but also on the characteristics of the traffic offered by the end users; clearly, for a given x, the loss probability is larger when the end users generate higher loads, or when the traffic rate fluctuates more fiercely in time.

The reasoning of the previous paragraph explains why it makes sense that the ISPs *measure* the traffic of their clients, so as to estimate the function $\delta(x)$ online and to insert this estimate in their utility function. We propose a procedure that estimates $\delta(x)$ in a (large-deviations-based) Bayesian way [108], which yields conservative estimates, and is therefore particularly useful for risk-averse ISPs. The analysis is based on the assumption of Gaussian traffic; as we have seen earlier in this book, this is justified as long as the traffic aggregation level (for instance in terms of the number of end users) is sufficiently high [102, 156, 211], which is typically the case for large ISPs.

The structure of the remainder of the chapter is as follows. In Section 16.2, we present our model and some preliminaries. In Section 16.3, we analyze the situation in which the utility curves are functions of the bandwidth allocated, and present a provably convergent iterative scheme for the prices and the allocations. Section 16.4 adapts this iterative scheme to the situation in which each ISP's utility is a function of the loss probability; in Section 16.6 it is explained in detail how the loss probability may be estimated from measurements (relying on the large-deviations theory described in Section 16.5). In Section 16.7, the procedures are illustrated through numerical experiments.

16.2 Model and preliminaries

In this section, we present our model of a profit-maximizing network that is used by utility-maximizing routes. The problem is stated in microeconomic terms, and a number of preliminaries and first observations are given. In the next section, we present an iterative scheme of price and bandwidth adaptation for this model. In the setting of Sections 16.2 and 16.3, utility is a function of bandwidth allocated; Section 16.4 then describes how the scheme should be adapted for the situation in which utility is a function of the loss probability.

Links. Let $\mathcal{L}$ denote the set of all physical links and assume each link $\ell \in \mathcal{L}$ has a finite service capacity (i.e., bandwidth) c_ℓ. Each link $\ell \in \mathcal{L}$ has the authority to set a price p_ℓ for each unit of bandwidth, with the aim of maximizing the total network profit.

Routes. A route is defined as a subset of the links, starting at an origin node and ending at a destination node. Let $\mathcal{R}$ denote the set of routes in the network. Each route $r \in \mathcal{R}$ is allocated the same amount of bandwidth at any link that it traverses, and when allocated x_r units, it derives a utility $U_r(x_r)$. Recall that routes can be identified with users (ISPs). The routes are 'at the mercy' of the network, in the sense that they request capacity given the prevailing prices; they choose the amount of capacity that maximizes their compensated utility (defined as utility less cost; a precise definition follows later). We assume that the routes cannot cooperate; in this way, we preclude the possibility that routes can collude to try and bring the prices down.

Identifying routes with ISPs, the utility optimization described earlier reflects the route's task to satisfy the SLA that it has agreed upon with the end users. This SLA could, for instance, contain guarantees on the packet loss probability. Clearly, the extent to which the ISP is able to meet the SLA of its customers decreases as the loss probability increases, whereas the loss fraction decreases in the amount of bandwidth allocated. Therefore, it is natural to assume that, for every ISP r, the utility curve $U_r(x_r)$ increases in x_r.

As is customary (cf. [150]), we assume that these utility functions are strictly increasing and concave in x_r. In addition, for reasons that will become clear below, we assume that the utility functions are differentiable, with derivatives that are one-to-one.

If the network sets prices that are too high, then the routes will request very low amounts of bandwidth. The profit derived from this allocation will not be large, and the network will not be satisfied. Conversely, if the prices are very low, the capacity requested will be larger, although possibly not large enough to offset the decreased prices. In addition, the routes' requests may exceed the finite link capacity. It is thus conceivable that there is an optimal set of prices that balance these opposing forces. Importantly, observe that the objectives of the various ISPs are not aligned with the network's objective. Let us now look in more detail at a formal description of these objectives.

User optimization problem. As stated above, routes (or users) occupy a subset of the physical links. As the route is meant to transport a stream of data packets, it requires the same amount of bandwidth from each link that it traverses. The network charges a price p_ℓ per unit of bandwidth on link ℓ. A route $r \in \mathcal{R}$ is then faced with the total price of $x_r \sum_{\ell \in r} p_\ell$ when acquiring x_r units of bandwidth. Since we assume that the routes cannot cooperate with each other, they make their decision solely based on their own utility function and the prevailing prices. Each user aims to maximize its own 'compensated utility', that is, the utility derived from being allocated a level of bandwidth less the cost of purchasing the bandwidth. The *user optimization problem* is therefore given by

$$\max_{x_r \geq 0} U_r(x_r) - x_r \cdot \sum_{\ell \in r} p_\ell. \tag{16.1}$$

Let $U_r'(x_r)$ denote the derivative of the utility function for route r with respect to x_r and $V_r(x_r)$ denote the inverse of this derivative. By the conditions imposed on the utility functions, the solution to the user problem can be expressed as (with $p^r := \sum_{\ell \in r} p_\ell$)

$$x_r(p^r) := V_r(p^r), \tag{16.2}$$

assuming the right-hand side of Equation (16.2) is nonnegative. In essence, $x_r(p^r)$ defines a *demand function* for bandwidth for each route r, which is, as expected, decreasing in price. We make the added assumption that the utility functions induce demand functions that are twice continuously differentiable with respect to p^r.

Network optimization problem. Now that we know how user r reacts to price p^r, the next question is how the prices are set. The network is entitled to change the prices of the links p_ℓ so as to maximize its profit. If the routes, request capacities equal to $\vec{x} \equiv (x_r, r \in \mathcal{R})$, then the profit equals the sum of the revenues earned on each link: $\sum_{\ell \in \mathcal{L}} \left(p_\ell \cdot \sum_{r:\ell \in r} x_r \right)$; for ease, we leave out the cost to the network in providing bandwidth to the routes. Since the capacities $\vec{x}$ are defined by the demand curve (16.2), the profit can be expressed as a function of the price, with $\vec{p} \equiv (p_\ell,\ \ell \in \mathcal{L})$:

$$\pi(\vec{p}) = \sum_{\ell \in \mathcal{L}} \left(p_\ell \cdot \sum_{r:\ell \in r} x_r(p^r) \right). \tag{16.3}$$

While the links choose prices that maximize their objective (16.3), they must also adhere to a number of constraints. In the first place, there must be constraints requiring that the capacity requested by the routes on a specific link be no larger than the link capacity. In addition, we impose a cap on the prices that each link may charge, $p_{\max}$; for ease, we assume that this maximum price is the same for each link. Such a maximum price may be due to regulatory reasons, but we illustrate in Section 16.3 that it is useful from a mathematical perspective as well. We denote the price cap by $p_{\max}$. We thus obtain the following *network optimization problem*:

$$\begin{aligned} &\max_{\vec{p}} \quad \pi(\vec{p}) \\ &\text{subject to} \quad \sum_{r:\ell \in r} x_r(p^r) \leq c_\ell \quad \text{for } \ell \in \mathcal{L} \\ &\qquad\qquad\quad 0 \leq p_r \leq p_{\max} \quad \text{for } r \in \mathcal{R}. \end{aligned} \tag{16.4}$$

We assume that $\sum_{r:\ell \in r} x_r(\sum_{\ell \in r} p_{\max}) < c_\ell$; this guarantees the existence of a feasible solution.

The formulation given in Equation (16.4) is a static optimization problem. So, from the network's perspective, if it could be solved centrally, the profit-maximizing prices could be set and the problem solved. However, there are issues that preclude this from happening.

- First, we seek a distributed implementation. We assume that there is no central controller of the network, and hence any decision regarding the change in prices must be performed in a manner that does not require direct communication between the links.

- In addition, observe that the formulation given in Equation (16.4) combines the prices charged by the links and the demand functions of the routes. We assume that the demand functions are not known by the links; they can only 'ask' the routes to reveal how much bandwidth they want at posted prices. This implies that any solution of this problem requires iterative feedback between the links and the routes that traverse it.

  The idea of reaching the solution dynamically is even more pertinent when the routes experience a change in their utility functions (for example, because the traffic offered by the end users changes; when the end users' load increases, more capacity is needed to meet the SLAs, and therefore the utility assigned to a given capacity decreases).

The intuitive explanation of a distributed mechanism is that each link should learn the demand curves of the routes during some sort of bidding process: the network advertises a set of prices $\vec{p}_1$, and sees the responses $x_r(\vec{p}_1)$ for $r \in \mathcal{R}$; then the network changes the prices to $\vec{p}_2$, and sees the new responses $x_r(\vec{p}_2)$ for $r \in \mathcal{R}$, etc. This bidding process must be automated, that is, performed through agents, to be practical in a network setting. Importantly, the algorithm that will be presented in the next section allows a distributed implementation.

16.3 Single-link network

In this section, we consider the network as a single link of capacity c, shared by a set of ISPs, each denoted $r \in \mathcal{R}$. For this single-link model, we devise a distributed algorithm.

The reason for commencing with a simplified network is twofold. First, analyzing the single-link model allows immediate insight into the differences between social welfare maximization and the profit maximization scenario that we explore in this chapter; we detail these insights below. In the second place, conditions required for convergence of the devised distributed algorithm are simpler to check in the single link case. We make a remark regarding extension to the network setting at the conclusion of this section.

Because we concentrate on a single link, we may suppress the dependence on ℓ, and we denote the price of bandwidth by p. Thus, the optimization problem user r is faced with is $\max_{x_r} U_r(x_r) - x_r p$, and the demand curve is given by $x_r(p) = V_r(p)$. For the sake of completeness, we translate the formulation in Equation (16.4) to the single-link case. The network problem is given by

$$\begin{aligned} \max_{p} \quad & p \cdot \sum_{r \in \mathcal{R}} x_r(p) \\ \text{subject to} \quad & \sum_{r \in \mathcal{R}} x_r(p) \leq \mathrm{c} \\ & 0 \leq p \leq p_{\max}. \end{aligned} \tag{16.5}$$

Since each demand function is decreasing and continuous in p, so is the aggregate demand $f(p) := \sum_r x_r(p)$ is also. Therefore, the link capacity constraint defines a minimum price $p_{\min}$ below which the constraint is violated. Of course, $p_{\min}$ could be negative. If this is the case, the nonnegativity constraint makes the link capacity constraint redundant. We assume that $p_{\min} \neq 0$.

Since the feasible region is closed and $\pi(p)$ is continuous over it, there exists an optimal solution. If the restriction that $p \leq p_{\max}$ was omitted, it is possible that no finite optimum exists. Take for example the case where $U_r(x_r)$ is concave, but $x_r(p)$ is convex with $x_r(p) \to \kappa$, a finite value, when $p \to \infty$. Then the link can make the price arbitrarily high and $\pi(p) \to \infty$. This shows that the introduction of $p_{\max}$ serves a mathematical purpose, besides reflecting the regulatory environment.

System optimization versus profit optimization. Before describing the iterative scheme, we first consider, through two insightful lemmas, the differences between the equilibrium solutions of the system-optimizing (social-welfare) and profit-maximizing problems in the single-link case, respectively.

The optimization-based flow control that Kelly *et al.* [153] and Low and Lapsley [179] performed are examples of system optimization, where the network acts in such a way that the equilibrium maximizes aggregate utility, $\sum_r U_r(x_r)$. However, the equilibrium vector of bandwidths $\vec{x}^{(s)}$ resultant from these scenarios may not be the same as that attained via the profit-maximizing problem described above, denoted as $\vec{x}^{(\pi)}$. Under system optimization, *all* of the link capacity c is allocated to the routes, while under profit optimization this is not necessarily the case.

Lemma 16.3.1 *The equilibrium reached yields an aggregate utility*

$$\sum_r U_r\left(x_r^{(s)}\right) \geq \sum_r U_r\left(x_r^{(\pi)}\right),$$

where this inequality can be strict. In economic terms, the profit maximization scenario creates a possible welfare loss because the network acts as a monopolist.

Proof. Consider the objective function $\pi(p)$ of the link problem (16.5). Let $x_r'(p)$ denote the derivative of route r's demand function with respect to p. Suppose that there exists a maximizer $0 < p^\star < p_{\max}$, where $p^\star$ satisfies

$$\pi'(p^\star) = p^\star \sum_{r\in\mathcal{R}} x_r'(p^\star) + \sum_{r\in\mathcal{R}} x_r(p^\star) = 0.$$

Since this is independent of the total available bandwidth c, the link capacity constraint may or may not be active at the optimum and therefore there may be bandwidth on the link left unallocated at $p^\star$. If spare capacity exists, aggregate utility is strictly less than in the welfare optimization case, since aggregate utility could be increased by allocating the excess capacity to one of the ISPs.

We separately analyze the cases in which the utility functions do not admit a solution to the above equation. If the cap on prices $p_{\max}$ comes into effect, then by assumption $\sum_r x_r(p_{\max}) < \mathrm{c}$, and hence there is unallocated link capacity at equilibrium. If $p^\star = 0$, then the link capacity constraint is inactive by assumption. Hence, again, there is unallocated capacity at equilibrium. Therefore, by the same argument as above, aggregate utility is strictly less than in the welfare-maximizing scenario.

Finally, if $p^\star = p_{\min}$, then all link capacity is allocated to the ISPs. The aggregate utility can only be as great as $\sum_r U_r\left(x_r^{(s)}\right)$, but no larger. □

Lemma 16.3.2 *The allocated bandwidth to each route under profit maximization is not greater than the allocated bandwidth in the system (social-welfare) optimization case, that is,*

$$x_r^{(\pi)} \leq x_r^{(s)} \quad \textit{for all } r \in \mathcal{R}.$$

Proof. Consider the optimal allocations under the profit-maximizing and the aggregate utility-maximizing scenarios, that is, $x_r^{(\pi)}$ and $x_r^{(s)}$, respectively. The latter can be mapped to some price q for which $U_r'(x_r^{(s)}) = q$. The maximum profit to the link is

$$\pi(p^\star) = p^\star \sum_{r \in \mathcal{R}} x_r^{(\pi)} \geq q \sum_{r \in \mathcal{R}} x_r^{(s)},$$

since the left-hand side is optimal.

Since the total allocated bandwidth in the profit scenario is not greater than for social optimization, we must have that $p^\star \geq q$. Since each demand function is a decreasing function of price, $x_r^{(\pi)} = x_r(p^\star) \leq x_r(q) = x_r^{(s)}$. □

These results are intuitive; in economic terms, they highlight the difference between a monopoly market and perfect competition, and the welfare loss that results from the former.

Remark 16.3.3 It follows from Lemma 16.3.2 that if all link capacity is allocated, then it is done so in a system (socially) optimum way. ◇

Existence of a solution. As the single-link problem is one of constrained optimization with respect to a one-dimensional variable p, the constraints reduce to a feasible interval from $\max\{0, p_{\min}\}$ to $p_{\max}$. Since $p_{\min} \neq 0$, at any time at most one constraint is active. Therefore, trivially, there is no linear dependence between the constraint gradients of the active constraints and hence any local maximizer is regular [31, Section 3.3].

The concavity of the utility functions is not sufficient to guarantee a unique optimal solution of Equation (16.5). Recall that the objective function and constraints of the link problem are written in terms of the demand functions $x_r(p)$ of each route. The assumptions on the utility functions guarantee that the demand functions $x_r(p)$ exists, but do not imply that they are concave (or convex). With $\pi(p) = pf(p)$, we have that

$$\pi''(p) = 2f'(p) + pf''(p);$$

$f'(p)$ is negative, but $f''(p)$ could be both positive and negative. We conclude that $\pi(p)$ is not necessarily concave, and hence we cannot guarantee the existence of a unique optimum.

Let $\mathcal{L}(p, \eta, \gamma)$ denote the Lagrangian of the network problem, where η is the multiplier of the link capacity inequality constraint and γ is the vector of multipliers of the price inequality constraints:

$$\mathcal{L}(p, \eta, \gamma) := pf(p) - \eta(f(p) - \mathrm{c}) - \gamma_1(p - p_{\max}) + \gamma_2 p. \tag{16.6}$$

Recall that $p_{\min}$ was defined as the solution to $f(p) = \mathrm{c}$ and hence is implicit in the statement of the Lagrangian.

We proceed by assuming the *second-order sufficient conditions* (SOSC) for inequality constrained problems [31, Proposition 3.3.2], which can be described as follows, hold. Let ∇_p denote the derivative operator with respect to p. Then, the SOSC state that the triple $(p^\star, \eta^\star, \gamma^\star)$ describes a local maximum if it satisfies

$$\nabla_p \mathcal{L}(p^\star, \eta^\star, \gamma^\star) = 0, \quad f(p^\star) \le \mathrm{c}, \quad 0 \le p^\star \le p_{\max}, \quad \eta^\star \ge 0, \quad \gamma^\star \ge 0,$$

where

$\eta^\star = 0,\ \gamma^\star = 0$ if the corresponding constraint is inactive;
$\eta^\star > 0,\ \gamma^\star > 0$ if the corresponding constraint is active ('strict complementarity');

$$y^{\mathrm{T}}\, \nabla^2 \mathcal{L}(p^\star, \eta^\star, \gamma^\star)\, y < 0,$$

for all vectors $y \neq 0$ such that the active-constraint gradients at $p^\star$ are perpendicular to y.

Iterative scheme. As discussed earlier, distributed implementation requires that the optimum of Equation (16.5) be found without the link having complete knowledge of each ISP's demand function and each ISP being unaware of the demand functions of the other ISPs. Instead, updates to both the prices and the allocated bandwidths must be performed iteratively, as reactions to one another. We require that the information transferred between the ISP and the link be local.

Since there is a separation between the users and the link, a first-order method is more appropriate; second-order methods (such as Newton-type schemes) typically involve nonlocal information. In the general network setting, first-order methods also fit into the requirement that links do not communicate directly with one another. The most basic first-order method involves the Lagrangian defined above. Its convergence, however, relies on the Hessian being concave at the local maximum. This is stronger than the SOSC and hence may not be satisfied.

For the single-link network, we proceed by defining the *augmented* Lagrangian

$$\mathcal{L}_m(p, \eta) := \pi(p) - \frac{1}{2m}\Big(\big(\max\{0, \eta + m(f(p) - \mathrm{c})\}\big)^2 - \eta^2\Big),$$

where only the link capacity constraint is involved. The form of the second term (penalty function) is derived by appending slack variables to the inequality constraints, adding a penalty for violation and finally maximizing over the slack variables analytically [31, Section 4.2].

Assume that the SOSC hold. Then, by [180, Ch. 13, Prop. 1], there exists an $\overline{m}$ such that, for all $m > \overline{m}$, the augmented Lagrangian $\mathcal{L}_m(p^\star, \eta^\star)$ has a local maximum at $p^\star$ with penalty parameter m. First-order Lagrange methods, while normally described using the original Lagrangian (16.6), can be applied with the augmented version also, when $m > \overline{m}$. The augmented Lagrangian not only allows us to use first-order methods, but imposes a penalty m on infeasibility. This has the benefit of discouraging iterates to stray too far from the feasible region.

Let p_j denote the price per unit bandwidth at the jth iteration and η_j denote the multiplier value at the jth iteration. The first-order augmented Lagrangian method is defined through the updates

$$\eta_{j+1} := \max\Big\{0, \eta_j + \epsilon\Big(f_{j+1} - \mathrm{c}\Big)\Big\}, \tag{16.7}$$

$$p_{j+1} := \Big[p_j + \epsilon\Big(f_{j+1} + f'_{j+1}[p_j - \max\{0, \eta_j + m(f_{j+1} - \mathrm{c})\}]\Big)\Big]_0^{p_{\max}}, \tag{16.8}$$

where $f_{j+1} := \sum_r x_r(p_j)$ and $f'_{j+1} := \sum_r x'_r(p_j)$; also, $[x]_b^a$ is defined as $\min\{\max\{x, a\}, b\}$.

There exists a neighborhood around $(p^\star, \eta^\star)$ and an $\overline{\epsilon}$ such that for all $\epsilon \in (0, \overline{\epsilon}]$, $(p^\star, \eta^\star)$ is a point of attraction of the updates given by Equations (16.7) and (16.8). This follows from a variation of Proposition 4.4.2 in [31], which concerned a first-order method applied to the original Lagrangian for equality-constrained problems. Details of the proof of convergence for a first-order method applied to the augmented Lagrangian for inequality-constrained problems can be found in the appendix of [194].

The update formulae (16.7) and (16.8) involve the demand functions for all routes $r \in \mathcal{R}$, through the f_{j+1} and f'_{j+1} terms. However, as discussed earlier, there is a separation between the link and the ISPs. Hence, the network does not know the individual ISP's utility functions (and hence demand curves). Therefore, the function $f(p)$ must be learned by the link at each iteration, by combining the requests that each route submits. The price update p_{j+1} not only involves f_{j+1}, the current total demand for bandwidth at the prevailing price, but also its derivative f'_{j+1}. This could be implemented by the requirement that each ISP, in addition, submit an 'elasticity' $x'_r(p_j)$ at each iteration. Since each ISP knows its own utility function, this is not an unrealistic requirement. The algorithm for the single link network could operate as shown below.

Algorithm 16.3.4 Profit maximization.

1. At iteration j, the link sets its price p_j and broadcasts it to the routes that traverse it.

2. Each ISP $r \in \mathcal{R}$ sees the current price and responds with two pieces of information:

$$x_r(p_j) = V_r(p_j) \quad \textit{and} \quad x'_r(p_j) = \frac{d}{dp}V_r(p_j).$$

3. Through agents, these requests get aggregated and the link sees a total desired bandwidth of f_{j+1} and its corresponding derivative f'_{j+1}.

4. Using the updates (16.7) and (16.8), the link updates its price to p_{j+1} and its multiplier to η_{j+1}.

5. It rebroadcasts a price of p_{j+1}, and the process continues until a prespecified convergence criterion is met.

Remark 16.3.5 In the setting described in this (and the previous) section, it is assumed that the utility functions $U_r(x_r)$ are increasing and convex for $x_r \geq 0$. It can be questioned, however, whether this is a realistic assumption in practical situations. In fact, each ISP has to be capable of handling a certain average rate that is generated by the end users, say μ_r on route r; if the ISP is allocated less than μ_r, then the ISP fails to transmit all traffic, and therefore it is rather unlikely that SLAs would have been met. Hence, each route r requires at least bandwidth μ_r, and in addition some extra bandwidth x_r can be allocated. In other words, the total bandwidth allocated, say $\overline{x}_r$, is the sum of the average rate μ_r (needed to ensure that all incoming traffic can be handled) and 'excess bandwidth' x_r (needed to provide QoS; the higher the x_r, the better the QoS delivered). Therefore, it is natural to assume that in this context the utility function is strictly increasing and concave in the *excess* bandwidth x_r (rather than the total bandwidth $\overline{x}_r$).

It is easy to adapt the Algorithm 16.3.4 to the situation described above. The user maximizes the utility of the excess bandwidth x_r less the cost of the bandwidth $\overline{x}_r = \mu_r + x_r$,

$$\max_{x_r \geq 0} U_r(x_r) - (\mu_r + x_r)p,$$

whereas the network solves

$$\max_p p \cdot \sum_{r \in \mathcal{R}} (\mu_r + x_r(p)) \quad \text{under} \quad \sum_{r \in \mathcal{R}} (\mu_r + x_r(p)) \leq \mathrm{c},$$

and $p \in [0, p_{\max}]$; this yields the optimal price $p^\star$, so that the total assigned bandwidth to user r is $\mu_r + x_r(p^\star) = \overline{x}_r$.

Using the same approach as for the original problem, we should use the updates

$$\eta_{j+1} := \max\left\{0, \eta_j + \epsilon\Big(f_{j+1} + \overline{\mu} - \mathrm{c}\Big)\right\}, \tag{16.9}$$

$$p_{j+1} := \Big[p_j + \epsilon\Big(f_{j+1} + \overline{\mu} + f'_{j+1}\big[p_j - \max\{0, \eta_j + m(f_{j+1} + \overline{\mu} - \mathrm{c})\}\big]\Big)\Big]_0^{p_{\max}}, \tag{16.10}$$

where f_{j+1} and f'_{j+1} are the total demand and the derivative of the total demand for *excess* bandwidth, respectively, and $\overline{\mu} := \sum_{r \in \mathcal{R}} \mu_r$. ◇

Remark 16.3.6 The first-order method can be extended to the general network setting; then a vector of prices should be updated. Convergence to a local maximum $\vec{p}^\star$ can be guaranteed if the second-order sufficient conditions are satisfied for the larger problem, and it can be shown that $\vec{p}^\star$ is regular. As expected, this is more complicated in the network case. We do not include further details here. ◇

In this section, we have assumed that each ISP assigns utility to bandwidth. As argued in the introduction, it may be more natural to consider the situation where utility is a function of the loss probability, as this is what the ISP has negotiated with its end users. The next section addresses the optimization problem for this situation.

16.4 Gaussian traffic; utility as a function of loss

In this section, we briefly review the concept of Gaussian queues and provide some examples of possible utility functions. These examples show that the utility curve can be considered a function of the loss probability δ, rather than the bandwidth x_r allocated to route r. Finally we explain how to change Algorithm 16.3.4 to this situation.

Gaussian queues. As we have seen earlier in this book, it was argued in [102, 156, 211] that aggregate traffic streams in packet networks can be accurately approximated by Gaussian processes, in particular when the traffic aggregation level (for instance in terms of numbers of users) is sufficiently high. This class of models covers a broad range of correlation structures, including long-range dependence.

The total amount of traffic $A_r(t)$ offered to the rth ISP in an interval of length t is normally distributed with mean $\mu_r t$ and variance $v_r(t)$. Suppose this ISP is allocated an amount of bandwidth $\overline{x}_r = \mu_r + x_r$ (cf. Remark 16.3.5). With the assumption of Gaussianity, we focus on the performance measure

$$\delta(x_r) := \mathbb{P}(A_r(T) \geq (\mu_r + x_r)T), \tag{16.11}$$

that is, the probability that the total offered traffic over this chosen timescale exceeds the allocated bandwidth. The choice of an appropriate T is primarily a task of the network manager, and depends on the types of applications being used; see Section 14.1. We rescale time such that $T \equiv 1$; for brevity we write $v_r \equiv v_r(1)$. The quantity $\delta(x_r)$ can be approximated as in Section 14.1: the Chernoff bound suggests

$$\delta(x_r) \approx \exp\left(-\frac{x_r^2}{2v_r}\right), \tag{16.12}$$

for $x_r > 0$, and 1 otherwise.

As argued in Remark 16.3.5, one has to allocate to route r minimally the mean rate μ_r, and in addition an excess rate x_r that can be determined by the iterative algorithm of the previous section. In practice, however, ISPs will assign utility to the loss probability that they experience rather than to (excess) bandwidth as such; recall that this is because it is the loss probability that they agreed upon with their clients (in the SLA). A way to do so, is to define for every function $U_r(\cdot)$, describing the utility derived from excess bandwidth, a function $\overline{U}_r(\cdot)$ such that

$$U_r(x_r) = \overline{U}_r(\delta(x_r)) = \overline{U}_r\left(\exp\left(-\frac{x_r^2}{2v_r}\right)\right);$$

the last equality is based on Approximation (16.12).

Example 16.4.1 In this example, we show that natural choices for $\overline{U}_r(\cdot)$ lead to functions $U_r(\cdot)$ that satisfy the assumptions made in Section 16.2. Consider the following choices: (1) $\overline{U}_r(\delta) = \alpha(-\log\delta)^\beta$, for positive α and $\beta \in (0, \frac{1}{2})$. Then,

$$U_r(x_r) = \frac{\alpha}{(2v_r)^\beta} x_r^{2\beta},$$

which is indeed strictly increasing and concave in x_r.

(2) $\overline{U}_r(\delta) = \alpha - \beta/(1-\delta)$ for positive β. Now,

$$U_r(x_r) = \alpha - \beta\left(1 - \exp\left(-\frac{x_r^2}{2v_r}\right)\right)^{-1},$$

which is also strictly increasing and concave. ◇

We observe that we can rewrite the user optimization problem, for the single link, to $\max_{x_r \geq 0} \overline{U}_r(\delta(x_r)) - (x_r + \mu_r)p$. To perform this optimization, we should adapt Algorithm 16.3.4 slightly, as will be explained in the next section.

Characterization of the demand curves. In Algorithm 16.3.4, the routes' utilities depend on the bandwidth that is allocated to them. We now describe how the algorithm should be adapted to the situation where utility curves $\overline{U}_r(\cdot)$ depend on the loss probability.

Recall that Step 2 of the algorithm required each ISP to tell the link (through agents) its current demand at the prevailing price and its current derivative of demand at the prevailing price. We now discuss how the ISPs can use knowledge of the function $\delta(x_r)$ to calculate these quantities. As we have seen earlier, the demand curve is defined as the solution to the equation $U_r'(x_r) = p$. By assumption, we know a solution exists (see Section 16.2). Since $U_r(x_r) = \overline{U}_r(\delta(x_r))$, by the chain rule, we have

$$U_r'(x) = \overline{U}_r'(\delta(x_r))\frac{\mathrm{d}}{\mathrm{d}x_r}\delta(x_r). \tag{16.13}$$

By equating Expression (16.13) to p, we find the demand for excess bandwidth at price p (denoted $x_r(p)$); in the algorithm, it can be found numerically, for example, via a Golden Section search. The algorithm also requires $x_r'(p)$; we can approximate this by

$$\frac{\mathrm{d}}{\mathrm{d}p}x_r(p) \approx \frac{x_r(p+h) - x_r(p)}{h}, \tag{16.14}$$

for a small $h > 0$. These quantities can then be used in Step 2 of the algorithm, with updates (16.9) and (16.10) used to adapt the price.

The above approach clearly requires that the ISPs *know* their required mean rate μ_r and the function $\delta(x_r)$. In a practical setting, however, the ISPs do not *a priori* know the μ_r and v_r of their customers. Therefore, they may rely on measurements to estimate this loss probability. In the next section, we explain the large-deviations background of this estimation procedure (relying extensively on Sanov's theorem and its inverse), which leads, in Section 16.6, to two ways to estimate $\delta(x_r)$ from measurements.

16.5 Sanov's theorem and its inverse

This section reviews a number of key large-deviations results. Sanov's theorem can be considered as one of the classical large-deviations theorems [261]; it describes the likelihood of an empirical distribution being far away from the real distribution. Its inverse, which was derived about a decade ago, does the opposite: it analyzes the 'probability' of a true distribution, given that a certain empirical distribution is observed. Particularly 'Inverse Sanov' turns out to be useful when estimating $\delta(x_r)$, as we will see in Section 16.6.

Sanov's theorem. Suppose we observe a sequence of i.i.d. random variables $X_1, X_2, \ldots$, distributed as a generic random variable X, attains values in $\Sigma \subseteq \mathbb{R}$. Define the *empirical distribution* after n samples as

$$L_n(x) := \frac{1}{n}\sum_{i=1}^{n} 1\{X_i = x\},$$

where $1\{A\} = 1$ if A is true, and 0 otherwise. As expected, under rather general conditions, $L_n(\cdot)$ converges to the density of the X_i (notice that talking about 'convergence' requires that a notion of distance be defined – we come back to this issue below). Informally, Sanov's theorem says how likely it is that the observed empirical distribution is (close to) ν, given that its actual distribution is ν_0. In fact, 'Sanov' says that $L_n(\cdot)$ satisfies an ldp (see Definition 4.1.4); it also gives an explicit expression for the corresponding rate function.

We show below that for the special case of X having an 'alternative distribution' (i.e., Σ consists of two elements; we say $|\Sigma| = 2$), 'Sanov' follows directly from 'Cramér'. Interestingly, we can 'bootstrap' this result for $|\Sigma| = 2$ immediately to the situation of finite alphabets, i.e., $|\Sigma| = N$ for some $N \in \mathbb{N}$. We then present Sanov's theorem in its most general form ($\Sigma = \mathbb{R}$).

But let us start with $|\Sigma| = 2$; say, for ease, that $X = 1$ with probability $p \in (0, 1)$ and $X = 0$ with probability $1 - p$. Suppose we are interested in the probability $\mathbb{P}(L_n(1) \approx a)$ for some $a \in (0, 1)$ (we intentionally write '$\approx$' rather than '$=$', as na need not be an integer number). Observe that $\{L_n(1) = a\}$ is equivalent to $\{n^{-1}\sum_{i=1}^{n} X_i = a\}$. With 'Cramér' we obtain

$$\begin{aligned}\lim_{n\to\infty}\frac{1}{n}\log\mathbb{P}(L_n(1)\approx a) &= -\sup_{\theta}\left(\theta a - \log\mathbb{E}e^{\theta X}\right)\\ &= -a\log\left(\frac{a}{p}\right) - (1-a)\log\left(\frac{1-a}{1-p}\right).\end{aligned}$$

We now proceed with $|\Sigma| = N$. Suppose that the random variable X equals j with probability p_j, with $\sum_{j=1}^{N} p_j = 1$. Let us analyze, asymptotically in n,

$$q_n(a) := \mathbb{P}\left(L_n(1)\approx a_1, \ldots, L_n(N-1)\approx a_{N-1}\right), \tag{16.15}$$

for $a_j \geq 0$ with $\sum_{j=1}^{N-1} a_j \leq 1$; notice that this entails that we automatically have that $L_n(N) \approx a_N := 1 - \sum_{j=1}^{N-1} a_j$. We can rewrite the joint probability (16.15) as the product of $N - 1$ conditional probabilities:

$$\prod_{j=1}^{N-1}\mathbb{P}(L_n(j)\approx a_j \mid L_n(1)\approx a_1, \ldots, L_n(j-1)\approx a_{j-1}).$$

Define

$$\overline{a}_j := \sum_{k=j}^{N} a_k = 1 - \sum_{k=1}^{j-1} a_k, \quad \overline{p}_j := \sum_{k=j}^{N} p_k = 1 - \sum_{k=1}^{j-1} p_k.$$

Then, it is not hard to see that, conditional on $L_n(1) \approx a_1, \ldots, L_n(j-1) \approx a_{j-1}$, it holds that $n\,L_n(j)$ is distributed as $\sum_{i=1}^{n\overline{a}_j} Y_i$ with the Y_i i.i.d., alternatively distributed, where Y_i is 1 with probability $\overline{p}_j/p_j$ and is 0 otherwise. Our result for $|\Sigma| = 2$ now implies that

$$\begin{aligned}&\lim_{n\to\infty}\frac{1}{n}\log\mathbb{P}(L_n(j)\approx a_j \mid L_n(1)\approx a_1, \ldots, L_n(j-1)\approx a_{j-1})\\ &= -\overline{a}_j\cdot\left(\left(\frac{a_j}{\overline{a}_j}\right)\log\left(\frac{a_j/\overline{a}_j}{\overline{p}_j/p_j}\right) + \left(1 - \frac{a_j}{\overline{a}_j}\right)\log\left(\frac{1 - a_j/\overline{a}_j}{1 - \overline{p}_j/p_j}\right)\right).\end{aligned}$$

After some elementary calculations, this reduces to

$$\psi_j := -a_j\log\frac{a_j}{p_j} + \overline{a}_j\log\frac{\overline{a}_j}{\overline{p}_j} - \overline{a}_{j+1}\log\frac{\overline{a}_{j+1}}{\overline{p}_{j+1}}.$$

It now immediately follows that

$$\lim_{n\to\infty} \frac{1}{n} \log q_n(a) = \sum_{j=1}^{N-1} \psi_j = -H(a),$$

$$\text{where} \quad H(a) \equiv H(a \mid p) := \sum_{j=1}^{N} a_j \log \frac{a_j}{p_j}.$$

This result could be called *Sanov's theorem for finite alphabets.* It says that the probability $q_n(a)$, as defined in Equation (16.15), vanishes at an exponential rate in n, and it does so with decay rate $H(a)$. This could alternatively be proved by applying Stirling's formula to the multinomial distribution; cf. [72, Section 2.1.1]. It entails that one could use the proxy $q_n(a) \approx \exp(-nH(a))$.

As mentioned earlier, Sanov's result can be extended to the case $\Sigma = \mathbb{R}$. Then, the statement is that $L_1(\cdot), L_2(\cdot), \ldots$ obeys the ldp with rate function $H(\cdot)$, where

$$H(\nu) \equiv H(\nu \mid \nu_0) := \int_{\Sigma} \log\left(\frac{\mathrm{d}\nu(x)}{\mathrm{d}\nu_0(x)}\right) \mathrm{d}\nu(x) \tag{16.16}$$

if $\mathrm{d}\nu(x)/\mathrm{d}\nu_0(x)$ exists, and is ∞ otherwise; $\nu_0(\cdot)$ is, as before, the distribution of X. This function H is often referred to as the Kullback–Leibler distance of ν with respect to ν_0 [72, Section 6.2]. The concept of ldp requires that open and closed sets be defined, and thus a metric is needed; in the context of Sanov's theorem one could use the so-called Prohorov metric for probability measures, see for instance [42, Ch. III]. Notice also that $H(\nu_0) = 0$, as expected.

Thus, 'Sanov' can be used to find an approximation of the probability that the empirical distribution $L_n(\cdot)$ is close to some $\nu(\cdot)$, given that its 'true distribution', say L, is $\nu_0(\cdot)$:

$$\mathbb{P}(L_n(\cdot) \approx \nu(\cdot) \mid L = \nu_0(\cdot)) \approx \exp\left(-nH(\nu \mid \nu_0)\right).$$

Exercise 16.5.1 *Sanov for normally distributed random variables.* Suppose that the X_i are i.i.d. samples $\mathcal{N}(\mu_0, v_0)$. Determine the asymptotics of $\mathbb{P}(L_n(\cdot) \approx \nu(\cdot))$, where $\nu(\cdot)$ corresponds to $\mathcal{N}(\mu, v)$.

Solution. To this end, we need to evaluate

$$\begin{aligned} H(\mu, v) \equiv H(\mu, v \mid \mu_0, v_0) &= \int_{-\infty}^{\infty} \log\left(\frac{f_{\mu,v}(x)}{f_{\mu_0,v_0}(x)}\right) \cdot f_{\mu,v}(x)\,\mathrm{d}x \\ &= \int_{-\infty}^{\infty} \left(\frac{1}{2}\log\left(\frac{v_0}{v}\right) - \frac{1}{2}\frac{(x-\mu)^2}{v} + \frac{1}{2}\frac{(x-\mu_0)^2}{v_0}\right) f_{\mu,v}(x)\,\mathrm{d}x. \end{aligned}$$

Use of standard calculus now yields (use that $\int x^2 f_{\mu,v}(x)\mathrm{d}x = \mu^2 + v$)

$$H(\mu, v) = \frac{1}{2}\log\left(\frac{v_0}{v}\right) + \frac{(\mu-\mu_0)^2 + v - v_0}{2v_0}.$$

Note that indeed $H(\mu_0, v_0) = 0$, as desired. ◇

The inverse of Sanov's theorem. As argued above, 'Sanov' tells us how likely it is to observe that the empirical distribution $L_n(\cdot)$ is close to $\nu(\cdot)$, given a 'true distribution' $\nu_0(\cdot)$. In many applications, however, the 'reverse question' is more relevant: what is the 'probability' that the true distribution is $\nu_0(\cdot)$, given that we observe that the empirical distribution $L_n(\cdot)$ is close to $\nu(\cdot)$? Notice that this question inherently has a Bayesian flavor.

To study this type of question in more detail, let us look at an example. Again, let $X_1, X_2, \ldots$ be a sequence of i.i.d. random variables, distributed as a random variable X, with $\mathbb{P}(X = 1) = 1 - \mathbb{P}(X = 0) = p$, but now with unknown p. How would a Bayesian estimate p? Well, he would impose some *prior* distribution $f(\cdot)$ on p (defined on $[0, 1]$), and then compute the *posterior* density of p, given observations $X_1, \ldots, X_n$. Given that $\sum_{i=1}^n X_i = k$, this posterior would be, in self-evident (but rather imprecise) notation,

$$\mathbb{P}\left(p \;\middle|\; \sum_{i=1}^n X_i = k\right) = \kappa_n(k)\cdot\binom{n}{k}p^k(1-p)^{n-k}f(p).$$

Here $\kappa \equiv \kappa_n(k)$ is a normalizing constant; as the expression in the previous equation is a density, the integral over $p \in [0, 1]$ should be 1. Suppose for instance that we take a uniform prior, i.e., $f(p) = 1$ on $[0, 1]$. Using the formulae for the Beta-integral, we find that $\kappa = (n + 1)$ (in this case κ is independent of k, but this is not necessarily the case; in general κ will depend on k). We thus obtain

$$\mathbb{P}\left(p \;\middle|\; \sum_{i=1}^n X_i = k\right) = (n+1)\binom{n}{k}p^k(1-p)^{n-k}. \tag{16.17}$$

Suppose we parameterize $k = an$ and consider the decay rate (in n) of Equation (16.17). Applying the (rough) Stirling approximation $n! \sim n^n e^{-n}$ (for decay rates the 'nonexponential terms' in Stirling's approximation do not play a role), and noticing that $n^{-1}\log(n + 1) \to 0$, we obtain

$$\lim_{n\to\infty}\frac{1}{n}\log\mathbb{P}\left(p \;\middle|\; \sum_{i=1}^n X_i = k\right) = -a\log\left(\frac{a}{p}\right) - (1-a)\log\left(\frac{1-a}{1-p}\right). \tag{16.18}$$

This is a remarkable finding. Informally it says that, for the uniform prior, the 'probability' that the success probability equals p, given that the observations would suggest that the success probability is a, is about $\exp(-nH(a \mid p))$. It can be seen as a true 'inverse Sanov', as it measures the likelihood of a distribution, given the observations.

Ganesh and O'Connell [107] show that considerably more general results can be derived. In particular, Equation (16.18) applies not only for the uniform prior, but in fact for any prior. Also, the result is not restricted to 0–1 random variables; it can be derived for any sequence of i.i.d. random variables on a finite state space Σ. Informally, when observing that $L_n(\cdot)$ equals a measure $\nu(\cdot)$, one may wonder what the 'probability' is that the real distribution L is governed by a measure $\nu_0(\cdot)$. Then, the decay rate of this 'probability' is given by

$$\lim_{n\to\infty} \frac{1}{n} \log \mathbb{P}(L = \nu_0(\cdot) \mid L_n = \nu(\cdot)) = -H(\nu \mid \nu_0),$$

with $H(\nu \mid \nu_0)$ as defined in Equation (16.16).

For instance, if we know that the X_i are normally distributed, and suppose we observe that the measurements reveal a sample mean $\hat{\mu}_n$ and sample variance $\hat{v}_n$, one could be interested in the 'probability' that they actually stem from $\mathcal{N}(\mu_0, v_0)$. Although in this case the state space is $\mathbb{R}$ (and hence infinite), one can show that the following approximation can be used (cf. Exercise 16.5.1):

$$q_n(\mu_0, v_0 \mid \hat{\mu}_n, \hat{v}_n) := \exp\left(-n\left(\frac{1}{2}\log\left(\frac{v_0}{\hat{v}_n}\right) + \frac{(\hat{\mu}_n - \mu_0)^2 + \hat{v}_n - v_0}{2v_0}\right)\right); \tag{16.19}$$

see [83, Theorem 6].

Exercise 16.5.2 Let $X_1, X_2, \ldots$ i.i.d. Poisson random variables with unknown mean μ. Suppose we observe that $\sum_{i=1}^n X_i = k$. Let the prior distribution be exponential with mean ζ^{-1}, i.e., the density $f(\mu) = \zeta e^{-\zeta\mu}$ applies. Compute the posterior

$$\psi_n(\mu \mid k) := \mathbb{P}\left(\mu \,\middle|\, \sum_{i=1}^n X_i = k\right).$$

Compute also the decay rate of $\psi_n(\mu \mid \alpha n)$, and relate this to 'inverse Sanov' (IS).

Solution. It is immediately seen that

$$\psi_n(\mu \mid k) = \kappa_n(k) \cdot \frac{(n\mu)^k e^{-n\mu}}{k!} \cdot \zeta e^{-\zeta\mu}.$$

The normalizing constant follows from the Gamma integral:

$$\kappa^{-1} = \int_0^\infty \frac{(n\mu)^k e^{-n\mu}}{k!} \cdot \zeta e^{-\zeta\mu} \mathrm{d}\mu = \frac{\zeta}{n}\left(\frac{n}{n+\zeta}\right)^{k+1}.$$

Thus we obtain that the posterior has an Erlang$(k, n+\zeta)$ distribution:

$$\psi_n(\mu \mid k) = \frac{(n+\zeta)^{k+1}\mu^k}{k!}e^{-(n+\zeta)\mu},$$

see for instance [275, Appendix B]. Direct computations show that

$$\lim_{n\to\infty}\frac{1}{n}\log\psi_n(\mu \mid \alpha n) = \alpha\log\frac{\mu}{\alpha} - \mu + \alpha.$$

This is independent of the ζ of the prior distribution, as could be expected from 'IS'. The last expression can be interpreted as the Kullback–Leibler distance of a Poisson random variable with mean α with respect to a Poisson random variable with mean μ. ◇

16.6 Estimation of loss probabilities

In this section, we describe two approaches to estimate the loss probability from measurements. In the first place, one could estimate the mean and variance, continue as if these are the true parameter values, and insert them in the approximation; this approach is called *certainty equivalence* (*CE*). As will be argued, it does not take into account the inherent randomness of the measurements, and therefore we propose a more conservative, Bayesian alternative, based on the so-called *Inverse Sanov* theorem. We conclude this section by showing how each ISP can use the Bayesian estimator to calculate the quantities needed in Step 2 of the algorithm in the context of Gaussian queues.

Consider a single ISP; it is offered an amount of traffic per unit time, say $A(1)$, that is normally distributed with *unknown* mean μ and variance v. There are essentially two different ways to estimate $\Delta(x) := \mathbb{P}(A(1) > x)$. A 'frequentist approach' would be to first estimate the mean and variance of $A(1)$. This would be done in the classical way: measure traffic over nonoverlapping windows of length 1, sufficiently placed to make sure that they are just weakly dependent. We thus obtain observations $a_1, \ldots, a_n$. Then we compute the estimates $\hat{\mu}_n := n^{-1}\sum_{i=1}^n a_i$ and $\hat{v}_n := (n-1)^{-1}\sum_{i=1}^n (a_i - \hat{\mu}_n)^2$. Then one could proceed as if the estimates $\hat{\mu}_n$ and $\hat{v}_n$ are the real parameter values, and apply Equation (16.12):

$$\mathbb{P}(\mathcal{N}(\hat{\mu}_n, \hat{v}_n) \geq x) \approx \exp\left(-\frac{1}{2}\frac{(x-\hat{\mu}_n)^2}{\hat{v}_n}\right).$$

This idea leads to the following approximation.

Approximation 16.6.1 *The* certainty-equivalence *estimator of* $\Delta(x)$ *is*

$$\Delta_{\mathrm{CE}}(x) := \exp\left(-\frac{1}{2}\frac{(x-\hat{\mu}_n)^2}{\hat{v}_n}\right) =: q(x \mid \hat{\mu}_n, \hat{v}_n). \tag{16.20}$$

The term 'certainty-equivalence' (CE) is used, as it is based on the idea that the values $\hat{\mu}_n$ and $\hat{v}_n$ are the true ones, thus ignoring the inherent uncertainty of the estimates. This explains why one could expect that this approximation cannot be guaranteed to be conservative.

The Bayesian alternative would be the following. First, realize that the event $\{A(1) > x\}$ could be considered as the combination of two effects. In the first place, it could be that the estimates $\hat{\mu}_n$ and $\hat{v}_n$ of the mean and variance are too optimistic, i.e., lower than the true values μ_0 and v_0. In the second place, it could be that, even if the estimates coincide with the true values μ_0 and v_0, $A(1)$ is unusually high (a 'large deviation').

Whereas the second probability can be approximated by standard large-deviations techniques, the first probability can be estimated by applying the so-called *Inverse Sanov* theorem [107]. Suppose we know that the observations stem from a normal distribution and suppose their realizations reveal a sample mean of $\hat{\mu}_n$ and a sample variance of $\hat{v}_n$. We may be interested in the 'probability' $q_n(\mu_0, v_0 \mid \hat{\mu}_n, \hat{v}_n)$ that the observations actually stem from a normal distribution with mean μ_0 and variance v_0. As we have seen, we can use Approximation (16.19). Recall that this statement, relying on the IS theorem, in fact says that, the 'probability' of μ_0, v_0 deviating from $\hat{\mu}_n$, $\hat{v}_n$ vanishes exponentially in n. As mentioned earlier, $I = 0$ for $\mu_0 = \hat{\mu}_n$ and $v_0 = \hat{v}_n$ and positive elsewhere; I is a so-called 'relative entropy' or Kullback–Leibler distance.

We would thus obtain the following first approximation that 'conditions' on the values of μ_0 and v_0. With q_n and q as defined in Equations (16.19) and (16.20),

$$\Delta(x) \approx \int_{\mu_0, v_0} q(x \mid \mu_0, v_0) \cdot q_n(\mu_0, v_0 \mid \hat{\mu}_n, \hat{v}_n)\, \mathrm{d}\mu_0\, \mathrm{d}v_0.$$

To further approximate $\Delta(x)$, we can apply the principle of the largest term, cf. [109, p. 25], and replace the integral by the maximum of the integrand. We thus obtain

$$\Delta(x) \approx \exp\left(-\frac{1}{2} Q_n(x, \hat{\mu}_n, \hat{v}_n)\right),$$

where $Q_n(x, \hat{\mu}_n, \hat{v}_n) := \min_{\mu_0, v_0} Q_n(x, \mu_0, v_0 \mid \hat{\mu}_n, \hat{v}_n)$;

$$Q_n(x, \mu_0, v_0 \mid \hat{\mu}_n, \hat{v}_n) := \frac{(x - \mu_0)^2}{v_0} + n \left(\log\left(\frac{v_0}{\hat{v}_n}\right) + \frac{(\hat{\mu}_n - \mu_0)^2 + \hat{v}_n - v_0}{v_0}\right).$$

The optimizing μ_0, v_0 are found explicitly; the first order condition

$$\frac{\mathrm{d}}{\mathrm{d}\mu_0} Q_n(x, \mu_0, v_0 \mid \hat{\mu}_n, \hat{v}_n) = 2\left(\frac{x + n\hat{\mu}_n - (n+1)\mu_0}{v_0}\right) = 0$$

immediately leads to

$$\mu_0^{\star} = \hat{\mu}_n + \frac{x - \hat{\mu}_n}{n+1},$$

whereas, on the other hand,

$$\frac{\mathrm{d}}{\mathrm{d}v_0} Q_n(x, \mu_0, v_0 \mid \hat{\mu}_n, \hat{v}_n) = -\frac{(x-\mu)^2}{v_0^2} + \frac{n}{v_0} - n \cdot \frac{(\hat{\mu}_n - \mu_0)^2 + \hat{v}_n}{v_0^2} = 0$$

leads to (after considerable calculus)

$$v_0^\star = \hat{v}_n + \frac{(x - \hat{\mu}_n)^2}{n+1}.$$

Inserting these into the objective function Q_n eventually yields the following approximation.

Approximation 16.6.2 *The* Inverse Sanov *estimator of* $\Delta(x)$ *is*

$$\Delta_{\mathrm{IS}}(x) := \left(1 + \frac{1}{n+1} \cdot \frac{(x - \hat{\mu}_n)^2}{\hat{v}_n}\right)^{-n/2}. \tag{16.21}$$

It is interesting to see how this estimate behaves as a function of n. In the first place, we observe that, for small n, the estimate $\mu_0^\star$ has a bias toward x, whereas for larger n it will be close to $\hat{\mu}_n$. A similar effect can be observed for $v_0^\star$: the impact of x vanishes when n grows, and for larger n it will be approximately equal to $\hat{v}_n$. Along the same lines it can be argued that for large n the Approximation (16.21) decreases to

$$\exp\left(-\frac{1}{2}\frac{(x - \hat{\mu}_n)^2}{\hat{v}_n}\right).$$

In other words, when n grows large, the uncertainty of the estimates $\hat{\mu}_n$ and $\hat{v}_n$ decreases, and, as a result, the approximation $\Delta_{\mathrm{IS}}(x)$ converges to the CE-based estimate (16.20). So, in general the frequentist and Bayesian approaches lead to different approximations, where the Bayesian approach is more conservative.

Remark 16.6.3 In the context of loss-probability-based utility functions, notice that the performance measure of loss discussed in this section, $\Delta(x)$, is based on the *total* allocated bandwidth. This is in contrast to the loss probability $\delta(x)$ based on *excess* bandwidth (see Equation (16.11)). To employ our estimators in the iterative scheme for allocating excess bandwidth, we can use the following:

$$\delta_{\mathrm{CE}}(x) = \Delta_{\mathrm{CE}}(x + \hat{\mu}_n), \quad \text{and} \quad \delta_{\mathrm{IS}}(x) = \Delta_{\mathrm{IS}}(x + \hat{\mu}_n).$$

In doing this, the IS estimator has an element of certainty equivalence to it, in that we use the sample mean $\hat{\mu}_n$ as the estimate of the true mean. However, as we will see through our numerical example, using this IS estimator still yields a more conservative outcome.

Exercise 16.6.4 One could say that the above analysis relates to a *packet-switched* environment: users generate data packets at some variable rate, and we argued (see Chapter 3) that the aggregate of the resulting traffic streams can be modeled as a Gaussian process. Traditional communication networks, however, were *circuit-switched*: a client uses, while active, a *line* of fixed capacity.

This situation could be modeled as follows: clients arrive according to a Poisson process of rate λ, stay in the system for some random duration D (where the durations of the individual jobs constitute a sequence of i.i.d. random variables), and transmit traffic at rate, say, 1 while in the system. In this model, the distribution of the number of clients present in the system, say M, is Poisson with mean $\lambda\mathbb{E}D$.

Suppose however that we want to estimate $\mathbb{P}(M \geq m)$, but that we do not know $\mu_0 := \lambda\mathbb{E}D$. Instead we have an approximation $\hat{\mu}_n$, based on n observations (which we suppose to be independent) of the number of clients $a_1, \ldots, a_n$, i.e., $\hat{\mu}_n := n^{-1}\sum_{i=1}^n a_i$. Determine a Bayesian approximation of $\mathbb{P}(M \geq m)$.

Solution. In the spirit of the above reasoning for $\mathbb{P}(A(1) \geq x)$, we may write

$$\mathbb{P}(M \geq m) \approx \int_{\mu_0} r(m \mid \mu_0) \cdot r_n(\mu_0 \mid \hat{\mu}_n)\,\mathrm{d}\mu_0,$$

with

$$r(m \mid \mu_0) = \sum_{i \geq m} e^{-\mu_0}\frac{\mu_0^i}{i!} \approx \exp\left(m \log \frac{\mu_0}{m} - \mu_0 + m\right),$$

where the last approximation is due to 'Cramér', and

$$r_n(\mu_0 \mid \hat{\mu}_n) \approx \exp\left(n\hat{\mu}_n \log \frac{\mu_0}{\hat{\mu}_n} - n\mu_0 + n\hat{\mu}_n\right),$$

because of Exercise 16.5.2. Performing the maximization to μ_0, we obtain (as for the Gaussian case) $\mu_0^\star = \hat{\mu}_n + (m - \hat{\mu}_n)/(n+1)$, which, inserted in the objective function yields the approximation

$$\mathbb{P}(M \geq m) \approx \left(\frac{n + m/\hat{\mu}_n}{n+1}\right)^{m+n\hat{\mu}_n}\left(\frac{\hat{\mu}_n}{m}\right)^m.$$

For large n this is close to $\exp(m\log(\hat{\mu}_n/m) - \hat{\mu}_n + m)$, i.e., the CE expression. ◇

Remark 16.6.5 The IS methodology (in conjunction with the principle of the largest term) is a powerful tool in many other situations, and nicely captures the possible 'dangers' of using the CE-approach. Although it is not entirely in the scope of this chapter, we have decided to include here an example that relates to measurement-based admission control, as it nicely illustrates the advantages of the IS-approach.

Consider a link in a network, on which i.i.d. sources are active. Let each individual source transmit traffic at a rate that is distributed as a random variable X, which we assume to represent a normal random variable, but with *unknown* mean and variance. Suppose we are able to obtain n measurements of this traffic rate (which we assume to be independent, and distributed as the random variable X), say, $a_1, \ldots, a_n$. The link has service capacity $\mathrm{c} \equiv nc$, and we focus on the regime of large n (i.e., we have many observations at our disposal).

We first compute the number of sources that can be admitted for the CE-approach, under the requirement that the probability of the aggregate input rate exceeding c is below ϵ, based on $a_1, \ldots, a_n$. Denoting by the X_i the rates of the individual sources, we wish to determine the biggest m such that

$$\mathbb{P}\left(\sum_{i=1}^{m} X_i \geq nc\right) \leq \epsilon.$$

As we consider CE, we assume that the X_i follow a normal distribution with mean $\hat{\mu}_n$ and variance $\hat{v}_n$. Applying the usual large-deviations approximation, we see that to find m, we must solve

$$\exp\left(-\frac{1}{2}\frac{(nc - m\hat{\mu}_n)^2}{m\hat{v}_n}\right) = \epsilon.$$

Elementary calculus yields that

$$m \equiv m(n) = \left(\frac{c}{\hat{\mu}_n}\right) n - \left(\frac{\sqrt{\hat{v}_n}}{\hat{\mu}_n}\sqrt{\frac{c}{\hat{\mu}_n}} \cdot \sqrt{-\log \epsilon}\right)\sqrt{n} + o(\sqrt{n}). \tag{16.22}$$

We now approximate the probability that the aggregate input rate exceeds c is below ϵ under the CE-based admission control. Suppose that we accept these m sources. It is natural to ask whether the probability of the aggregate input rate exceeding c is indeed in the order of ϵ; if it is, then the CE-approach is a reliable method. Given our discussion on CE, one might expect that the probability is larger than ϵ. To verify this, we follow the Bayesian approach:

$$\mathbb{P}\left(\sum_{i=1}^{m} X_i \geq nc\right) \approx \int_{\mu_0, v_0} \exp\left(-\frac{1}{2}\frac{(nc - m\mu_0)^2}{mv_0}\right.$$
$$\left. - \frac{n}{2}\left(\log\left(\frac{v_0}{\hat{v}_n}\right) + \frac{(\hat{\mu}_n - \mu_0)^2 + \hat{v}_n - v_0}{v_0}\right)\right) \mathrm{d}\mu_0 \, \mathrm{d}v_0.$$

Again applying the principle of the largest term, we find that the optimizing μ_0 and v_0 are

$$\mu_0^\star = \frac{n}{m+n}(c + \hat{\mu}_n); \quad v_0^\star = \frac{1}{m} \cdot \frac{1}{n+m}(nc - m\hat{\mu}_n)^2 + \hat{v}_n,$$

which yields the approximation, with $m \equiv m(n)$ as given in Equation (16.22),

$$\mathbb{P}\left(\sum_{i=1}^{m} X_i \geq nc\right) \approx \left(1 + \frac{1}{m} \cdot \frac{1}{m+n} \cdot \frac{(nc - m\hat{\mu}_n)^2}{\hat{v}_n}\right)^{-n/2}.$$

According to Equation (16.22), it holds that $(nc - m\hat{\mu}_n)^2 = \hat{v}_n \cdot (c/\hat{\mu}_n) \cdot \sqrt{-\log \epsilon} \cdot n + o(n)$, i.e., essentially linear in n. With $m \approx nc/\hat{\mu}_n$, we obtain that for n large

$$\mathbb{P}\left(\sum_{i=1}^{m} X_i \geq nc\right) \approx \exp\left(-\frac{1}{2} \cdot \frac{\hat{\mu}_n}{\hat{\mu}_n + c} \cdot (-\log \epsilon)\right) = (\sqrt{\epsilon})^{\hat{\mu}_n/(\hat{\mu}_n + c)}$$
$$= (\sqrt{\epsilon})^{\hat{\mu}_n/(\hat{\mu}_n + \mathrm{c}/n)}.$$

Since for large n we have that $\hat{\mu}_n/(\hat{\mu}_n + \mathrm{c}/n) \approx 1$, we observe that this probability is approximately $\sqrt{\epsilon}$, which is considerably larger than ϵ.

We may draw the following conclusions from the above computations. In the first place, consider CE-based admission control, i.e., admit m connections, with m as in Equation (16.22). The performance target is that the probability of exceeding c is below ϵ, but it turns out that this probability is rather in the order of $\sqrt{\epsilon}$, i.e., considerably too high. Apparently, the uncertainty of the estimates $\hat{\mu}_n$ and $\hat{v}_n$ is so strong that the CE-based admission control is too optimistic.

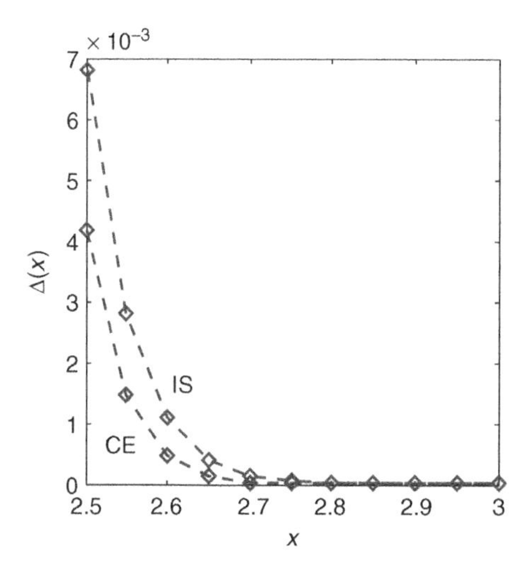

Figure 16.1: Difference between CE and the IS loss estimators with $N = 50$; System parameters: $\mu = 2$, $\sigma = 0.15$, $T = 1$.

In the second place, we observe that, in order to meet the performance target, one may still use the CE-method, but now with an adjusted ϵ, i.e., $\epsilon' := \epsilon^2$. These observations are in line with the findings of [121, Section 2]; notice that the approaches used are entirely different. We conclude that

$$m(n) = \left(\frac{c}{\hat{\mu}_n}\right) n - \left(\frac{\sqrt{\hat{v}_n}}{\hat{\mu}_n}\sqrt{\frac{c}{\hat{\mu}_n}} \cdot \sqrt{-2\log\epsilon}\right)\sqrt{n} + o(\sqrt{n})$$

is the resulting 'correct' admission criterion. ◇

Examples of IS estimator. Figure 16.1 shows the difference between the two loss estimators. Using a sample size of $N = 50$, we consider a system with traffic source parameters $\mu = 2$, $\sigma = 0.15$, and $T = 1$. The graph is generated by taking 20 samples from a $\mathcal{N}(2, 0.15^2)$ distribution, constructing the estimators $\Delta_{\mathrm{CE}}(x)$ and $\Delta_{\mathrm{IS}}(x)$, respectively, and plotting the means of the calculated quantities.

The graph shows the two estimators of loss as the allocated capacity x increases. The upper curve is the result of the IS estimator, confirming our previous observation that it was the more cautious estimator of the loss probability.

16.7 Numerical example

We illustrate the performance of the two loss estimators via a numerical example. Consider a network of four links of capacity $c = 20$, shared by five ISPs, of the form given in Figure 16.2. The routes are denoted by dashed lines, and traverse at least one link between an origin and a destination node.

One ISP traverses 3 links, two ISPs traverse 2 links and the remaining ISPs traverse a single link. We assume each ISP has a utility function of the form

$$\overline{U}(\delta) = \alpha - \frac{\beta}{(1-\delta)}. \tag{16.23}$$

As seen in Example 16.4.1, this utility function induces a utility function in terms of excess bandwidth that is strictly increasing and concave. Given the form

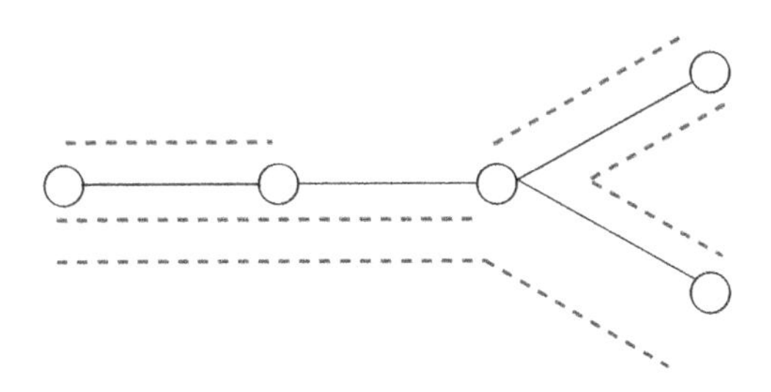

Figure 16.2: Network of four links shared by five routes.

Network Parameter	Parameter Value
Links	$c = [20, 20, 20, 20]$
Utility function	$\alpha = 1, \beta = 10$
True mean	$\mu = [3, 3, 3, 3, 3]$
True standard deviation	$\sigma = [0.2, 0.2, 0.2, 0.2, 0.2]$
Price cap	$p_{\max} = 1$ for all links
Step size	$\epsilon = 0.02$

Table 16.1: Network parameters for numerical example.

of Equation (16.23), the demand for excess bandwidth tends to 0 as the price $p \to \infty$. This implies that in this example we know *a priori* that $p^\star = p_{\max}$. The numerical example uses network parameters given in Table 16.1.

Updates to the link prices and multipliers are made according to Equations (16.9) and (16.10). From these equations, it can be seen that, at each link, not only the demand for excess bandwidth and the corresponding derivative should be known, but also the mean traffic rate. In the measurement-based context, the ISPs must estimate these quantities based on observations of the system. At each iteration, an independent sample of size n is taken. From this sample, the empirical mean can be calculated, and the ISPs are first allocated this quantity as an estimate of the mean rate (the minimum amount required to carry their traffic). This is aggregated on each link ℓ to estimate $\overline{\mu}_\ell$, the total mean rate on link ℓ, leaving a remaining $c_\ell - \hat{\overline{\mu}}_\ell$ units of bandwidth available.

The amount of this remaining bandwidth that gets allocated to the ISPs is derived from the demand for excess bandwidth. This depends on how δ is evaluated, which brings us to our two loss estimators of Section 16.6. As explained, the IS estimator $\Delta_{\mathrm{IS}}(x)$ is more conservative than the CE estimator $\Delta_{\mathrm{CE}}(x)$, in the sense that for a given bandwidth x, the estimated loss probability is higher using $\Delta_{\mathrm{IS}}(x)$.

Since we know *a priori* that in this example $p^\star = p_{\max}$ for all links, it follows that the excess bandwidth demanded by the ISPs is larger when they use $\Delta_{\mathrm{IS}}(x)$ as the estimator of the loss probability. The final profit $\pi(p^\star)$ is made up of the link prices multiplied by the total allocated bandwidth, where the latter is made up of the mean rate plus the demand for excess bandwidth. Hence, the previous discussion implies that the profit will also be larger when $\Delta_{\mathrm{IS}}(x)$ is used.

Figure 16.3 shows the profit derived as the iterative procedure progresses, when a sample size of $n = 20$ is used. Profit is calculated as the current price multiplied by the estimated mean rate plus the demand for excess bandwidth at the prevailing price. The solid straight lines indicate the average behavior of the IS and CE estimators, respectively. As expected, on average, the profit derived using $\Delta_{\mathrm{IS}}(x)$ is larger than when using $\Delta_{\mathrm{CE}}(x)$. The difference between the profits can be thought of as a premium that the network earns from being used by risk-averse ISPs.

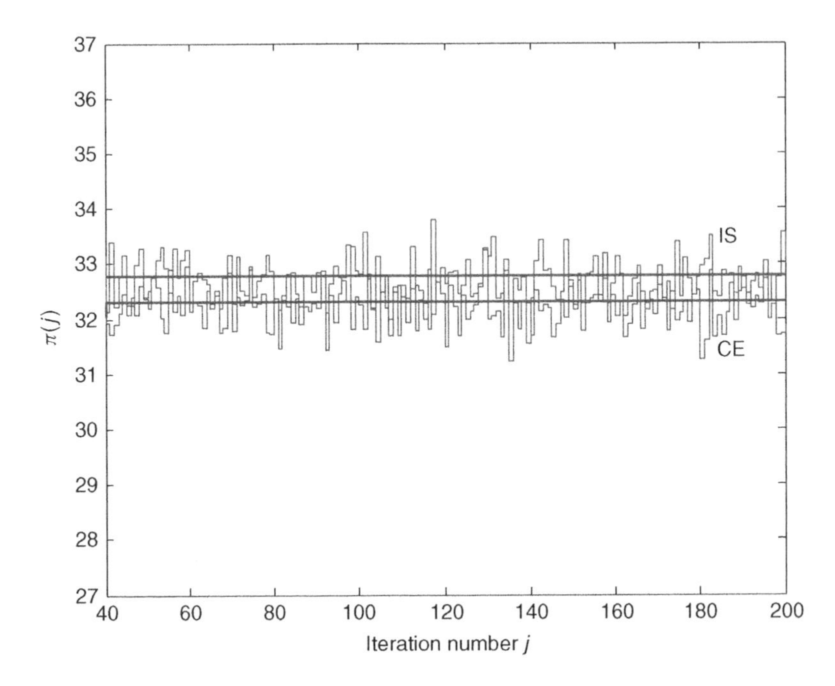

Figure 16.3: Derived profit using two estimators of loss and the first-order iterative method.

The difference between the IS and CE estimators is further highlighted by Figure 16.4. It shows the total allocated capacity for the three-link route as the price update procedure progresses. To aid comparison between the scenarios, we have removed the fluctuations in bandwidth as the iterative scheme progresses and only shown the 'average' behavior (achieved by assuming that the ISP views a sample mean $\hat{\mu}_n$ and sample variance $\hat{v}_n$ equal to the true mean and variance at each iteration). The benefit of this is the ease with which we can see how the sample size affects the impact of the IS estimator. The bottom-most curve is the total allocated bandwidth when the ISP uses the CE estimator. When using measurements, the $\Delta_{\text{CE}}(x)$-based total allocation fluctuates about this curve. So, quite frequently, the CE-based allocation is below the true optimal allocation. The remaining curves are the total allocated bandwidth when using the IS estimator for various sample sizes. The uppermost is when the smallest sample size $n = 10$ is used. Here, the ISP mistrusts the observed mean and variance, and hence compensates by demanding more bandwidth. This mistrust diminishes as n is made larger (here $n = 20, 50, 100$ are shown), which translates to a corresponding decrease in the profit of the network.

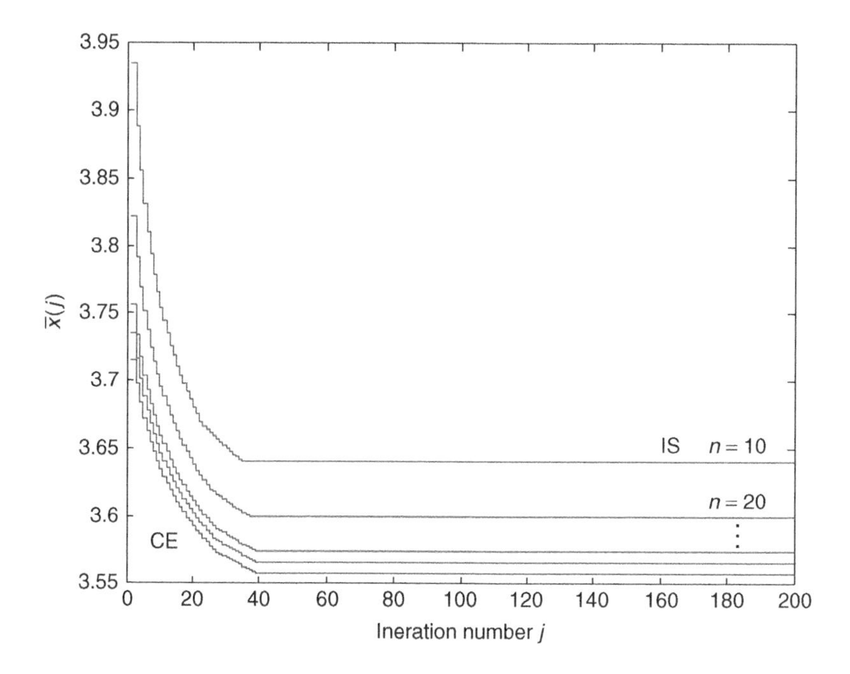

Figure 16.4: Total allocated capacity of ISP 1 in the network – includes mean rate and demand for excess bandwidth.

Bibliographical notes

This chapter is based on [194]; see also [248]. Some interesting references on bandwidth trading are [11, 12, 169]. Our framework with a network, service providers and end users is in line with the one sketched in [105].

The analysis of bandwidth trading algorithms falls in the area of *network economics*, which is a field of research that is in the intersection of performance analysis, microeconomics, and engineering. Microeconomic methods [278] turned out to be extremely useful when analyzing the competition for scarce resources (bandwidth, in this chapter), or negative externalities [181, 182, 266]. It was recognized that pricing could be used to provide the users with certain incentives; in this way differentiation of Quality of Service could be achieved [114, 184, 214, 229]. Interestingly, this could be achieved by marking packets during congestion periods [8, 111, 112]; the relation with large deviations is explained in [290]. Some of the textbooks on network economics are [56, 129, 209, 269].

It was observed in [23, 113, 120, 121] that a CE-approach, in which measured quantities are used as if they were the real ones, may lead to performance estimates that are considerably too optimistic. Reference [106] describes the use of the Bayesian approach in a couple of general network problems. Such an approach was also used in [82, 83], with emphasis on the use of on-line measurements when performing admission control; for an overview and background on measurement-based admission control algorithms, see for instance [26] and the introductions of [44, 140, 141]. The IS result was derived in [107]; for interesting related results see [108]. Standard textbooks on Bayesian analysis are by Berger [28] and Winkler [289].

Bibliography

[1] Abate, J., and Whitt, W. Transient behavior of regulated Brownian motion, I: starting at the origin. *Adv. Appl. Probab. 19* (1987), 560–598. [*p. 195, 196*]

[2] Abate, J., and Whitt, W. Transient behavior of regulated Brownian motion, II: non-zero initial conditions. *Adv. Appl. Probab. 19* (1987), 599–631. [*p. 196*]

[3] Abate, J., and Whitt, W. Asymptotics for M/G/1 low-priority waiting-time tail probabilities. *Queueing Syst. 25* (1997), 173–233. [*p. 143*]

[4] Abry, P., Baraniuk, R., Flandrin, P., Riedi, R., and Veitch, D. Multiscale nature of network traffic. *IEEE Signal Proc. Mag. 19* (2002), 28–46. [*p. 30*]

[5] Addie, R., Mannersalo, P., and Norros, I. Most probable paths and performance formulae for buffers with Gaussian input traffic. *Eur. Trans. Telecomm. 13* (2002), 183–196. [*p. 24, 48, 51, 76, 78, 90, 120, 220*]

[6] Adler, R. *An Introduction to Continuity, Extrema, and Related Topics for General Gaussian Processes. Lecture Notes-Monograph Series, Vol. 12*. Institute of Mathematical Statistics, Hayward, CA, USA, 1990. [*p. 24, 48*]

[7] Agrawal, R., Makowski, A., and Nain, P. On a reduced load equivalence for fluid queues under subexponentiality. *Queueing Syst. 33* (1999), 5–41. [*p. 67*]

[8] Alvarez, J., and Hajek, B. On the use of packet classes in communication networks to enhance congestion pricing based on marks. *IEEE Trans. Autom. Contr. 47* (2002), 1020–1026. [*p. 301*]

[9] Anick, D., Mitra, D., and Sondhi, M. Stochastic theory of a data handling system with multiple resources. *Bell Syst. Techn. J. 61* (1982), 1871–1894. [*p. 220*]

[10] Applebaum, D. *Lévy Processes and Stochastic Calculus*. Cambridge University Press, Cambridge, UK, 2004. [*p. 206*]

[11] Arvidsson, A., Chiera, B., Krzesinski, A., and Taylor, P. A distributed scheme for value-based bandwidth reconfiguration. *Submitted* (2006). [*p. 301*]

[12] Arvidsson, A., de Kock, J., Krzesinski, A., and Taylor, P. Cost-effective deployment of bandwidth partitioning in broadband networks. *Telecomm. Syst. 25* (2004), 33–49. [*p. 301*]

[13] Asmussen, S. *Applied Probability and Queues*, 2nd edition. Springer, New York, NY, USA, 2003. [*p. 5, 17*]

[14] Asmussen, S., and Rubinstein, R. Steady state rare events simulation in queueing models and its complexity properties. In *Advances in Queueing*, J. Dshalalow, Ed. CRC Press, 1995, pp. 429–461. [*p. 118*]

[15] Avi-Itzhak, B. A sequence of service stations with arbitrary input and regular service times. *Manage. Sci. 11* (1965), 565–571. [*p. 121, 133*]

[16] Azencott, R. *Ecole d'Eté de Probabilités de Saint-Flour VIII-1978*, vol. 774 of *Lecture Notes in Mathematics*. Springer, Berlin, 1980, ch. Grandes déviations et applications, pp. 1–176. [*p. 51*]

[17] Baccelli, F., and Brémaud, P. *Elements of Queuing Theory*, 2nd edition. Springer, New York, NY, USA, 2002. [*p. 5*]

[18] Bahadur, R., and Rao, R. On deviations of the sample mean. *Ann. Math. Stat. 31* (1960), 1015–1027. [*p. 40*]

[19] Bahadur, R., and Zabell, S. Large deviations of the sample mean in general vector spaces. *Ann. Probab. 7* (1979), 587–621. [*p. 51*]

[20] Baldi, P., and Pacchiarotti, B. Importance sampling for the ruin problem for general Gaussian processes. *Submitted* (2004). [*p. 118*]

[21] Barakat, C., Thiran, P., Iannaccone, G., Diot, C., and Owezarski, P. Modeling Internet backbone traffic at the flow level. *IEEE Trans. Signal Proc. 51* (2003), 2111–2114. [*p. 30, 248*]

[22] Baxter, G., and Donsker, M. On the distribution of the supremum functional for the processes with stationary independent increments. *Trans. Am. Math. Soc. 85* (1957), 73–87. [*p. 197*]

[23] Bean, N. Robust connection acceptance control for ATM networks with incomplete source information. *Ann. Oper. Res. 48* (1994), 357–379. [*p. 301*]

[24] Ben Azzouna, N., Clérot, F., Fricker, C., and Guillemin, F. Modeling ADSL traffic on an IP backbone link. *Ann. Telecomm. 59* (2004), 1260–1314. [*p. 30*]

[25] Beran, J. *Statistics for Long-Memory Processes*. Chapman & Hall, New York, NY, USA, 1994. [*p. 24, 27, 29, 251*]

[26] van den Berg, H., and Mandjes, M. Admission control in integrated networks: overview and evaluation. In *Proceedings of the 8th International Conference on Telecommunication Systems*. Nashville, (2000), pp. 132–151. [*p. 301*]

[27] van den Berg, H., Mandjes, M., van de Meent, R., Pras, A., Roijers, F., and Venemans, P. QoS-aware bandwidth provisioning of IP links. *Comp. Netw. 50* (2006), 631–647. [*p. 235, 243*]

[28] Berger, J. *Statistical Decision Theory and Bayesian Statistics*, 2nd edition. Springer, New York, NY, USA, 1999. [*p. 301*]

[29] Berger, A., and Whitt, W. Extending the effective bandwidth concept to networks with priority classes. *IEEE Comm. Mag.* August (1998), 78–83. [*p. 140*]

[30] Bertoin, J. *Lévy Processes*. Cambridge University Press, Cambridge, UK, 1996. [*p. 206*]

[31] Bertsekas, D. *Nonlinear Programming*, 2nd edition. Athena Scientific, Belmont, MA, USA, 1999. [*p. 119, 212, 281, 282, 283*]

[32] Bertsekas, D., and Gallager, R. *Data Networks*. Prentice-Hall, Englewood Cliffs, NJ, USA, 1987. [*p. 5*]

[33] Bertsimas, D., Paschalidis, I., and Tsitsiklis, J. Large deviations analysis of generalized processor sharing policy. *Queueing Syst. 32* (1999), 319–349. [*p. 167*]

[34] Bingham, N., Goldie, C., and Teugels, J. *Regular Variation*. Cambridge University Press, Cambridge, UK, 1987. [*p. 33, 34, 110*]

[35] de Boer, P.-T., Kroese, D., Mannor, S., and Rubinstein, R. A tutorial on the cross-entropy method. *Ann. Oper. Res. 134* (2005), 19–67. [*p. 118*]

[36] Bonald, T., Olivier, P., and Roberts, J. Dimensioning high speed IP access networks. *Proc. ITC 18* (2003), pp. 241–251. [*p. 248*]

[37] Boots, N.-K., and Mandjes, M. Fast simulation of a queue fed by a superposition of many(heavy-tailed) sources. *Probab. Eng. Inform. Sci. 16* (2002), 205–232. [*p. 116*]

[38] Borst, S., Mandjes, M., and van Uitert, M. GPS queues with heterogeneous traffic classes. *Proc. IEEE Infocom* (2002), pp. 74–83. [*p. 167*]

[39] Borst, S., Mandjes, M., and van Uitert, M. Generalized processor sharing queues with light-tailed and heavy-tailed input. *IEEE/ACM Trans. Netw. 11* (2003), 821–834. [*p. 167*]

[40] Botvich, D., and Duffield, N. Large deviations, the shape of the loss curve, and economies of large scale multiplexers. *Queueing Syst. 20* (1995), 293–320. [*p. 87, 90, 174*]

[41] Bratley, P., Fox, B., and Schrage, L. *A Guide to Simulation*, 2nd edition. Springer, New York, NY, USA, 1987. [*p. 117*]

[42] Bucklew, J. *Large Deviation Techniques in Decision, Simulation and Estimation*. Wiley, New York, NY, USA, 1990. [*p. 5, 289*]

[43] Bucklew, J. *Introduction to Rare Event Simulation*. Springer, New York, NY, USA, 2004. [*p. 117*]

[44] Casetti, C., Kurose, J., and Towsley, D. A new algorithm for measurement-based admission control in integrated services packet networks. In *Proceedings of the Fifth International Workshop on Protocols for High-Speed Networks*. Antipolis, France, (1996), pp. 13–28. [*p. 301*]

[45] Chang, C.-S. Sample path large deviations and intree networks. *Queueing Syst. 20* (1995), 7–36. [*p. 143*]

[46] Chang, C.-S. *Performance Guarantees in Communication Networks*. Springer, London, UK, 2000. [*p. 5*]

[47] Chang, C.-S., Heidelberger, P., Juneja, S., and Shahabuddin, P. Effective bandwidth and fast simulation of ATM intree networks. *Perf. Eval. 20* (1994), 45–65. [*p. 143*]

[48] Chernoff, H. A measure of asymptotic efficiency for test of hypothesis based on the sum of observations. *Ann. Math. Stat. 23* (1952), 493–507. [*p. 39*]

[49] Choe, J., and Shroff, N. A central-limit-theorem-based approach for analyzing queue behavior in high-speed networks. *IEEE/ACM Trans. Netw. 6* (1998), 659–671. [*p. 69*]

[50] Choe, J., and Shroff, N. On the supremum distribution of integrated stationary Gaussian processes with negative linear drift. *Adv. Appl. Probab. 31* (1999), 135–157. [*p. 69*]

[51] Choe, J., and Shroff, N. Use of the supremum distribution of Gaussian processes in queueing analysis with long-range dependence and selfsimilarity. *Stoch. Models 16* (2000), 209–231. [*p. 69*]

[52] Cohen, J. *The Single Server Queue*, 2nd edition. North Holland, Amsterdam, Netherlands, 1982. [*p. 5*]

[53] Cohen, J. On the effective bandwidth in buffer design for the multi-server channels. *CWI Report BS-R9406*, `http://ftp.cwi.nl/CWIreports/BS/BS-R9406.pdf`, (1994). [*p. 247*]

[54] Cottrell, M., Fort, J.-C., and Malgouyres, G. Large deviations and rare events in the study of stochastic algorithms. *IEEE Trans. Automat. Contr. 28* (1983), 907–920. [*p. 117*]

[55] Courcoubetis, C., and Weber, R. Buffer overflow asymptotics for a buffer handling many traffic sources. *J. Appl. Probab. 33* (1996), 886–903. [*p. 90*]

[56] Courcoubetis, C., and Weber, R. *Pricing Communication Networks: Economics, Technology and Modelling*. Wiley, Chichester, 2003. [*p. 301*]

[57] Cramér, H. Sur un nouveau théorème limite de la théorie des probabilités. *Actual. Sci. Ind. 736* (1938), 5–23. [*p. 39*]

[58] Crovella, M., and Bestavros, A. Self-similarity in World Wide Web traffic: evidence and possible causes. *IEEE/ACM Trans. Netw. 5* (1997), 835–847. [*p. 26, 35*]

[59] Davies, R., and Harte, D. Tests for Hurst effect. *Biometrika 74* (1987), 95–101. [*p. 115*]

[60] Dębicki, K. A note on LDP for supremum of Gaussian processes over infinite horizon. *Stat. Probab. Lett. 44* (1999), 211–220. [*p. 66, 69*]

[61] Dębicki, K. Some properties of generalized Pickands constants. *Prob. Th. Appl. 50* (2006), 290–298. [*p. 67, 69*]

[62] Dębicki, K., Dieker, T., and Rolski, T. Quasi-product forms for Lévy-driven fluid networks. *Math. Oper. Res.* (2007), in press. [*p. 206*]

[63] Dębicki, K., and Mandjes, M. Exact overflow asymptotics for queues with many Gaussian inputs. *J. Appl. Probab. 40* (2003), 702–720. [*p. 105*]

[64] Dębicki, K., and Mandjes, M. Traffic with an FBM limit: convergence of the stationary workload process. *Queueing Syst. 46* (2003), 113–127. [*p. 32, 59*]

[65] Dębicki, K., and Mandjes, M. A note on large-buffer asymptotics for generalized processor sharing with Gaussian inputs. *Queueing Syst.* (2007), in press. [*p. 167*]

[66] Dębicki, K., Mandjes, M., and van Uitert, M. A tandem queue with Lévy input: a new representation of the downstream queue length. *Probab. Eng. Inform. Sci. 21* (2007), 83–107. [*p. 143, 197, 206*]

[67] Dębicki, K., and Palmowski, Z. Heavy traffic Gaussian asymptotics of on-off fluid model. *Queueing Syst. 33* (1999), 327–338. [*p. 30, 32, 35, 59*]

[68] Dębicki, K., and Rolski, T. Gaussian fluid models; a survey. *Symposium on Performance Models for Information Communication Networks*, Sendai, Japan. Available from: `http://www.math.uni.wroc.pl/~rolski/publications.html`, 2000. [*p. 67, 69*]

[69] Dębicki, K., and Rolski, T. A note on transient Gaussian fluid models. *Queueing Syst. 41* (2002), 321–342. [*p. 69, 101*]

[70] Dębicki, K., and van Uitert, M. Large buffer asymptotics for generalized processor sharing queues with Gaussian inputs. *Queueing Syst. 54* (2006), 111–120. [*p. 167*]

[71] Delas, S., Mazumdar, R., and Rosenberg, C. Tail asymptotics for HOL priority queues handling a large number of independent stationary sources. *Queueing Syst. 40* (2002), 183–204. [*p. 143*]

[72] Dembo, A., and Zeitouni, O. *Large Deviations Techniques and Applications*, 2nd edition. Springer, New York, NY, USA, 1998. [*p. 5, 39, 45, 289*]

[73] Demers, A., Keshav, S., and Shenker, S. Analysis and simulation of a fair queueing algorithm. *Proc. ACM Sigcomm* (1989), pp. 1–12. [*p. 145*]

[74] Dempster, A. Generalized D_n^+ statistics. *Ann. Math. Stat. 30* (1959), 593–597. [*p. 60*]

[75] Deuschel, J.-D., and Stroock, D. *Large Deviations*. Academic Press, Boston, MA, USA, 1989. [*p. 5, 51*]

[76] Dieker, T. Tools for simulating fractional Brownian motion traces; *note: these tools are available from:* `http://www.proba.ucc.ie/~td3/fbm/index.html`, 2003. [*p. 256*]

[77] Dieker, T. *Simulation of Fractional Brownian Motion*. Master's Thesis, Vrije Universiteit, Amsterdam, Netherlands, 2002. [*p. 118*]

[78] Dieker, T. Extremes of Gaussian processes over an infinite horizon. *Stoch. Proc. Appl. 115* (2005), 207–248. [*p. 67, 69*]

[79] Dieker, T. Reduced-load equivalence for queues with Gaussian input. *Queueing Syst. 49* (2005), 405–414. [*p. 67*]

[80] Dieker, T., and Mandjes, M. On spectral simulation of fractional Brownian motion. *Probab. Eng. Inform. Sci. 17* (2003), 417–434. [*p. 118*]

[81] Dieker, T., and Mandjes, M. Fast simulation of overflow probabilities in a queue with Gaussian input. *ACM Trans. Model. Comp. Simul. 16* (2006), 119–151. [*p. 115, 117, 118*]

[82] Duffield, N. Asymptotic sampling properties of effective bandwidth estimation for admission control. *Proc. IEEE Infocom* (1999), pp. 1532–1538. [*p. 301*]

[83] Duffield, N. A large deviation analysis of errors in measurement based admission control to buffered and bufferless resources. *Queueing Syst. 34* (2000), 131–168. [*p. 291, 301*]

[84] Duffield, N., and O'Connell, N. Large deviations and overflow probabilities for general single-server queue, with applications. *Math. Proc. Camb. Phil. Soc. 118* (1995), 363–374. [*p. 69*]

[85] Dukkipati, N., Kuri, J., and Jamadagni, H. Optimal call admission control for generalized processor sharing (GPS) schedulers. *Proc. IEEE Infocom* (2001), pp. 468–477. [*p. 224*]

[86] Duncan, T., Yan, Y., and Yan, P. Exact asymptotics for a queue with fractional Brownian input and applications in ATM networks. *J. Appl. Probab. 38* (2001), 932–945. [*p. 69*]

[87] Dupuis, P., and Ellis, R. *A Weak Convergence Approach to the Theory of Large Deviations.* Wiley, New York, NY, USA, 1997. [*p. 5*]

[88] Dupuis, A., and Guillemin, F. Simulation-based analysis of weighted fair queueing algorithms for ATM networks. *Telecomm. Syst. 12* (1999), 149–166. [*p. 224*]

[89] Dupuis, P., and Wang, H. Importance sampling, large deviations, and differential games. *Stoch. Stoch. Rep. 76* (2004), 481–508. [*p. 117*]

[90] Dupuis, P., and Wang, H. Dynamic importance sampling for uniformly recurrent Markov chains. *Ann. Appl. Probab. 15* (2005), 1–38. [*p. 117*]

[91] Ellis, R. *Entropy, Large Deviations, and Statistical Mechanics.* Springer, New York, NY, USA, 1985. [*p. 5, 39*]

[92] Elwalid, A., and Mitra, D. Effective bandwidth of general Markovian traffic sources and admission control of high speed networks. *IEEE/ACM Trans. Netw. 1* (1993), 329–343. [*p. 235, 247*]

[93] Elwalid, A., and Mitra, D. Analysis, approximations and admission control of a multi-service multiplexing system with priorities. *Proc. IEEE Infocom* (1995), pp. 463–472. [*p. 138*]

[94] Elwalid, A., and Mitra, D. Design of generalized processor sharing schedulers which statistically multiplex heterogeneous QoS classes. *Proc. IEEE Infocom* (1999), pp. 1220–1230. [*p. 146, 209, 224*]

[95] Elwalid, A., Mitra, D., and Wentworth, R. A new approach for allocating buffers and bandwidth to heterogeneous regulated traffic in an ATM node. *IEEE J. Sel. Areas Comm. 13* (1995), 1115–1127. [*p. 84, 235*]

[96] Erlang, A. The theory of probabilities and telephone conversations. *Nyt. Tidsskr. Mat. B 20* (1909), 33–39. [*p. 246*]

[97] Erramilli, A., Narayan, O., and Willinger, W. Experimental queueing analysis with long-range dependent packet traffic. *IEEE/ACM Trans. Netw. 4* (1996), 209–223. [*p. 30*]

[98] Erramilli, A., Roughan, M., Veitch, D., and Willinger, W. Self-similar traffic and network dynamics. *Proc. IEEE 90* (2002), 800–819. [*p. 30*]

[99] Fayolle, G., Mitrani, I., and Iasnogoroski, R. Sharing a processor among many job classes. *J. ACM 27* (1980), 519–532. [*p. 167*]

[100] Fiedler, M., and Arvidsson, A. A resource allocation law to satisfy QoS demands on ATM burst and connection level. *COST257 TD(99)06* (1998). [*p. 231*]

[101] Fraleigh, C. *Provisioning Internet Backbone Networks to Support Latency Sensitive Applications*. PhD Thesis, Stanford University, USA, 2002. [*p. 248*]

[102] Fraleigh, C., Tobagi, F., and Diot, C. Provisioning IP backbone networks to support latency sensitive traffic. *Proc. IEEE Infocom* (2003). [*p. 25, 29, 30, 229, 230, 231, 237, 248, 275, 285*]

[103] Fricker, C., and Jaibi, R. On the fluid limit of the M/G/∞ server queue. *Submitted* (2006). [*p. 35*]

[104] Friedman, H. Reduction methods for tandem queueing systems. *Oper. Res. 13* (1965), 121–131. [*p. 121, 133*]

[105] Fulp, E., and Reeves, D. Optimal provisioning and pricing of internet differentiated services in hierarchical markets. In *International Conference of Networking, ICN 2001, Colmar, France. Lecture Notes in Computer Science (LNCS) Series, 2093*, P. Lorenz, Ed. Springer, 2001, pp. 409–418. [*p. 301*]

[106] Ganesh, A., Green, P., O'Connell, N., and Pitts, S. Bayesian network management. *Queueing Syst. 28* (1998), 267–282. [*p. 301*]

[107] Ganesh, A., and O'Connell, N. An inverse of Sanov's theorem. *Stat. Probab. Lett. 42* (1999), 201–206. [*p. 291, 293, 301*]

[108] Ganesh, A., and O'Connell, N. A large deviation principle for Dirichlet posteriors. *Bernoulli 6* (2000), 1021–1034. [*p. 275, 301*]

[109] Ganesh, A., O'Connell, N., and Wischik, D. *Big Queues, Lecture Notes in Mathematics, Vol. 1838*. Springer, Berlin, Germany, 2004. [*p. 5, 39, 69, 86, 90, 293*]

[110] Gibbens, R., and Hunt, P. Effective bandwidths for the multi-type UAS channel. *Queueing Syst. 9* (1991), 17–28. [*p. 247*]

[111] Gibbens, R., and Kelly, F. Resource pricing and the evolution of congestion control. *Automatica 35* (1999), 1969–1985. [*p. 301*]

[112] Gibbens, R., and Kelly, F. On packet marking at priority queues. *IEEE Trans. Automat. Contr. 47* (2002), 1016–1020. [*p. 301*]

[113] Gibbens, R., Kelly, F., and Key, P. A decision-theoretic approach to call admission control in ATM networks. *IEEE J. Sel. Areas Comm. 13* (1995), 1102–1114. [*p. 301*]

[114] Gibbens, R., Mason, R., and Steinberg, R. Internet service classes under competition. *IEEE J. Sel. Areas Comm. 18* (2000), 2490–2498. [*p. 301*]

[115] Giordano, S., Gubinelli, M., and Pagano, M. Bridge Monte-Carlo: a novel approach to rare events of Gaussian processes. In *Proceedings of the 5th Workshop on Simulation*. St. Petersburg, (2005), pp. 281–286. [*p. 118*]

[116] Glasserman, P., Heidelberger, P., Shahabuddin, P., and Zajic, T. Multilevel splitting for estimating rare event probabilities. *Oper. Res. 47* (1999), 585–600. [*p. 118*]

[117] Glasserman, P., and Wang, Y. Counterexamples in importance sampling for large deviations probabilities. *Ann. Appl. Probab. 7* (1997), 731–746. [*p. 118*]

[118] Glynn, P. Efficiency improvement techniques. *Ann. Oper. Res. 53* (1994), 175–197. [*p. 118*]

[119] Glynn, P., and Whitt, W. Logarithmic asymptotics for steady-state tail probabilities in a single-server queue. In *Studies in Applied Probability, Papers in Honour of Lajos Takács*, J. Galambos and J. Gani, Eds. Applied Probability Trust, 1994, pp. 131–156. [*p. 191*]

[120] Grossglauser, M., Keshav, S., and Tse, D. RCBR: A simple and efficient service for multiple time-scale traffic. *IEEE/ACM Trans. Netw. 5* (1997), 741–755. [*p. 301*]

[121] Grossglauser, M., and Tse, D. A framework for robust measurement-based admission control. *IEEE/ACM Trans. Netw. 7* (1999), 293–309. [*p. 297, 301*]

[122] Guérin, R., Ahmadi, H., and Naghshineh, M. Equivalent capacity and its application to bandwidth allocation in high-speed networks. *IEEE J. Sel. Areas Comm. 9* (1991), 968–981. [*p. 235, 247*]

[123] Guillemin, F., Mazumdar, R., Dupuis, A., and Boyer, J. Analysis of the fluid weighted fair queueing system. *J. Appl. Probab. 40* (2003), 180–199. [*p. 167*]

[124] Guillemin, F., and Pinchon, D. Analysis of generalized processor-sharing systems with two classes of customers and exponential services. *J. Appl. Probab. 41* (2004), 832–858. [*p. 167*]

[125] Guo, L., and Matta, I. The war between mice and elephants. In *Proceedings of the International Conference on Network Protocols* (2001), pp. 180–191. [*p. 35*]

[126] Hammersley, J., and Handscomb, D. *Monte Carlo Methods*. Methuen, London, 1964. [*p. 117*]

[127] Hansen, M., and Pitts, S. Nonparametric inference from the M/G/1 workload. *Bernoulli 12* (2006), 737–759. [*p. 255*]

[128] Harrison, J. *Brownian Motion and Stochastic Flow Systems*. Wiley, New York, NY, USA, 1985. [*p. 195*]

[129] Hassin, R., and Haviv, M. *To Queue or Not to Queue*. Kluwer, Dordrecht, Netherlands, 2003. [*p. 301*]

[130] Heath, D., Resnick, S., and Samorodnitsky, G. Heavy tails and long range dependence in on/off processes and associated fluid models. *Math. Oper. Res. 23* (1998), 145–165. [*p. 33*]

[131] Heidelberger, P. Fast simulation of rare events in queueing and reliability models. *ACM Trans. Model. Comp. Simul. 5* (1995), 43–85. [*p. 118*]

[132] den Hollander, F. *Large Deviations. Fields Institute Monographs 14*. American Mathematical Society, Providence, RI, USA, 2000. [*p. 5*]

[133] Hopmans, A., and Kleijnen, J. Importance sampling in systems simulation: A practical failure? *Math. Comp. Simul. 21* (1979), 209–220. [*p. 117*]

[134] Huang, C., Devetsikiotis, M., Lambadaris, I., and Kaye, A. Fast simulation of queues with long-range dependent traffic. *Stoch. Models 15* (1999), 429–460. [*p. 118*]

[135] Huang, X., Ozdağlar, A., and Acemoğlu, D. Efficiency and Braess' paradox under pricing in general networks. *IEEE J. Sel. Areas Comm. 24* (2006), 977–991. [*p. 275*]

[136] Hui, J. Resource allocation for broadband networks. *IEEE J. Sel. Areas Comm. 6* (1988), 1598–1608. [*p. 235, 247*]

[137] Humblet, P., Bhargava, A., and Hluchyj, M. Ballot theorems applied to the transient analysis of nD/D/1 queues. *IEEE/ACM Trans. Netw. 1* (1993), 81–95. [*p. 60*]

[138] Hüsler, J., and Piterbarg, V. Extremes of a certain class of Gaussian processes. *Stoch. Proc. Appl. 83* (1999), 257–271. [*p. 67, 69*]

[139] Jacod, J., and Shiryaev, A. *Limit Theorems for Stochastic Processes*, 2nd edition. Springer, Berlin, 2003. [*p. 35*]

[140] Jamin, S., Danzig, P., Shenker, S., and Zhang, L. A measurement-based admission control algorithm for integrated service packet networks. *IEEE/ACM Trans. Netw. 5* (1997), 56–70. [*p. 301*]

[141] Jamin, S., Shenker, S., and Danzig, P. Comparison of measurement-based call admission control algorithms for controlled-load service. *Proc. IEEE Infocom* (1997), pp. 973–980. [*p. 301*]

[142] Johari, R., and Tan, D. End-to-end congestion control for the Internet: delays and stability. *IEEE/ACM Trans. Netw. 9* (2001), 818–832. [*p. 274*]

[143] Juva, I., Susitaival, R., Peuhkuri, M., and Aalto, S. Traffic characterization for traffic engineering purposes: analysis of Funet data. In *Proceedings of the 1st EuroNGI Conference on Next Generation Internet Networks–Traffic Engineering* (2005). [*p. 30*]

[144] Kaj, I., and Taqqu, M. Convergence to fractional Brownian motion and to the Telecom process: the integral representation approach. *Department of Mathematics, Uppsala University, U.U.D.M. 2004:16*. Submitted (2004). [*p. 35*]

[145] Kaufman, J. Blocking in a shared resource environment. *IEEE Trans. Comm. 29* (1981), 1474–1481. [*p. 246*]

[146] Kella, O. Parallel and tandem fluid networks with dependent Lévy inputs. *Ann. Appl. Probab. 3* (1993), 682–695. [*p. 206*]

[147] Kella, O., and Whitt, W. A tandem fluid network with Lévy input. In *Queueing and Related Models*, U. Bhat and I. Basawa, Eds. Oxford University Press, 1992, pp. 112–128. [*p. 133, 206*]

[148] Kelly, F. Loss networks. *Ann. Appl. Probab. 1* (1991), 319–378. [*p. 246*]

[149] Kelly, F. Notes on effective bandwidths. In *Stochastic Networks: Theory and Applications*, F. P. Kelly, S. Zachary, and I. Ziedins, Eds. Oxford University Press, 1996, pp. 141–168. [*p. 235, 247*]

[150] Kelly, F. Charging and rate control for elastic traffic. *Eur. Trans. Telecomm. 8* (1997), 33–37. [*p. 274, 277*]

[151] Kelly, F. Models for a self-managed Internet. *Phil. Trans. R. Soc. A358* (2000), 2335–2348. [*p. 230, 274*]

[152] Kelly, F. Mathematical modelling of the Internet. In *Mathematics Unlimited – 2001 and Beyond*, B. Engquist and W. Schmid, Eds. Springer, 2001, pp. 685–702. [*p. 274*]

[153] Kelly, F., Maulloo, A., and Tan, D. Rate control in communication networks: shadow prices, proportional fairness and stability. *J. Oper. Res. Soc. 49* (1998), 237–252. [*p. 274, 280*]

[154] Kesidis, G., Walrand, J., and Chang, C.-S. Effective bandwidths for multiclass Markov fluids and other ATM sources. *IEEE/ACM Trans. Netw. 1* (1993), 424–428. [*p. 247*]

[155] Kilkki, K. *Differentiated Services for the Internet*. Macmillan Technical Publishing, Indianapolis, IN, USA, 1999. [*p. 119, 229*]

[156] Kilpi, J., and Norros, I. Testing the Gaussian approximation of aggregate traffic. In *Proceedings of the 2nd Internet Measurement Workshop* (2002), pp. 49–61. [*p. 25, 26, 28, 29, 30, 231, 261, 275, 285*]

[157] Kobelkov, S. The ruin problem for the stationary Gaussian process. *Theory Probab. Appl. 49* (2005), 155–163. [*p. 69*]

[158] Konstant, D., and Piterbarg, V. Extreme values of the cyclostationary Gaussian random processes. *J. Appl. Probab. 30* (1993), 82–97. [*p. 99, 104*]

[159] Kosten, L. Stochastic theory of data-handling systems with groups of multiple sources. In *Performance of Computer-Communication Systems*, H. Rudin and W. Bux, Eds. Elsevier, 1984, pp. 321–331. [*p. 224*]

[160] Kulkarni, V. *Frontiers in Queueing*. CRC Press, Boca Raton, FL, USA, 1997, ch. Fluid models for single buffer systems, pp. 321–338. [*p. 224*]

[161] Kulkarni, V., and Rolski, T. Fluid model driven by an Ornstein-Uhlenbeck process. *Probab. Eng. Inform. Sci. 8* (1994), 403–417. [*p. 59*]

[162] Kumaran, K., and Mandjes, M. The buffer-bandwidth trade-off curve is convex. *Queueing Syst. 38* (2001), 471–483. [*p. 90, 224*]

[163] Kumaran, K., and Mandjes, M. Multiplexing regulated traffic streams: design and performance. *Proc. IEEE Infocom* (2001), pp. 527–536. [*p. 84*]

[164] Kumaran, K., Mandjes, M., and Stolyar, A. Convexity properties of loss and overflow functions. *Oper. Res. Lett. 31* (2003), 95–100. [*p. 90*]

[165] Kumaran, K., Margrave, G., Mitra, D., and Stanley, K. Novel techniques for the design and control of generalized processor sharing schedulers for multiple QoS classes. *Proc. IEEE Infocom* (2000), pp. 932–941. [*p. 146, 224*]

[166] Kumaran, K., and Mitra, D. Performance and fluid simulations of a novel shared buffer management system. *Proc. IEEE Infocom* (1998), pp. 1449–1461. [*p. 214*]

[167] Kurose, J., and Ross, K. *Computer Networking: A Top-Down Approach Featuring the Internet.* Addison-Wesley Longman, Boston, MA, USA, 2002. [*p. 5*]

[168] Kvols, K., and Blaabjerg, S. Bounds and approximations for the periodic on-off queue with application to ATM traffic control. *Proc. IEEE Infocom* (1992), pp. 487–494. [*p. 84*]

[169] Lanning, S., Massey, W., Rider, B., and Wang, Q. Optimal pricing in queueing systems with quality of service constraints. *Proc. ITC 16* (1999), pp. 747–756. [*p. 301*]

[170] Law, A., and Kelton, W. *Simulation Modeling and Analysis.* 3rd edition. McGraw-Hill, New York, NY, USA, 2000. [*p. 117*]

[171] L'Ecuyer, P. Efficiency improvement via variance reduction. In *Proceedings of the Winter Simulation Conference* (1994), pp. 122–132. [*p. 118*]

[172] Leland, W., Taqqu, M., Willinger, W., and Wilson, D. On the self-similar nature of Ethernet traffic (extended version). *IEEE/ACM Trans. Netw. 2* (1994), 1–15. [*p. 30, 247*]

[173] Lieshout, P., and Mandjes, M. Generalized processor sharing: characterization of the admissible region and selection of optimal weights. *Comp. Oper. Res.* (2007), in press. [*p. 219, 223, 224*]

[174] Lieshout, P., and Mandjes, M. A note on the delay distribution in generalized processor sharing. *Oper. Res. Lett.* (2007), in press. [*p. 169*]

[175] Lieshout, P., and Mandjes, M. Tandem Brownian queues. *Math. Meth. Oper. Res.* (2007), in press. [*p. 202, 206*]

[176] Lieshout, P., Mandjes, M., and Borst, S. GPS scheduling: selection of optimal weights and comparison with strict priorities. *Proc. ACM Sigmetrics Perform. 34* (2006), pp. 75–86. [*p. 219, 224*]

[177] Likhanov, N., and Mazumdar, R. Cell loss asymptotics in buffers fed with a large number of independent stationary sources. *J. Appl. Probab. 36* (1999), 86–96. [*p. 104*]

[178] LoPresti, F., Zhang, Z., Towsley, D., and Kurose, J. Source time scale optimal buffer/bandwidth trade-off for regulated traffic in an ATM node. *Proc. IEEE Infocom* (1997), pp. 675–682. [*p. 84*]

[179] Low, S., and Lapsley, D. Optimization flow control, I: Basic algorithm and convergence. *IEEE/ACM Trans. Netw. 7* (1999), 861–875. [*p. 274, 280*]

[180] Luenberger, D. *Linear and Nonlinear Programming*, 2nd edition. Addison Wesley, Reading, MA, USA, 1984. [*p. 283*]

[181] MacKie-Mason, J., and Varian, H. Pricing congestible network resources. *IEEE J. Sel. Areas Comm. 13* (1995), 1141–1149. [*p. 301*]

[182] MacKie-Mason, J., and Varian, H. Pricing the Internet. In *Public Access to the Internet*, B. Kahin and J. Keller, Eds. MIT Press, 1995, pp. 269–314. [*p. 301*]

[183] Mandjes, M. Asymptotically optimal importance sampling for tandem queues with Markov fluid input. *AEÜ Int. J. Elect. Comm. 52* (1998), 152–161. [*p. 143*]

[184] Mandjes, M. Pricing strategies under heterogeneous service requirements. *Comp. Netw. 42* (2003), 231–249. [*p. 301*]

[185] Mandjes, M. A note on the benefits of buffering. *Stoch. Models 20* (2004), 43–54. [*p. 24, 90*]

[186] Mandjes, M. Packet models revisited: tandem and priority systems. *Queueing Syst. 47* (2004), 363–377. [*p. 197*]

[187] Mandjes, M. Large deviations for complex buffer architectures: the short-range dependent case. *Stoch. Models 22* (2006), 99–128. [*p. 191*]

[188] Mandjes, M., and Borst, S. Overflow behavior in queues with many long-tailed inputs. *Adv. Appl. Probab. 32* (2000), 1150–1167. [*p. 34, 90*]

[189] Mandjes, M., and Kim, J.-H. An analysis of the phase transition phenomenon in packet networks. *Adv. Appl. Probab. 33* (2001), 260–280. [*p. 86*]

[190] Mandjes, M., and Kim, J.-H. Large deviations for small buffers: an insensitivity result. *Queueing Syst. 37* (2001), 349–362. [*p. 77, 90*]

[191] Mandjes, M., Mannersalo, P., and Norros, I. Priority queues with Gaussian input: a path-space approach to loss and delay asymptotics. *Proc. ITC 19* (2005), pp. 1135–1144. [*p. 143*]

[192] Mandjes, M., Mannersalo, P., and Norros, I. Gaussian tandem queues with an application to dimensioning of switch fabrics. *Comp. Netw. 51* (2007), 781–797. [*p. 141, 143*]

[193] Mandjes, M., Mannersalo, P., Norros, I., and van Uitert, M. Large deviations of infinite intersections of events in Gaussian processes. *Stoch. Proc. Appl. 116* (2006), 1269–1293. [*p. 142*]

[194] Mandjes, M., and Ramakrishnan, M. Bandwidth trading under misaligned objectives: decentralized measurement-based control. *Submitted* (2006). [*p. 283, 301*]

[195] Mandjes, M., and Ridder, A. Optimal trajectory to overflow in a queue fed by a large number of sources. *Queueing Syst. 31* (1999), 137–170. [*p. 90*]

[196] Mandjes, M., and van de Meent, R. Inferring traffic burstiness by sampling the buffer occupancy. In *Networking Technologies, Services, and Protocols; Performance of Computer and Communication Networks; Mobile and Wireless Communication Systems. Fourth International IFIP-TC6 Networking Conference (Networking 2005), Waterloo, Canada. Lecture Notes in Computer Science (LNCS) Series, 3462*, R. Boutaba, K. Almeroth, R. Puigjaner, S. Shen, and J. Black, Eds. Springer, 2005, pp. 303–315. [*p. 271*]

[197] Mandjes, M., Saniee, I., and Stolyar, A. Load characterization, overload prediction, and load anomaly detection for voice over IP traffic. *IEEE Trans. Neural Netw. 16* (2005), 1019–1028. [*p. 18*]

[198] Mandjes, M., and van de Meent, R. Resource provisioning through buffer sampling. *Submitted* (2006). [*p. 271*]

[199] Mandjes, M., and van Uitert, M. Sample-path large deviations for tandem queues with Gaussian inputs. *Proc. ITC 18* (2003), pp. 521–530. [*p. 141*]

[200] Mandjes, M., and van Uitert, M. Sample-path large deviations for Generalized processor sharing queues with Gaussian inputs. *Perf. Eval. 61* (2005), 225–256. [*p. 158, 160, 166, 224*]

[201] Mandjes, M., and van Uitert, M. Sample-path large deviations for tandem and priority queues with Gaussian inputs. *Ann. Appl. Probab. 15* (2005), 1193–1226. [*p. 123, 129, 131, 141, 149, 158*]

[202] Mandjes, M., and Weiss, A. Sample-path large deviations of a multiple time-scale queueing model. *Submitted* (2004). [*p. 78*]

[203] Mankiw, G. *Macro-economics*, 2nd edition. Worth Publishers, New York, NY, USA, 1994. [*p. 89*]

[204] Mannersalo, P., and Norros, I. Approximate formulae for Gaussian priority queues. *Proc. ITC 17* (2001), pp. 991–1002. [*p. 120, 140, 141, 143, 146*]

[205] Mannersalo, P., and Norros, I. GPS schedulers and Gaussian traffic. *Proc. IEEE Infocom* (2002), pp. 1660–1667. [*p. 120, 146, 166, 216, 224*]

[206] Mannersalo, P., and Norros, I. A most probable path approach to queueing systems with general Gaussian input. *Comp. Netw. 40* (2002), 399–412. [*p. 51, 86, 120, 141*]

[207] Massoulié, L. Large deviations estimates for polling and weighted fair queueing service systems. *Adv. Perf. Anal. 2* (1999), 103–127. [*p. 167*]

[208] Massoulié, L., and Simonian, A. Large buffer asymptotics for the queue with FBM input. *J. Appl. Probab. 36* (1999), 894–906. [*p. 69*]

[209] McKnight, L., and Bailey, J. *Internet Economics*. MIT Press, Cambridge, MA, USA, 1998. [*p. 301*]

[210] van de Meent, R., and Mandjes, M. Evaluation of 'user-oriented' and 'black-box' traffic models for link provisioning. In *Proceedings of the 1st EuroNGI Conference on Next Generation Internet Networks – Traffic Engineering* (2005). [*p. 29*]

[211] van de Meent, R., Mandjes, M., and Pras, A. Gaussian traffic everywhere? In *Proceedings of the IEEE International Conference on Communications* (2006). [*p. 25, 29, 30, 231, 248, 275, 285*]

[212] van de Meent, R., Mandjes, M., and Pras, A. Smart dimensioning of IP network links. *Submitted* (2007). [*p. 271*]

[213] van de Meent, R., Pras, A., Mandjes, M., H. van den Berg, and Nieuwenhuis, L. Traffic measurements for link dimensioning – a case study. In *Self-managing distributed systems. 14th IFIP/IEEE International Workshop on Distributed Systems Operations and Management, DSOM 2003. Heidelberg, Germany. Lecture Notes in Computer Science (LNCS) Series, 2867*, M. Brunner and A. Keller, Eds. Springer, 2003, pp. 106–117. [*p. 250*]

[214] Mendelson, H., and Whang, S. Optimal incentive-compatible priority pricing for the M/M/1 queue. *Oper. Res. 38* (1990), 870–883. [*p. 301*]

[215] Michna, Z. On tail probabilities and first passage times for fractional Brownian motion. *Math. Meth. Oper. Res. 49* (1999), 335–354. [*p. 118*]

[216] Mikosch, T., Resnick, S., Rootzén, H., and Stegeman, A. Is network traffic approximated by stable Lévy motion or fractional Brownian motion? *Ann. Appl. Probab. 12* (2002), 23–68. [*p. 33, 35*]

[217] Mitra, D. Stochastic theory of a fluid model of producers and consumers coupled by a buffer. *Adv. Appl. Probab. 20* (1988), 646–676. [*p. 224*]

[218] Mitrinović, D. *Analytic Inequalities*. Springer, Berlin, Germany, 1970. [*p. 109*]

[219] Mo, J., and Walrand, J. Fair end-to-end window-based congestion control. *IEEE/ACM Trans. Netw. 8* (2000), 556–567. [*p. 274*]

[220] Narayan, O. Exact asymptotic queue length distribution for fractional Brownian traffic. *Adv. Perf. Anal. 1* (1998), 39–63. [*p. 67, 69*]

[221] Norros, I. A storage model with self-similar input. *Queueing Syst. 16* (1994), 387–396. [*p. 69*]

[222] Norros, I. On the use of fractional Brownian motion in the theory of connectionless networks. *IEEE J. Sel. Areas Comm. 13* (1995), 953–962. [*p. 69*]

[223] Norros, I. Busy periods of fractional Brownian storage: a large deviations approach. *Adv. Perf. Anal. 2* (1999), 1–20. [*p. 123, 142*]

[224] Norros, I. Most probable paths in Gaussian priority queues. *COST257 TD(99)16* (1999). [*p. 183, 184*]

[225] Norros, I. Large deviations of queues with long-range dependent input. In *Theory and Applications of Long Range Dependence*, P. Doukhan, G. Oppenheim, and M. Taqqu, Eds. Birkhäuser, 2002, pp. 409–416. [*p. 141*]

[226] Norros, I. Most probable path techniques for Gaussian systems. In *Proc. Networking 2002* (2002), pp. 86–104. [*p. 141*]

[227] Norros, I., Mannersalo, P., and Wang, J. Simulation of fractional Brownian motion with conditionalized random midpoint displacement. *Adv. Perf. Anal. 2* (1999), 77–101. [*p. 115, 118*]

[228] Norros, I., Roberts, J., Simonian, A., and Virtamo, J. The superposition of variable bit rate sources in an ATM multiplexer. *IEEE J. Sel. Areas Comm. 9* (1991), 378 –387. [*p. 60*]

[229] Odlyzko, A. Paris metro pricing for the internet. In *Proceedings of the ACM Conference on Electronic Commerce* (1999), pp. 140–147. [*p. 301*]

[230] Oetiker, T. MRTG: Multi Router Traffic Grapher; *note*: this tool is available from: `http://people.ee.ethz.ch/oetiker/webtools/mrtg/`, 2007. [*p. 230, 245*]

[231] Onvural, R. *Asynchronous Transfer Mode Networks: Performance Issues*, 2nd edition. Artech House, Norwood, MA, USA, 1995. [*p. 246, 247, 275*]

[232] Ozdağlar, A. Price competition with elastic traffic. *Submitted* (2006).

[233] Panagakis, A., Dukkipati, N., Stavrakakis, I., and Kuri, J. Optimal call admission control on a single link with a GPS scheduler. *IEEE/ACM Trans. Netw. 12* (2004), 865–878. [*p. 224*]

[234] Papagiannaki, K. *Provisioning IP Backbone Networks Based on Measurements*. PhD Thesis, University of London, UK, 2003. [*p. 248*]

[235] Parekh, A., and Gallager, R. A generalized processor sharing approach to flow control in integrated services network: the single node case. *IEEE/ACM Trans. Netw. 1* (1993), 344–357. [*p. 145*]

[236] Parekh, A., and Gallager, R. A generalized processor sharing approach to flow control in integrated services networks: The multiple node case. *IEEE/ACM Trans. Netw. 2* (1994), 137–150. [*p. 145*]

[237] Paschalidis, Y. Class-specific quality of service guarantees in multimedia communication networks. *Automatica 12* (1999), 1951–1969. [*p. 169*]

[238] Paxson, V. Fast, approximate synthesis of fractional Gaussian noise for generating self-similar network traffic. *Comp. Comm. Rev. 27* (1997), 5–18. [*p. 115*]

[239] Paxson, V., and Floyd, S. Wide area traffic: the failure of Poisson modeling. *IEEE/ACM Trans. Netw. 3* (1995), 226–244. [*p. 247*]

[240] Pickands III, J. Asymptotic properties of the maximum in a stationary Gaussian process. *Trans. Am. Math. Soc. 145* (1969), 75–86. [*p. 69*]

[241] Pigou, A. *The Economics of Welfare*, 3rd edition. Macmillan, London, UK, 1929. [*p. 275*]

[242] Piterbarg, V. *Asymptotic Methods in the Theory of Gaussian Processes*. American Mathematical Society, Providence, RI, USA, 1995. [*p. 69*]

[243] Piterbarg, V., and Fatalov, V. The Laplace method for probability measures in banach spaces. *Russ. Math. Surv. 50* (1995), 1151–1239. [*p. 69*]

[244] Piterbarg, V. Large deviations of a storage process with fractional Brownian motion as input. *Extremes 4* (2001), 147–164. [*p. 69*]

[245] Piterbarg, V., and Prisyazhnyuk, V. Asymptotic behavior of the probability of a large excursion for a nonstationary Gaussian processes. *Theory Probab. Math. Stat. 18* (1978), 121–133. [*p. 99*]

[246] Prabhu, N. *Stochastic Storage Processes: Queues, Insurance Risk, Dams and Data Communication*, 2nd edition. Springer, New York, NY, USA, 1998. [*p. 5, 206*]

[247] Pyke, R. The supremum and infimum of the Poisson process. *Ann. Math. Stat. 30* (1959), 568–576. [*p. 60*]

[248] Ramakrishnan, M. *Distributed Approaches to Capacity Reallocation in Networks*. PhD Thesis, University of Melbourne, Australia, 2006. [*p. 301*]

[249] Reich, E. On the integrodifferential equation of Takács I. *Ann. Math. Stat. 29* (1958), 563–570. [*p. 57, 122, 148, 149*]

[250] Robert, P. *Stochastic Networks and Queues*. Springer, New York, NY, USA, 2003. [*p. 5*]

[251] Roberts, J. A service system with heterogeneous service requirements – applications to multi-service telecommunications systems. In *Proceedings of Performance of Datacommunications Systems and their Applications*, G. Pujolle, Ed. North Holland, Amsterdam, The Netherlands, 1981, pp. 423–431. [*p. 246*]

[252] Roberts, J. Traffic theory and the Internet. *IEEE Comm. Mag. 39* (2001), 94–99. [*p. 215*]

[253] Roberts, J., Mocci, U., and Virtamo, J. *Broadband Network Teletraffic, Final Report of European Action COST 242*. Springer, Berlin, Germany, 1996. [*p. 5, 246, 247*]

[254] Roberts, J., and Virtamo, J. The superposition of periodic cell arrival streams in an ATM multiplexer. *IEEE Trans. Comm. 39* (1991), 298–303. [*p. 60*]

[255] Rockafellar, R. *Convex Analysis*. Princeton University Press, Princeton, NJ, USA, 1970. [*p. 95*]

[256] Rogers, L. Fluid models in queueing theory and Wiener-Hopf factorization of Markov chains. *Ann. Appl. Probab. 4* (1994), 390–413. [*p. 224*]

[257] Roughgarden, T. *Selfish Routing and the Price of Anarchy*. MIT Press, Cambridge, MA, USA, 2004. [*p. 275*]

[258] Rubinstein, R. Optimization of computer simulation models with rare events. *Eur. J. Oper. Res. 99* (1997), 89–112. [*p. 118*]

[259] Rubinstein, R., and Kroese, D. *The Cross-Entropy Method: A Unified Approach to Combinatorial Optimization, Monte-Carlo Simulation, and Machine Learning*. Springer, New York, NY, USA, 2004. [*p. 118*]

[260] Sadowsky, J., and Bucklew, J. On large deviations theory and asymptotically efficient Monte Carlo estimation. *IEEE Trans. Inf. Theory 36* (1990), 579–588. [*p. 117*]

[261] Sanov, I. *Selected Translations in Mathematical Statistics and Probability*, vol. 1. American Mathematical Society, Providence, RI, USA, 1961, ch. On the probability of large deviations of random variables, pp. 213–244. [*p. 287*]

[262] Sato, K. *Lévy Processes and Infinitely Divisible Distributions*. Cambridge University Press, Cambridge, UK, 1999. [*p. 206*]

[263] Shakkottai, S., and Srikant, R. Many-sources delay asymptotics with applications to priority queues. *Queueing Syst. 39* (2001), 183–200. [*p. 143*]

[264] Sharma, V. Estimating traffic intensities at different nodes in networks via a probing stream. *Proc. IEEE Globecom* (1999), pp. 374–380. [*p. 255*]

[265] Sharma, V., and Mazumdar, R. Estimating the traffic parameters in queueing systems with local information. *Perf. Eval. 32* (1998), 217–230. [*p. 255*]

[266] Shenker, S. Fundamental design issues for the future Internet. *IEEE J. Sel. Areas Comm. 13* (1995), 1176–1188. [*p. 301*]

[267] Shwartz, A., and Weiss, A. *Large Deviations for Performance Analysis. Queues, Communications, and Computing*. Chapman & Hall, London, UK, 1995. [*p. 5*]

[268] Siegmund, D. Importance sampling in the Monte Carlo study of sequential tests. *Ann. Stat. 4* (1976), 673–684. [*p. 117*]

[269] Songhurst, D. *Charging Communication Networks – from Theory to Practice*. Elsevier, Amsterdam, the Netherlands, 1999. [*p. 301*]

[270] Sriram, K., and Whitt, W. Characterizing superposition arrival processes in packet multiplexers for voice and data. *IEEE J. Sel. Areas Comm. 4* (1986), 833–846. [*p. 220*]

[271] Szczotka, W. Central limit theorem in $D[0;\infty)$ for breakdown processes. *Probab. Math. Stat. 1* (1980), 49–57. [*p. 35*]

[272] Takács, L. *Combinatorial Methods in the Theory of Stochastic Processes*. Wiley, New York, NY, USA, 1967. [*p. 60*]

[273] Taqqu, M., Teverovsky, V., and Willinger, W. Estimators for long-range dependence: an empirical study. *Fractals 3* (1995), 785–798. [*p. 29*]

[274] Taqqu, M., Willinger, W., and Sherman, R. Proof for a fundamental result in self-similar traffic modeling. *Comp. Comm. Rev. 27* (1997), 5–23. [*p. 33, 35*]

[275] Tijms, H. *A First Course in Stochastic Models*. Wiley, New York, NY, USA, 2003. [*p. 236, 292*]

[276] van Uitert, M., and Borst, S. Generalised processor sharing networks fed by heavy-tailed traffic flows. *Proc. IEEE Infocom* (2001), pp. 269–278. [*p. 167*]

[277] van Uitert, M., and Borst, S. A reduced-load equivalence for generalised processor sharing networks with long-tailed traffic flows. *Queueing Syst. 41* (2002), 123–163. [*p. 167*]

[278] Varian, H. *Microeconomic Analysis*, 3rd edition. Norton, New York, NY, USA, 1992. [*p. 301*]

[279] de Veciana, G., Kesidis, G., and Walrand, J. Resource management in wide-area ATM networks using effective bandwidths. *IEEE J. Sel. Areas Comm. 13* (1995), 1081–1090. [*p. 247*]

[280] Villen-Altamirano, M., and Villen-Altamirano, J. Restart: A straightforward method for fast simulation of rare events. In *Proceedings of the Winter Simulation Conference* (1994), pp. 282–289. [*p. 118*]

[281] Vinnicombe, G. On the stability of networks operating TCP-like congestion control. In *Proceedings of the 15th IFAC World Congress on Automatic Control* (2002). [*p. 274*]

[282] Weber, R. The interchangeability of ·/M/1 queues in series. *J. Appl. Probab. 16* (1979), 690–695. [*p. 121*]

[283] Weiss, A. A new technique for analyzing large traffic systems. *Adv. Appl. Probab. 18* (1986), 506–532. [*p. 78, 90*]

[284] Weiss, W. QoS with differentiated services. *Bell Labs Techn. J. 3* (1998), 48–62. [*p. 119, 229*]

[285] Whitt, W. Tail probabilities with statistical multiplexing and effective bandwidth in multi class queues. *Telecomm. Syst. 2* (1993), 71–107. [*p. 247*]

[286] Whitt, W. *Stochastic-Process Limits: An Introduction to Stochastic-Process Limits and their Application to Queues*. Springer, New York, NY, USA, 2002. [*p. 32, 33, 35*]

[287] Whittle, P. *Optimization under Constraints.* Wiley, Chichester, UK, 1971. [*p. 212*]

[288] Willinger, W., Taqqu, M., Sherman, R., and Wilson, D. Self-similarity through high-variability: statistical analysis of Ethernet LAN traffic at the source level. *IEEE/ACM Trans. Netw. 5* (1997), 71–86. [*p. 35*]

[289] Winkler, R. *Introduction to Bayesian Inference and Decision*, 2nd edition. Probabilistic Publishing, Sugar Land, TX, USA, 2003. [*p. 301*]

[290] Wischik, D. *Large deviations and Internet Congestion.* PhD Thesis, University of Cambridge, UK, 1999. [*p. 301*]

[291] Wischik, D. The output of a switch, or, effective bandwidths for networks. *Queueing Syst. 32* (1999), 383–396. [*p. 143*]

[292] Wischik, D. Sample path large deviations for queues with many inputs. *Ann. Appl. Probab. 11* (2001), 389–404. [*p. 90, 120, 140, 141, 143*]

[293] Wischik, D., and Ganesh, A. The calculus of Hurstiness. *Queueing Syst.* (2007), in press. [*p. 143*]

[294] Wroclawski, J. Specification of the controlled load network element service. *Internet Engineering Task Force RFC 2211* (1997). [*p. 229*]

[295] Zeevi, A., and Glynn, P. On the maximum workload in a queue fed by fractional Brownian motion. *Ann. Appl. Probab. 10* (2000), 1084–1099. [*p. 69*]

[296] Zhang, Z.-L. Large deviations and the Generalized processor sharing scheduling for a two-queue system. *Queueing Syst. 26* (1997), 229–254. [*p. 151, 152, 167, 168, 191*]

[297] Zhang, Z.-L. Large deviations and the Generalized processor sharing scheduling for a multiple-queue system. *Queueing Syst. 28* (1998), 349–376. [*p. 167*]

[298] Zhang, Z.-L., Liu, Z., Kurose, J., and Towsley, D. Call admission control schemes under Generalized processor sharing scheduling. *Telecomm. Syst.* (1997), 125–152. [*p. 224*]

[299] Zhang, Z.-L., Ribeiro, V., Moon, S., and Diot, C. Small-time scaling behaviors of Internet backbone traffic: An empirical study. *Proc. IEEE Infocom* (2003). [*p. 252*]

[300] Zwart, B., Borst, S., and Dębicki, K. Reduced-load equivalence for Gaussian processes. *Oper. Res. Lett. 33* (2005), 502–510. [*p. 67*]

Index

Large deviations for Gaussian queues M. Mandjes

Printed and bound by CPI Group (UK) Ltd, Croydon, CR0 4YY

16/04/2025

14658545-0004